Springer Proceedings in Physics

Volume 153

For further volumes:
http://www.springer.com/series/361

Stefano Bellucci
Editor

Breaking of Supersymmetry and Ultraviolet Divergences in Extended Supergravity

Proceedings of the INFN-Laboratori Nazionali di Frascati School 2013

Springer

Editor
Stefano Bellucci
Laboratori Nazionali di Frascati
National Institute of Nuclear Physics
Frascati RM
Italy

ISSN 0930-8989 ISSN 1867-4941 (electronic)
ISBN 978-3-319-03773-8 ISBN 978-3-319-03774-5 (eBook)
DOI 10.1007/978-3-319-03774-5
Springer Cham Heidelberg New York Dordrecht London

Library of Congress Control Number: 2014936032

Printed on acid-free paper

Springer is part of Springer Science+Business Media (www.springer.com)

Preface

The AdS/CFT correspondence, which states that $D = 4, N = 4$ super Yang-Mills is equivalent to Type IIB superstring on $AdS_5 \times S^5$, is a remarkable manifestation of the holographic principle. This correspondence has given rise to a plethora of trends, approaches, techniques, and developments in nowadays physics. They range from "standard" pure theoretical aspects to those having interesting experimental applications. This correspondence is extremely useful for stringy interpretation of large N gauge theory with a subsequent application to problems of gravity physics such as entropy characterization of black hole physics. It also turns out useful to understand strongly coupled high-energy systems such as RHIC and LHC experiments, and intricate condensed matter systems.

Inspired by this success, one could try to involve another far reaching idea which is partially already present in the AdS/CFT correspondence. This idea is a duality which generalizes the notion of electromagnetic duality in Maxwell theory. Combined with supersymmetry, it leads to intriguing developments.

Namely, $N = 2$ supergravity, deformed by a genuine supersymmetric completion of the λR^4 term, makes the previously "silent" ingredient play an active role. This $(R\ldots)^4$ counterterm in supergravity had its ups and downs since the time it was first proposed as a candidate for the UV divergence in $N = 1$ supergravity. For $N = 2$, a linearized version of the candidate for the UV divergence was proposed a year later and its $N = 8$ version was constructed 4 years later. Its gravitational part is the square of the Bel–Robinson tensor and it also has a term quartic in graviphotons. Its 3-loop UV finiteness was explained via the $E_{7(7)}$ duality discovered by Cremmer and Julia.

This book is based upon lectures held on 25–28 March, 2013 at the INFN-Laboratori Nazionali di Frascati Breaking of supersymmetry and Ultraviolet Divergences in extended Supergravity Workshop BUDS 2013, directed by Stefano Bellucci, with the participation of prestigious lecturers, including E. Bergshoeff, M. Cederwall, T. Dennen, P. Di Vecchia, A. Karlsson, M. Koehn, B. Ovrut, G. Ruppeiner, A. Van Proeyen, R. Kallosh, P. Aschieri, and S. Ferrara; a special attention is devoted to discuss topics related to the cancelation of ultraviolet divergences in extended supergravity and Born-Infeld like actions.

All talks were followed by extensive discussions and related reworking of the various, contributions, a feature which reflects itself in the specific "flavor" of this volume.

Superconformal symmetry and higher derivative Lagrangians were discussed. It was stressed that the interest in higher derivative terms has several motivations:

- they appear as α' terms in the effective action of string theory,
- they yield corrections to the black hole entropy,
- they provide higher orders in the AdS/CFT correspondence, and
- they give counterterms for UV divergences of quantum loops.

The current knowledge about general supergravity/supersymmetry theories was reviewed. The superconformal method (and in which supergravity theories we can use it) was discussed. Higher derivative supergravity actions and supergravity loop results were extensively re-examined. The Dirac-Born-Infeld—Volkov-Akulov actions were analyzed and the deformation of supersymmetry was provided as an example of an all order higher derivative supersymmetry action. Quantum calculations show that there are unknown relevant properties of supergravity theories. An interesting question is whether (broken) superconformal symmetry can be such an extra quantum symmetry. The nonexistence of (broken) superconformal-invariant counterterms and anomalies in $\mathrm{N} = 4, \mathrm{D} = 4$ could in that case explain 'miraculous' vanishing results.

Progress toward determining the UV Behavior of Maximal Supergravity was then discussed extensively. After 35 years of supergravity, we can only now make very precise statements about the $\mathrm{D} = 4$ ultraviolet structure. No $\mathrm{D} = 4$ divergence of pure supergravity has been found to date. Supersymmetry forbids 1, 2 loop divergences, and pure gravity was found to be 1-loop finite, 2-loop divergent by Goroff and Sagnotti. Including matter, the theory becomes 1-loop divergent as it was demonstrated by 't Hooft and Veltman. Naively, supersymmetry allows for a 3-loop divergence. $\mathrm{N} = 8$ SG and $\mathrm{N} = 4$ SG are found to be 3-loop finite! In $\mathrm{N} = 8$ supergravity no divergence can be there before 7 loops. A 7-loop divergence in $\mathrm{D} = 4$ implies a 5-loop divergence in $\mathrm{D} = 24/5$, a calculation currently in progress, which the groups involved reported at the Workshop.

If $\mathrm{N} = 8$ supergravity is perturbatively finite, the interest will lie in the reason behind the finiteness. Several possibilities arise for such a reason, a hidden new symmetry, for example. Understanding the mechanism might open a host of possibilities. Potential indications of hidden structures include the following fascinating options:

- Gravity is a double copy of gauge theories.
- Color-Kinematics according to which kinematics is a Lie algebra.
- Constraints from electric–magnetic duality.
- Hidden superconformal $\mathrm{N} = 4$ supergravity.

Let us recall that $\mathrm{N} = 4$ and $\mathrm{N} = 8$ supergravities arise as the low-energy limit of strings. String theory provides a consistent ultraviolet finite theory of quantum gravity. One could wonder if one can remove the string massive modes and address the question of ultraviolet behavior of pure supergravity. In the Workshop, also the String theory approach to UV divergences in supergravity was discussed, in order to elucidate the role of supersymmetry in perturbative computation, as well as the role of non-perturbative duality symmetries in string theory.

This is the seventh volume in a series of books on the general topics of supersymmetry, supergravity, black holes, and the attractor mechanism. Indeed, based on previous meetings, six volumes were already published:

BELLUCCI S. (2006). Supersymmetric Mechanics—Vol. 1: Supersymmetry, Noncommutativity and Matrix Models. (vol. 698, pp. 1–229). ISBN: 3-540-33313-4. BERLIN HEIDELBERG: Springer Verlag (GERMANY). Springer Lecture Notes in Physics Vol. 698.

BELLUCCI S., S. FERRARA, A. MARRANI. (2006). Supersymmetric Mechanics—Vol. 2: The Attractor Mechanism and Space Time Singularities. (vol. 701, pp. 1–242). ISBN 13: 9783540341567. BERLIN HEIDELBERG: Springer Verlag (GERMANY). Springer Lecture Notes in Physics Vol. 701.

BELLUCCI S. (2008). Supersymmetric Mechanics—Vol. 3: Attractors and Black Holes in Supersymmetric Gravity. (vol. 755, pp. 1–373). ISBN-13: 9783540795223. BERLIN HEIDELBERG: Springer Verlag (GERMANY). Springer Lecture Notes in Physics Vol. 755.

BELLUCCI S. (2010). The Attractor Mechanism. Proceedings of the INFN-Laboratori Nazionali di Frascati School 2007. ISSN 0930-8989, ISBN 978-3-642-10735-1, e-ISBN 978-3-642-10736-8. DOI 10.1007/978-3-642-10736-8. Springer Heidelberg Dordrecht London New York. Springer Proceedings in Physics Vol. 134.

BELLUCCI S. (2013). Supersymmetric Gravity and Black Holes. Proceedings of the INFN-Laboratori Nazionali di Frascati School on the Attractor Mechanism 2009. ISSN 0930-8989, ISBN 978-3-642-31379-0, ISBN 978-3-642-31380-6 (eBook), DOI 10.1007/978-3-642-31380-6, Springer Heidelberg New York Dordrecht London. Springer Proceedings in Physics Vol. 142.

BELLUCCI S. (2013). Black Objects in Supergravity. Proceedings of the INFN-Laboratori Nazionali di Frascati School 2011. ISBN 978-3-319-00214-9, Springer Heidelberg New York Dordrecht London. Springer Proceedings in Physics Vol. 144.

I wish to thank all speakers and participants to the Workshop for contributing to the success of the Workshop, which prompted the realization of this volume. I wish to thank my wife Gloria and our beloved daughters Costanza, Eleonora, Annalisa, Erica, and Maristella for love and inspiration, in want of which I would have never had the strength to complete this effort.

October 2013 Stefano Bellucci

Contents

Chapter 1
Superconformal Symmetry and Higher-Derivative Lagrangians

Antoine Van Proeyen

Superconformal methods are useful to build invariant actions in supergravity. We have a good insight in the possibilities of actions that are at most quadratic in spacetime derivatives, but insight in general higher-derivative actions is missing. Recently higher-derivative actions got more attention for several applications. One of these is the understanding of finiteness of loop computations in supergravities. Divergences can only occur if invariant counterterms or anomalies exist. One can wonder whether conformal symmetry might also play a role in this context. In order to construct higher-derivative supergravities with the conformal methods, one should first get more insight in such rigid supersymmetric actions with extra fermionic symmetries. We show how Dirac–Born–Infeld actions with Volkov–Akulov supersymmetries can be constructed in all orders.

1.1 Introduction

In the last 35 years, supergravity actions with terms that are at most quadratic in spacetime derivatives have been studied a lot. But recently higher-derivative terms in supergravity actions got more interest. There are different reasons for this. They appear as order α' terms in the effective action of string theory. It has also been realized that they lead to corrections to the black hole entropy. Furthermore, they can give higher order results in the AdS/CFT correspondence. In this talk, we will also consider them as counterterms for UV divergences of quantum loops.

In Sect. 1.2, we will review what we know about general sugra (supergravity) and susy (supersymmetry) theories. Our preferred method to obtain such theories uses the superconformal method, which we review in Sect. 1.3. We will also discuss

A. Van Proeyen (✉)
Instituut voor Theoretische Fysica, KU Leuven,
Celestijnenlaan 200D, 3001 Leuven, Belgium
e-mail: Antoine.VanProeyen@fys.kuleuven.be

S. Bellucci (ed.), *Breaking of Supersymmetry and Ultraviolet Divergences in Extended Supergravity*, Springer Proceedings in Physics 153,
DOI: 10.1007/978-3-319-03774-5_1,

there in which sugra theories these can be used. Then, in Sect. 1.4 we will turn to higher-derivative sugra actions and explain the relation with sugra loop results. We will see that we miss a lot of insight in the possibilities for higher-derivative actions. In view of this, we studied Dirac–Born–Infeld actions for vector multiplets, obtaining closed expressions and exhibiting extra Volkov–Akulov type supersymmetries. They are examples of all order higher-derivative susy actions. They are the deformation of the well-known lowest order supersymmetry action, and can be considered also perturbatively in a bottom-up construction. We will summarize this result in Sect. 1.5, before giving conclusions in Sect. 1.6.

1.2 General Sugra/Susy Theories

An overview of possible actions with supersymmetry and supergravity has been given in Chap. 12 of the book [1], starting from the basics. The theories considered there are 'ordinary' supersymmetry and supergravity theories, which means that the bosonic terms in the action are at most quadratic in spacetime derivatives, while the terms with fermions are at most linear in spacetime derivatives. In 4 dimensions they typically contain the frame field e_μ^a, gauge fields A_μ^A, with field strengths $F_{\mu\nu}^A$, scalars φ^u, gravitinos ψ_μ^i, and spin-1/2 fermions λ^m and a Lagrangian of the form

$$\begin{aligned} e^{-1}\mathcal{L} = {} & \tfrac{1}{2}R + \tfrac{1}{4}(\mathrm{Im}\,\mathcal{N}_{AB})F_{\mu\nu}^A F^{\mu\nu B} - \tfrac{1}{8}(\mathrm{Re}\,\mathcal{N}_{AB})e^{-1}\varepsilon^{\mu\nu\rho\sigma}F_{\mu\nu}^A F_{\rho\sigma}^B \\ & -\tfrac{1}{2}g_{uv}D_\mu\varphi^u D^\mu\varphi^v - V(\varphi) \\ & \left\{-\tfrac{1}{2}\bar{\psi}_{\mu i}\gamma^{\mu\nu\rho}D_\nu\psi_\rho{}^i - \tfrac{1}{2}g_A{}^B\bar{\lambda}^A \not{D}\lambda_B + \text{h.c.}\right\} + \cdots , \end{aligned} \tag{1.1}$$

where $\mathcal{N}_{AB}$, g_{uv} and $g_A{}^B$ are functions of the scalars φ. In general, the possibilities for susy theories depend on the properties of irreducible spinors in each dimension. For theories with Minkowski signature, these can be summarised in Table 1.1. For each spacetime dimension it is indicated whether Majorana (M), Majorana–Weyl (MW), symplectic (S) or symplectic Weyl (SW) spinors can be defined as the 'minimal spinor', and the number of real components of this minimal spinor is given. To make a complete list, we further use the information of what is the maximal number of susy generators in such theories. This is based on an analysis of representations of susy in 4 dimensions, which leads to maximal $\mathcal{N} = 8$ for sugra, and maximal $\mathcal{N} = 4$ for susy. This thus translates to maximal 32 real generators for sugra and 16 for susy. This information is based on an analysis of particle states i.e. states with momentum, spin and helicity $|p^\mu, s, h\rangle$. One needs that susy generators transform a boson state to a fermion state and that they square to translations, which is an invertible operator. Considering these operators as acting from bosonic states to fermionic states or the inverse, leads to the conclusion that there are an equal number of bosonic and fermionic states (number of degrees of freedom), and to the possible particle representations [2]. The information of the maximal number of susy

Table 1.1 Irreducible spinors, number of components and symmetry properties

Dim	Spinor	Min # components
2	MW	1
3	M	2
4	M	4
5	S	8
6	SW	8
7	S	16
8	M	16
9	M	16
10	MW	16
11	M	32

generators can also be used in dimensions higher than 4, since any higher-dimensional theory can be reduced on tori to $D = 4$, keeping the same number of susy generators. We recalled the essential elements of the proofs here, in order to distinguish supersymmetries of this kind, to the Volkov–Akulov supersymmetries. The latter do not transform between such bosonic and fermionic states and should thus not be included in the relevant counting of the number of supersymmetry generators. Using this information leads to Table 1.2. An entry in the table represents the possibility to have supergravity theories in a specific dimension D with the number of (real) supersymmetries indicated in the top row. We first repeat for every dimension the type of spinors that can be used. Theories with up to 16 (real) supersymmetry generators allow 'matter' multiplets. Considering the on-shell states of the free theories we distinguish different kinds of such multiplets. Those that contain a gauge field A_μ are called vector multiplets or gauge multiplets, and are indicated in Table 1.2 with $\heartsuit$. Tensor multiplets in $D = 6$ contain an antisymmetric tensor $T_{\mu\nu}$, are are indicated by $\diamondsuit$. Multiplets with only scalars and spin-1/2 fields are indicated with $\clubsuit$. They are the hypermultiplets in case of 8 supersymmetry generators, or the Wess–Zumino chiral multiplets for $\mathcal{N} = 1$, $D = 4$. At the bottom is indicated whether these theories exist only in supergravity (SG), or also with just global supersymmetry (SUSY).[1]

For each entry in Table 1.2 there are basic supergravities and 'deformations'. Basic supergravities have only gauged supersymmetry and general coordinate transformations (and U(1)s of vector fields). There is no potential for the scalars, and there are only Minkowski vacua. A deformation means that, without changing the kinetic terms of the fields, the couplings are changed. Many deformations are 'gauged supergravities'. That means that a Yang–Mills group is gauged, introducing a potential. Such supergravities are produced by fluxes on branes in string theory. There are also other deformations (e.g. massive deformations, the superpotential in $\mathcal{N} = 1$ supersymmetry, ...).

[1] Some exotic possibilities, like (4, 0), (2, 1) theories, for which no full action exists, are omitted here.

Table 1.2 Supersymmetry and supergravity theories in dimensions 4 to 11

D	SUSY	32	24	20	16	12	8	4
11	M	M						
10	MW	IIA IIB			I ♡			
9	M	$\mathcal{N}=2$			$\mathcal{N}=1$ ♡			
8	M	$\mathcal{N}=2$			$\mathcal{N}=1$ ♡			
7	S	$\mathcal{N}=4$			$\mathcal{N}=2$ ♡			
6	SW	(2, 2)	(2, 1)		(1, 1) ♡ (2, 0) ◇		(1, 0) ♡, ◇, ♣	
5	S	$\mathcal{N}=8$	$\mathcal{N}=6$		$\mathcal{N}=4$ ♡		$\mathcal{N}=2$ ♡, ♣	
4	M	$\mathcal{N}=8$	$\mathcal{N}=6$	$\mathcal{N}=5$	$\mathcal{N}=4$ ♡	$\mathcal{N}=3$ ♡	$\mathcal{N}=2$ ♡, ♣	$\mathcal{N}=1$ ♡, ♣
			SG		SG/SUSY	SG	SG/SUSY	

The embedding tensor formalism offers a way to classify the gauged supergravities. It defines the gauge group as a subgroup of the isometry group G, as can be seen from the covariant derivative $\left(\partial_\mu - A_\mu{}^M \Theta_M{}^\alpha \delta_\alpha\right)\phi$. Here, α labels all the rigid symmetries, while M labels those that are gauged. The 'embedding tensor' $\Theta_M{}^\alpha$ determines which symmetries are gauged and in which amount they contribute. E.g. the coupling constants are part of this tensor. The tensor should satisfy a number of constraints, whose solutions determine the possible gaugings [3–5]. This thus allows to get a complete picture of supergravities with at most two spacetime derivatives in Lagrangian, though it still needs more work to get all the explicit solutions of the constraints.

For higher-derivative actions there is no such systematic knowledge. There are various constructions of higher derivative terms, e.g. using supersymmetric Dirac–Born–Infeld actions, but there is no systematic construction or classification of possibilities; certainly not for supergravity, but even not for supersymmetry.

1.3 The Superconformal Method

There are various ways to construct supergravity actions. A basic way is the order-by-order Noether method: starting from a globally symmetric action, next order terms in the gravitational coupling constant are added using the concepts of Noether currents. This is in fact the only possibility for the theories with more than 16 susy generators. The superspace method is very useful for rigid $\mathcal{N} = 1$ supersymmetry. However, it becomes very difficult for supergravity. One needs many fields and many gauge transformations to get to a supergravity action. There is also the (super)group

Table 1.3 Conformal construction of Poincaré gravity

Conformal gauge multiplet coupled to a scalar	local conformal
⇓ gauge fix non-conformal symmetries	
Poincaré gravity	local Poincaré symmetry

manifold approach, where optimal use is made of the symmetries using constraints on the curvatures. We adhere to the method of superconformal tensor calculus whenever possible. This method has the advantage that it uses the nice features of superspace, like the the structure of multiplets, but it avoids its immense number of unphysical degrees of freedom. The extra symmetries that are used in this method often lead to insight in the structure of a supergravity theory.

Superconformal symmetry is the maximal extension of spacetime symmetries according to the Coleman–Mandula theorem. What we have in mind, is not the construction of the supersymmetric completion of Weyl gravity, $\int \mathrm{d}^4x\sqrt{g}\left[R^2_{\mu\nu\rho\sigma} - 2R^2_{\mu\nu} + \frac{1}{3}R^2\right]$, but the construction of Poincaré gravity,

$$S_{\text{Poinc}} = \int \mathrm{d}^4x\, \frac{1}{2\kappa^2}\sqrt{g}\, R, \tag{1.2}$$

using conformal methods, where the dimensionful gravitational coupling constant κ signals a breaking of the conformal symmetry. Thus, we use the conformal symmetry as a tool for the construction of actions. It allows us to use multiplet calculus similar to superspace, and it makes hidden symmetries explicit.

We first explain the strategy for the construction of pure gravity in a conformal way. One starts with a conformal coupling of a scalar field, which will act as 'compensator':

$$\mathcal{L} = -\tfrac{1}{2}\sqrt{g}\,\phi\,\square^C\phi = -\tfrac{1}{2}\sqrt{g}\,\phi\square\phi + \tfrac{1}{12}\sqrt{g}\,R\phi^2. \tag{1.3}$$

This action has local scale transformations $\delta\phi(x) = \lambda_{\mathrm{D}}(x)\phi(x)$. These can be gauge-fixed by choosing a value

$$\phi = \sqrt{6}/\kappa. \tag{1.4}$$

This introduces the scale κ, indicating the breaking of conformal symmetry. Using (1.4) in (1.3) leads to (1.2). The mechanism thus starts with a conformal invariant action, and has a Poincaré invariant action as a result after gauge fixing. This is systematically indicated in Table 1.3.

For the supersymmetric theories, a similar construction allows to get more insight in the structure of supergravity actions. A main difference between supersymmetry and supergravity is that multiplets have a clear structure in supersymmetry, but after coupling to supergravity they often get mixed, and they are not clearly identifiable in the final action. In another language: superfields are an easy conceptual tool for globally supersymmetric theories. With the similar method as described above for gravity, supergravity can also be obtained by starting with an action with superconformal symmetry and gauge fixing the superfluous symmetries. This is especially useful for matter-coupled supergravities. Before the gauge fixing, everything looks like in global supersymmetry, just adding covariantizations for the superconformal symmetries. Only after the gauge fixing, the multiplets get mixed.

To elucidate the superconformal symmetry, it is useful to consider it in the way of transformations of supermatrices of the form

$$\begin{pmatrix} \text{conformal algebra} & Q, S \\ Q, S & R\text{−symmetry} \end{pmatrix}. \tag{1.5}$$

Q is the ordinary supersymmetry and S is the extra, 'special' supersymmetry. The R-symmetry depends on the dimension and extension of supersymmetry. It is clarifying to order the generators according to their weight under dilatations (here for the $\mathcal{N} = 1$ superconformal algebra)

$$\begin{aligned} 1 &: P_a \\ \tfrac{1}{2} &: Q \\ 0 &: D,\ M_{ab},\ T \\ -\tfrac{1}{2} &: S \\ -1 &: K_a. \end{aligned} \tag{1.6}$$

P_a, D, M_{ab} and K_a are the conformal generators. The R-symmetry is in this case just U(1), whose generator is indicated by T. The weights in the first column of (1.6) determine the commutators involving D, for example

$$[D, Q] = \tfrac{1}{2}Q, \qquad [D, S] = -\tfrac{1}{2}S. \tag{1.7}$$

As we discussed above, T is an R-symmetry. All (anti)commutators are consistent with the weights, e.g.

$$\begin{aligned} \left\{Q_\alpha, Q^\beta\right\} &= -\tfrac{1}{2}(\gamma^a)_\alpha{}^\beta P_a, \qquad \left\{S_\alpha, S^\beta\right\} = -\tfrac{1}{2}(\gamma^a)_\alpha{}^\beta K_a, \\ \left\{Q_\alpha, S^\beta\right\} &= -\tfrac{1}{2}\delta_\alpha{}^\beta D - \tfrac{1}{4}(\gamma^{ab})_\alpha{}^\beta M_{ab} + \tfrac{1}{2}\mathrm{i}(\gamma_*)_\alpha{}^\beta T. \end{aligned} \tag{1.8}$$

The strategy for the superconformal construction of $\mathcal{N} = 1$ supergravity is analogous as for gravity in Table 1.3. It is depicted in Table 1.4.

A similar scheme holds for $\mathcal{N} = 4$ supergravity [6, 7] as shown in Table 1.5. The special feature is that the gauge compensating multiplets are on-shell multiplets. Remember that in any case the action should be invariant without use of the field equations, but the algebra of the symmetries may close only modulo field equations. However, the problem is that in this way there is no flexibility in the field equations.

Table 1.4 Superconformal construction of pure $\mathcal{N} = 1$ supergravity

Superconformal gauge multiplet ($\mathcal{N} = 1$ Weyl multiplet) coupled to a chiral multiplet
$\Downarrow$ gauge fix dilatations, special conformal transformations, local U(1)-symmetry and special supersymmetry
pure $\mathcal{N} = 1$ supergravity

Table 1.5 Superconformal construction of pure $\mathcal{N} = 4$ supergravity

Superconformal gauge multiplet ($\mathcal{N} = 4$ Weyl multiplet) coupled to 6 gauge compensating multiplets (on-shell)
$\Downarrow$ gauge fix dilatations, special conformal transformations, local SU(4), local U(1) and special supersymmetry
pure $\mathcal{N} = 4$ Cremmer-Scherk-Ferrara supergravity

They are already fixed by the supersymmetry transformation laws. This gives thus a problem when we want to modify the action with higher-derivative terms, since then the field equations will change. Therefore, higher-derivative terms cannot be added to $\mathcal{N} = 4$ supergravity without a modification of the field equations. The hypermultiplets of $\mathcal{N} = 2$ supergravity already have this feature of an 'on-shell algebra' (at least for a generic hyper-Kähler manifold). The $\mathcal{N} = 4$ gauge multiplets also share this property. This is especially relevant since they are compensating multiplets. It implies that the supersymmetry transformations of the $\mathcal{N} = 4$ super-Poincaré theory can only close modulo field equations. But one can apply the superconformal method.

In which supergravity theories can we use the superconformal methods? There are two necessary ingredients. First, one should have a superconformal algebra. Second, there should be compensating multiplets. Which theories allow superconformal algebras was already analysed by Nahm [8]. He analysed in which simple superalgebras the conformal algebra $\mathfrak{so}(D, 2)$ is a factor in the bosonic subalgebra. This lead to Table 1.6 (also a long list of superconformal algebras exist for $D = 2$). In each case the bosonic subgroup contains the covering group[2] of SO$(D, 2)$, such that spinor representations are possible, and a compact R-symmetry group. The last column gives the number of real supersymmetry generators. Other superconformal algebras have been considered where the conformal algebra is not a factor, but still a subalgebra of the bosonic part of the superalgebra. E.g. SO(11, 2) $\subset$ Sp(64) $\subset$ OSp(1|64) [9, 10]. However, these have not been successfully applied for constructing actions. Thus, the superconformal methods are restricted to the dimensions and extensions that

[2] The equality sign in the 'conf' column of Table 1.6 is only valid at the level of the algebra.

Table 1.6 Superconformal algebras

D	Supergroup	Conf.	R	Ferm.
3	$\mathrm{OSp}(N\|4)$	$\mathrm{SO}(3,2)=\mathrm{Sp}(4)$	$\mathrm{SO}(N)$	$4N$
4	$\mathrm{SU}(2,2\|N)$	$\mathrm{SO}(4,2)=\mathrm{SU}(2,2)$	$\mathrm{U}(N)$ for $N\neq 4$ $\mathrm{SU}(4)$ for $N=4$	$8N$
5	$F^2(4)$	$\mathrm{SO}(5,2)$	$\mathrm{SU}(2)$	16
6	$\mathrm{OSp}(8^*\|2N)$	$\mathrm{SO}(6,2)=\mathrm{SO}^*(8)$	$\mathrm{USp}(2N)$	$16N$

Table 1.7 Supergravity theories for which superconformal methods can be used

D	SUSY	32	24	20	16	12	8	4
11	M	M						
10	MW	IIA IIB			I			
9	M	$\mathcal{N}=2$			$\mathcal{N}=1$			
8	M	$\mathcal{N}=2$			$\mathcal{N}=1$			
7	S	$\mathcal{N}=4$			$\mathcal{N}=2$			
6	SW	(2, 2)	(2, 1)		(1, 1) $\boxed{(2,0)}$		$\boxed{(1,0)}$	
5	S	$\mathcal{N}=8$	$\mathcal{N}=6$		$\mathcal{N}=4$		$\boxed{\mathcal{N}=2}$	
4	M	$\mathcal{N}=8$	$\mathcal{N}=6$	$\mathcal{N}=5$	$\boxed{\mathcal{N}=4}$	$\boxed{\mathcal{N}=3}$	$\boxed{\mathcal{N}=2}$	$\boxed{\mathcal{N}=1}$

appear in Table 1.6 and furthermore to a number of supersymmetries ≤16, such that compensating multiplets exist.[3] This leads to those indicated in boxes in Table 1.7.

1.4 Higher Derivative Sugra Actions and Sugra Loop Results

For many years it was believed that supergravity could not be a finite theory. However, since the calculations of [12] revealed the 3-loop finiteness of $\mathcal{N}=8$, $D=4$ supergravity, we realize that quantum supergravity has more surprising features than we understood so far. In [13] the result was extended to 4 loops and even to $D=5$. But then, also $\mathcal{N}=4$ supergravity in $D=4$ turned out to be finite up to 3 loops [14] (and further results followed for $D=5$). This brings us to reflections on the nature of supergravity and possible counterterms. Divergences would imply that supersymmetric counterterms should exist (or there should be supersymmetric anomalies). But our present knowledge on higher-derivative terms in supergravity is not sufficient to be sure about which invariants can be consistently defined.

[3] For $D=10$ with 16 supersymmetries, a superconformal formulation, not based on a Lie superalgebra but rather on a soft algebra has been found in [11].

1.4.1 Superconformal Methods for the $\mathcal{N} = 2$ Example

Superconformal methods have been used to construct higher-derivative supergravities, starting with the work of Cecotti and Ferrara [15]. Especially for $\mathcal{N} = 2$ supergravity, the tensor calculus allows us to construct various terms [16]. The constructions use tensor calculus with chiral multiplets, which are similar to chiral superfields. The multiplets contain fields

$$S = \{X, \Omega_i, \ldots, C\}. \tag{1.9}$$

Any sum and product of these gives another chiral multiplets. These manipulations allow 'tensor calculus'. A useful tool is the kinetic multiplet of a chiral multiplet (which is also chiral) and starts with the complex conjugate of the highest component of a chiral multiplet:

$$\mathbb{T}(\bar{S}) = \{\bar{C}, \ldots\}. \tag{1.10}$$

To construct higher-derivative terms, one needs also another chiral multiplet, formed from the $\mathcal{N} = 2$ Weyl multiplet

$$W^2 = \{T^-_{ab}T^{ab\,-}, \ldots\}. \tag{1.11}$$

It starts from the square of an auxiliary field (antisymmetric tensor) of the Weyl multiplet. One can then use tensor calculus on these multiplets to construct new chiral multiplets, of which the highest components defines actions. In order to be able to define these in the superconformal framework, one has to take into account the dilatation symmetry. This implies that the function of chiral multiplets that is used to construct actions should satisfy homogeneity properties. Using such homogeneous functions of the chiral multiplets, one obtains supergravity theories using superconformal covariantization of the expressions used for global supersymmetry. Hence this leads to many possibilities, which are invariants contributing to the entropy and central charges of black holes.

In order to see how these actions lead to DBI theories, R^4 actions are considered in [17], using the above-mentioned constructions with

$$\left[S^2 + \lambda \frac{W^2}{S^2}\mathbb{T}\left(\frac{\bar{W}^2}{\bar{S}^2}\right)\right]_C. \tag{1.12}$$

It uses the action formula 'C', which means in global supersymmetry the highest component of th chiral multiplet. In superconformal calculus, there are some correction terms involving the gravitino, to obtain local conformal symmetry. S is the chiral compensating multiplet (which due to constraints is in fact a vector multiplet). Using just the first term in (1.12) would lead to pure supergravity.[4] The second term

[4] In fact, a second compensating multiplet is necessary in $\mathcal{N} = 2$, but we do not discuss this here, since this can be neglected for the present purposes.

in (1.12) uses the multiplet (1.11) and the construction of a kinetic multiplet (1.10). The powers of S are chosen in order to satisfy the homogeneity properties leading to conformal-invariant actions. That second term is taken with a coupling constant λ, in which an expansion will be considered.

Apart from a term of the form $\lambda C^4_{....}$, where $C_{....}$ is the Weyl tensor, and thus creating terms of the form R^4, the action formula in (1.12) produces also terms of the type $\lambda(\partial T)^4$, where T stands for the auxiliary field of the Weyl multiplet. In the standard supergravity action, the field equations imply that T is on-shell proportional to the graviphoton. For the action (1.12), we get, symbolically

$$T_{ab} = \frac{2}{X} F_{ab} + \lambda(\partial^4 T^3)_{ab}, \tag{1.13}$$

where X is the scalar of the compensating multiplet, which is in the Poincaré theory dependent on κ similar to (1.4). This equation is solved recursively, and we thus get an expression with an infinite number of higher derivative terms with higher and higher powers of the graviphoton F:

$$T_{ab} = \frac{2}{X} F_{ab} + \lambda(\partial^4 F^3)_{ab} + \lambda^2 \partial^4 F^2 \partial^4 F^3 + \cdots. \tag{1.14}$$

The action with auxiliary field eliminated leads to a DBI-type action with higher derivatives

$$S_{\text{deformed}} = -\tfrac{1}{4} F^2 + \lambda(\partial F)^4 + \lambda^2 \partial^8 F^6 + \cdots. \tag{1.15}$$

Note that before the elimination of the auxiliary field, this action has a finite number of terms with auxiliary fields. The infinite series is produced by the elimination of the auxiliary fields. They lead thus to a deformation of the lowest order action in powers of λ. At the same time also the transformation laws are deformed. Again, the transformation laws are finite expressions before the elimination of the auxiliary fields. E.g. for the gravitino transformation

$$\delta\psi^i_\mu = D_\mu \epsilon^i - \tfrac{1}{16}\gamma^{ab} T^-_{ab} \varepsilon^{ij} \gamma_\mu \epsilon_j - \gamma_\mu \eta^i, \tag{1.16}$$

where the covariant derivative uses the superconformal connections, and S-supersymmetry with parameter η^i is included. Then the on-shell value of the auxiliary fields is used as a power series in λ:

$$\phi_{\text{aux}} = \phi^{(0)}_{\text{aux}} + \Delta\phi_{\text{aux}}, \qquad \Delta\phi_{\text{aux}} = \sum_{n=1} \lambda^n \phi^{(n)}_{\text{aux}}. \tag{1.17}$$

This leads, with (1.14), to deformations in the supersymmetry transformation law of the gravitino of the form [17]

$$\Delta\psi^i_\mu = -4\lambda[\partial^4 F^3]_\mu{}^\nu \gamma_\nu \epsilon^i + \cdots. \tag{1.18}$$

Here also contributions have been used that originate from the 'decomposition law' expressing the parameter η^i in terms of ϵ^i after gauge fixing of S-supersymmetry.

We conclude that the tensor calculus allows us to obtain higher-derivative terms, determined first off-shell, which can lead to deformations of the action and transformation laws on-shell. They are obtained from (broken) superconformal actions. For pure gravity, the 3-loop counterterm that contains R^4 is obtained from the local conformal expression

$$\int d^4\sqrt{g}\,\phi^{-4}(C_{\mu\nu\rho\sigma}C^{\mu\nu\rho\sigma})^2, \tag{1.19}$$

where ϕ is the compensating scalar and $C_{\mu\nu\rho\sigma}$ is the Weyl tensor. For $\mathcal{N}=2$ a superconformal R^4 counterterm can be obtained from the λ-term in (1.12). What do we know about $\mathcal{N}=4$, where miraculous cancelations have been found?

1.4.2 Problem and Conjecture for $\mathcal{N}=4$ Supergravity

The problem is that it is not easy to construct counterterms for $\mathcal{N}=4$ supergravity. We cannot multiply the compensating multiplets to suitable powers, and thus we cannot make constructions as those for $\mathcal{N}=2$. The essential problem is that the algebra of supersymmetry holds only on shell. When we would like to write a modified action, then it implies modified field equations, and thus the transformations have to be modified (or in other words: the structure of the multiplets). For $\mathcal{N}=2$, deformed transformations could be found due to the possibility to work first with auxiliary fields. The field equations for the latter lead to deformed transformation laws on shell. For $\mathcal{N}=4$ we do not know auxiliary fields. How can we then establish the the existence or non-existence of the consistent order by order deformation of $\mathcal{N}=4$ supergravity?

This question lead to the conjecture made in [18]. If such counterterms do not exist, this may explain finiteness results (if meanwhile the explicit calculations do not find that $\mathcal{N}=4$, $D=4$ is divergent at higher loops). Until invariant counterterms are constructed we have no reason to expect UV divergences. We can also conjecture that such counterterms should be broken superconformal expressions, if conformal symmetry is more than a classical symmetry. Thus there are two points of view. The first one is that legitimate counterterms are not available yet, and we still have to construct them. The second one is that legitimate counterterms are not available, and cannot be constructed, offering an explanation of finiteness.

In fact, if the UV finiteness will persist in higher loops, one would like to view this as an opportunity to test some new ideas about gravity. One possible idea is that superconformal symmetry, used in the classical theory as a tool to construct actions, is more fundamental and has also a quantum significance. As mentioned in Sect. 1.3, the classical theory can be obtained from gauge fixing a superconformal action. In that way, the Planck mass appears only in the gauge-fixing procedure. This looks analogous to the appearance of the masses of W and Z vector mesons in the standard model. They are not present in the gauge-invariant action, and show up when the gauge symmetry is spontaneously broken. In the unitary gauge these masses give the impression of being fundamental. In the renormalizable gauge, where the UV properties are analysed, they are absent. One may hope that a similar understanding can be obtained in the future to give a more fundamental significance to the superconformal symmetry. The possible non-existence of (broken) superconformal-invariant counterterms and anomalies in $\mathcal{N} = 4$, $D = 4$ supergravity could then explain the 'miraculous' results of the quantum calculations.

Such ideas would give a simple explanation of the 3-loop finiteness and predict perturbative UV finiteness in higher loops. The same conjecture applies to higher-derivative superconformal invariants and to the existence of a consistent superconformal anomaly. Also for the latter, one may either say that we still have to understand how to construct such an anomaly, or maybe it does not exist. Therefore, the conjecture is economical, sparing in the use of resources: either the local $\mathcal{N} = 4$ superconformal symmetry is a good symmetry, or it is not. The conjecture is falsifiable by the $\mathcal{N} = 4$ 4-loop computations (which are already underway, as we heard during the conference). If the conjecture survives these computations (if they show further UV finiteness), then this gives a further hint that the models with superconformal symmetry serve as a basis for constructing a consistent quantum theory where the Planck mass appears only in the process of gauge fixing the superconformal symmetry. However, it is also falsifiable by our own calculations: if we find a way to construct (non-perturbative) superconformal invariants that can serve as counterterms, then this conjecture is circumvented. We will start to search in that direction, following a quote of R. Feynman: "We are trying to prove ourselves wrong as quickly as possible, because only in that way can we find progress."

1.5 Dirac–Born–Infeld-Volkov–Akulov and Deformation of Supersymmetry

The main problem for the superconformal construction of counterterms in $\mathcal{N} = 4$ supergravity is thus that the compensating multiplets have only been defined with transformations that close on-shell using the field equations of the 2-derivative action. These compensating multiplets are vector multiplets. In our recent work [19] we search for deformations of vector multiplet actions such that higher-derivative terms occur. We will find all-order higher derivative globally supersymmetric invariant actions. They are of the Dirac–Born–Infeld (DBI) type, and have extra symmetries, of Volkov–Akulov (VA) type. The latter are not yet S-supersymmetry transformations that we would like in the context of the superconformal programme mentioned above, but we will comment on this at the end.

We will consider vector multiplets with a gauge vector and a spinor field. We want that the supersymmetry algebra is closed, but not necessary off-shell, since the main problems that we want to address are theories with only an on-shell closed algebra.

A gauge vector in D dimensions has $D-2$ on-shell degrees of freedom,[5] while a spin-1/2 fermion has on-shell half the number of degrees of freedom of the number of components of the spinor. Considering Table 1.1 shows that one can have an equal number of bosonic and fermionic degrees of freedom for these fields in the cases $D=10$ with Majorana–Weyl spinors, $D=6$ with symplectic Majorana–Weyl spinors, $D=4$ with Majorana spinors, and even $D=3$ with Majorana spinors. Comparing with Table 1.2 shows that these are the maximal dimensions to have vector multiplets for supersymmetries with 16, 8, 4 and 2 generators. Other vector multiplets are obtained from these by dimensional reduction, which generates also scalar on-shell degrees of freedom.

These theories are described by an action of the form[6]

$$S=\int \mathrm{d}^D x\left\{-\tfrac{1}{4}(F_{\mu\nu})^2-\tfrac{1}{2}\bar{\lambda}\partial\!\!\!/\lambda\right\}. \tag{1.20}$$

They are invariant under supersymmetry transformations[7]

$$\delta_\epsilon A_\mu=-\tfrac{1}{2}\bar{\epsilon}\Gamma_\mu\lambda,\qquad \delta_\epsilon\lambda=\tfrac{1}{4}\Gamma^{\mu\nu}F_{\mu\nu}\epsilon, \tag{1.21}$$

where the spinors are of the appropriate type mentioned before, and for the case of symplectic Majorana-Weyl spinors also the extension index $i=1,2$ has been suppressed with the understanding that e.g.

$$\bar{\epsilon}\Gamma_\mu\lambda=\bar{\epsilon}^i\Gamma_\mu\lambda_i=\varepsilon^{ij}\bar{\epsilon}_j\Gamma_\mu\lambda_i. \tag{1.22}$$

The action has also an extra trivial (global) fermionic shift symmetry

$$\delta_\eta A_\mu=0,\qquad \delta_\eta\lambda=-\tfrac{1}{2\alpha}\eta, \tag{1.23}$$

where the normalization with a constant α has been used in order to match with formulas that will follow below.

1.5.1 The Bottom-Up Approach

We first attempt a 'bottom-up' approach. This means that we define a deformation of the action with terms proportional to a parameter α, and adapt simultaneously the transformation laws. In this we follow [20], where this was considered for $D=6$,

---

[5] All these ingredients are well defined and discussed in [1].

[6] With respect to [19] all barred spinors are multiplied with a factor $-1/2$ in order to agree with the normalizations as in (1.8) and [1].

[7] We use here Γ rather than γ for the gamma matrices in the D, to distinguish them later from the 4-dimensional matrices, see (1.45).

and an action was obtained of the form

$$\begin{aligned}S=\int \mathrm{d}^Dx\,\{&-\tfrac14 F^2-\tfrac12\bar\lambda\partial\!\!\!/\lambda\}+\alpha c_4F^{\mu\nu}\bar\lambda\Gamma_\mu\partial_\nu\lambda\\&+\tfrac18\alpha^2\Big[\mathrm{Tr}\,F^4-\tfrac14\left(F^2\right)^2-2(1+4c_4^2)\left(F^2\right)^{\mu\nu}\bar\lambda\Gamma_\mu\partial_\nu\lambda\\&-\tfrac12(1-4c_4^2)F_\mu{}^\lambda(\partial_\lambda F_{\nu\rho})\bar\lambda\Gamma^{\mu\nu\rho}\lambda-\tfrac14(c_1+8c_4^2)F^2\bar\lambda\partial\!\!\!/\lambda\\&+\tfrac14c_2F_{\mu\nu}(\partial_\lambda F^\lambda{}_\rho)\bar\lambda\Gamma^{\mu\nu\rho}\lambda+\tfrac14(c_3+4c_4^2)F_{\mu\nu}F_{\rho\sigma}\bar\lambda\Gamma^{\mu\nu\rho\sigma}\partial\!\!\!/\lambda\Big]\\&+\mathcal{O}(\alpha^2\lambda^4)+\mathcal{O}(\alpha^3).\end{aligned}\qquad(1.24)$$

The parameters λ_i are undetermined. However, they are all related to field redefinitions

$$\begin{aligned}A_\mu(0)&=A_\mu+\tfrac{1}{32}\alpha^2c_2F^{\nu\rho}\bar\lambda\Gamma_{\mu\nu\rho}\lambda,\\\lambda(0)&=\lambda+\tfrac12\alpha c_4F_{\mu\nu}\Gamma^{\mu\nu}\lambda+\tfrac{1}{32}\alpha^2c_1F^2\lambda-\tfrac{1}{32}\alpha^2c_3F_{\mu\nu}F_{\rho\sigma}\Gamma^{\mu\nu\rho\sigma}\lambda,\end{aligned}\qquad(1.25)$$

where on the right-hand side are the fields corresponding to $c_i=0$, and on the left-hand side those for arbitrary c_i. Hence, up to these redefinitions, the answer is unique up to this order. Remark e.g. that it contains in the bosonic part the unique combination

$$\mathrm{Tr}\ F^4-\tfrac14\left(F^2\right)^2,\qquad \mathrm{Tr}\ F^4\equiv F_{\mu\nu}F^{\nu\rho}F_{\rho\sigma}F^{\sigma\mu},\qquad F^2=F_{\mu\nu}F^{\mu\nu}.\qquad(1.26)$$

Also the transformation laws are deformed with respect to (1.21). As well ordinary supersymmetry transformations (parameter ϵ) as the extra supersymmetry (1.23) can be defined. E.g. for the latter we have now

$$\begin{aligned}\delta_\eta A^\mu&=-\tfrac{\alpha}{8}\bar\eta F^{\nu\mu}\Gamma_\nu\lambda-\tfrac{\alpha}{16}\bar\eta\Gamma^{\mu\nu\rho}F_{\nu\rho}\lambda+\tfrac{1}{32}\alpha c_2F_{\nu\rho}\bar\eta\Gamma^{\mu\nu\rho}\lambda+\mathcal{O}(\alpha\eta\lambda^3)+\mathcal{O}(\alpha^2),\\\delta_\eta\lambda&=-\tfrac{1}{2\alpha}\eta+\alpha\left[-\tfrac{1}{32}F^2-\tfrac{1}{64}\Gamma^{\mu\nu\rho\sigma}F_{\mu\nu}F_{\rho\sigma}\right]\eta\\&\quad+\tfrac14c_4F_{\mu\nu}(c)\Gamma^{\mu\nu}\left[\eta-\tfrac12\alpha c_4F_{\rho\sigma}(c)\Gamma^{\rho\sigma}\eta\right]\\&\quad+\tfrac{1}{64}\alpha c_1F^2\eta-\tfrac{1}{64}\alpha c_3F_{\mu\nu}F_{\rho\sigma}\Gamma^{\mu\nu\rho\sigma}\eta+\mathcal{O}(\alpha\eta\lambda^2)+\mathcal{O}(\alpha^2).\end{aligned}\qquad(1.27)$$

It turns out that we can write this for all $D=10,6,4,3$ with the appropriate spinors types (Majorana, Majorana–Weyl, symplectic Majorana–Weyl) as mentioned above. The only spinor properties that we need are the Majorana flip relations, like

$$\bar\lambda_1\Gamma^\mu\lambda_2=-\bar\lambda_2\Gamma^\mu\lambda_1,\qquad \bar\lambda_1\Gamma^{\mu\nu\rho}\lambda_2=\bar\lambda_2\Gamma^{\mu\nu\rho}\lambda_1,\qquad(1.28)$$

and the cyclic Fierz identity

$$\Gamma_\mu\lambda_1\bar\lambda_2\Gamma^\mu\lambda_3+\Gamma_\mu\lambda_2\bar\lambda_3\Gamma^\mu\lambda_1+\Gamma_\mu\lambda_3\bar\lambda_1\Gamma^\mu\lambda_2=0.\qquad(1.29)$$

These are valid for all these cases. Note that all bilinears in spinors contain odd-rank gamma matrices, as is consistent with the fact that the spinors are all of the same chirality in $D = 10$ and $D = 6$. But this property holds also for e.g. $D = 4$.

The results look very complicated and it seems hopeless to continue this to all orders in α and adding higher order spinor terms.

1.5.2 The Top-Down Approach

We [19] found a solution to the problem of the construction of the infinite series of deformations starting from the κ-symmetric action for Dp branes. This action is of the form

$$S_{\rm DBI} + S_{\rm WZ} = -\frac{1}{\alpha^2}\int \mathrm{d}^{p+1}\sigma\,\sqrt{-\det(G_{\mu\nu}+\alpha\mathcal{F}_{\mu\nu})} + \frac{1}{\alpha^2}\int \Omega_{p+1}, \qquad (1.30)$$

where the first term is a DBI action, and κ-supersymmetry implies that it should be complemented with a Wess–Zumino (WZ) term in terms of an appropriate $(p+1)$-form Ω_{p+1} (see e.g. (45) in [21]). In the DBI term appear

$$\begin{aligned} G_{\mu\nu} &\equiv \eta_{mn}\Pi^m_\mu \Pi^n_\nu\,, \qquad \Pi^m_\mu \equiv \partial_\mu X^m + \tfrac{1}{2}\bar\theta\Gamma^m\partial_\mu\theta, \\ \mathcal{F}_{\mu\nu} &\equiv F_{\mu\nu} + \alpha^{-1}\bar\theta\sigma_3\Gamma_m\partial_{[\mu}\theta\left(\partial_{\nu]}X^m + \tfrac{1}{4}\bar\theta\Gamma^m\partial_{\nu]}\theta\right). \end{aligned} \qquad (1.31)$$

We consider these actions in the context of the IIB theory, and thus X^m with $m = 0,\ldots,9$ denote the spacetime coordinates of the $D = 10$ theory. The coordinates on the brane are indicated by $\mu = 0,\ldots,p$, and p should be odd. θ is a doublet of Majorana–Weyl spinors, of which we omit again the extension index. $F_{\mu\nu}$ is an Abelian field strength.

This action has the following symmetries. First, there is a rigid supersymmetry doublet parameter ϵ^1, ϵ^2. There is also rigid Poincaré symmetry in $D = 10$. Furthermore, there are local symmetries on the brane. On the bosonic side these are the worldvolume general coordinate transformations. Furthermore there is the κ-supersymmetry doublet. Effectively only half of these are present, since they this is a reducible symmetry, which means that it appears only in the form

$$\delta_\kappa\theta = (1+\Gamma)\kappa, \qquad (1.32)$$

where Γ is a matrix such that $(1+\Gamma)$ is a projection on half of the spinor space.

Though this has been obtained from IIB superstring theory in $D = 10$, it turns out that the action (1.30) has also the same symmetries when we consider $D = 6$, just changing the index range to $m = 0,\ldots,5$ and using symplectic Majorana–Weyl spinors. This implies that we consider the $(2,0)$ theory in the $D = 6$, 16 supersymmetries entry of Table 1.2. This theory is often called iib. The action has then also a brane interpretation, (using again odd p) [22] as has been clarified in the talk of E. Bergshoeff in this conference. Moreover, we can also consider it solutions of $D=4, \mathcal{N}=2$ supergravity with worldvolume action as in (1.30) (thus $m=0,\ldots,3$ and $p=3$ or 1).

We then gauge-fix local symmetries imposing for a p-brane (describing here the embedding in $D=10$, but the other cases are obtained by changing the range of indices)

$$\begin{aligned} X^m &= \{\delta_\mu^{m'}\sigma^\mu, \phi^I\}, \quad m'=0,1,\ldots,p, \quad I=1,\ldots,9-p \\ \theta &= (\theta^1=0,\ \theta^2\equiv\alpha\lambda). \end{aligned} \tag{1.33}$$

The first line fixes the worldvolume general coordinate transformations by identifying the coordinates in the embedding spacetime with the worldvolume coordinates. This leaves $9-p$ scalars. In the second line, the effective κ-symmetry is fixed, and the remaining coordinate is renamed λ in order to make the connection with the down-up approach. These gauges lead to decomposition laws, implying that the parameters of the worldvolume general coordinate transformations and κ-symmetry become functions of the remaining (global) symmetries. There are thus two, deformed, fermionic symmetries ϵ^1 and ϵ^2. Two combinations of these symmetries are called ϵ and ζ, and can be related to the ϵ and η symmetries of the bottom-up approach.

We first consider the action for the case $p=9$ in this gauge, which reduces (1.30) to

$$S=-\frac{1}{\alpha^2}\int \mathrm{d}^{10}x\left\{\sqrt{-\det(G_{\mu\nu}+\alpha\mathcal{F}_{\mu\nu})}-1\right\}, \tag{1.34}$$

where

$$\begin{aligned} G_{\mu\nu} &= \eta_{mn}\Pi_\mu^m\Pi_\nu^n\,, \qquad \Pi_\mu^m=\delta_\mu^m+\tfrac12\alpha^2\bar\lambda\Gamma^m\partial_\mu\lambda, \\ \mathcal{F}_{\mu\nu} &\equiv F_{\mu\nu}-\alpha\bar\lambda\Gamma_{[\nu}\partial_{\mu]}\lambda, \qquad \mu=0,1,\ldots,9, \qquad m=0,1,\ldots,9. \end{aligned} \tag{1.35}$$

This action possesses 16 ϵ transformations, which are deformations of the Maxwell supermultiplet supersymmetries:

$$\begin{aligned} \delta_\epsilon\lambda &= -\tfrac{1}{2\alpha}(\mathbb{1}-\beta)\,\epsilon+\tfrac14\alpha\partial_\mu\lambda\bar\lambda\Gamma^\mu(\mathbb{1}+\beta)\,\epsilon, \\ \delta_\epsilon A_\mu &= \tfrac14\bar\lambda\Gamma_\mu(\mathbb{1}+\beta)\epsilon \\ &\quad +\tfrac18\alpha^2\bar\lambda\Gamma_m(\tfrac13\mathbb{1}+\beta)\epsilon\bar\lambda\Gamma^m\partial_\mu\lambda+\tfrac14\alpha\bar\lambda\Gamma^\rho(\mathbb{1}+\beta)\,\epsilon F_{\rho\mu}, \end{aligned} \tag{1.36}$$

where β is a matrix ($\hat\Gamma^\mu=\Pi_m^\mu\Gamma^m$)

$$\begin{aligned} \beta &= \left[\det\left(\delta_\mu{}^\nu+\alpha\mathcal{F}_{\mu\rho}G^{\rho\nu}\right)\right]^{-1/2}\sum_{k=0}^{5}\frac{\alpha^k}{2^k k!}\hat\Gamma^{\mu_1\nu_1\ldots\mu_k\nu_k}\mathcal{F}_{\mu_1\nu_1}\ldots\mathcal{F}_{\mu_k\nu_k} \\ &= 1+\mathcal{O}(\alpha). \end{aligned} \tag{1.37}$$

Furthermore, there are 16 ζ transformations:

$$\delta_\zeta \lambda = \alpha^{-1}\zeta - \tfrac{1}{2}\alpha\partial_\mu\lambda\bar{\lambda}\Gamma^\mu\zeta,$$
$$\delta_\zeta A_\mu = -\tfrac{1}{2}\bar{\lambda}\Gamma_\mu\zeta - \tfrac{1}{2}\alpha\bar{\lambda}\Gamma^\rho\zeta F_{\rho\mu} - \tfrac{1}{12}\alpha^2\bar{\lambda}\Gamma_m\zeta\bar{\lambda}\Gamma^m\partial_\mu\lambda. \tag{1.38}$$

Note that these transformations do not transform states of a fermion field to states of a bosonic field, and are thus not regular supersymmetries. They are transformations of the Volkov–Akulov (VA)-type. To stress this difference, we say that the theory has $16 + 16$ supersymmetries.

When we expand the action in orders of α, we find that the action (1.34) agrees with (1.24) when we choose the coefficients

$$c_1 = 2, \qquad c_2 = 0, \qquad c_3 = -1, \qquad c_4 = -\tfrac{1}{2}. \tag{1.39}$$

This eliminates in fact all ∂F terms from (1.24).

Also the transformation laws (1.36) and (1.38) can be identified with those in the bottom-up approach, modulo a 'zilch symmetry', i.e. a trivial on-shell symmetry. To complete the identification, ζ is recognized as a linear combination of ϵ and η.

This proves that our all-order result is indeed the full deformed theory that we were looking for.

The theories that we can obtain in this way are schematically indicated in Fig. 1.1. The supergravities with each a doublet of local symmetries from which one starts are indicated as open yellow boxes. The branes type DBI actions are the D9, D7, D5, D3 when we start from $D = 10$, and are indicated as V5 and V3 when we start from $D = 6$. The V stands for vector branes as explained in the talk of Eric Bergshoeff. This thus shows that we can construct deformed super-Maxwell theories for various dimensions and supersymmetry extensions, including $\mathcal{N} = 4$, $\mathcal{N} = 2$ and $\mathcal{N} = 1$ in 4 dimensions.

1.5.3 D = 4, $\mathcal{N}$ = 4 Gauge Multiplet

Let us in particular consider the D3 case, i.e. the $D = 4$, $\mathcal{N} = 4$ theory that we discussed in previous sections. The full action of the deformed theory is

$$S = -\frac{1}{\alpha^2}\int \mathrm{d}^4x \left\{\sqrt{-\det(G_{\mu\nu} + \alpha\mathcal{F}_{\mu\nu})} - 1\right\}, \qquad \mu = 0, 1, 2, 3, \tag{1.40}$$

with

$$G_{\mu\nu} = \eta_{mn}\Pi^m_\mu\Pi^n_\nu = \eta_{m'n'}\Pi^{m'}_\mu\Pi^{n'}_\nu + \delta_{IJ}\Pi^I_\mu\Pi^J_\nu, \quad m' = 0, 1, 2, 3,$$
$$\Pi^{m'}_\mu = \delta^{m'}_\mu + \tfrac{1}{2}\alpha^2\bar{\lambda}\Gamma^{m'}\partial_\mu\lambda\,, \qquad \Pi^I_\mu = \partial_\mu\phi^I + \tfrac{1}{2}\alpha^2\bar{\lambda}\Gamma^I\partial_\mu\lambda, \quad I = 1, \dots, 6,$$
$$\mathcal{F}_{\mu\nu} \equiv F_{\mu\nu} + \alpha\bar{\lambda}\Gamma_{[\mu}\partial_{\nu]}\lambda + \alpha\bar{\lambda}\Gamma_I\partial_{[\mu}\lambda\partial_{\nu]}\phi^I. \tag{1.41}$$

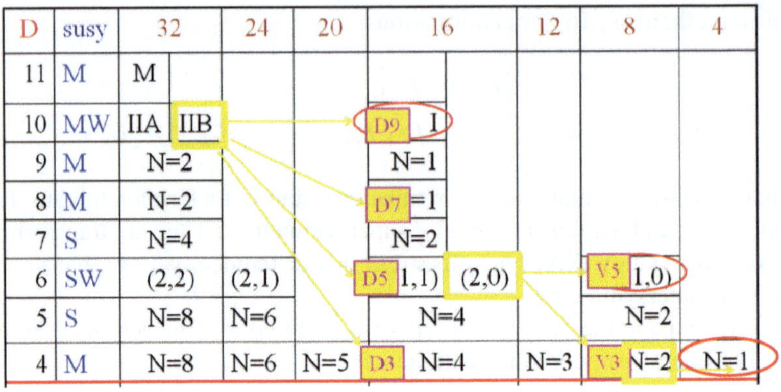

Fig. 1.1 The IIB supergravities have solutions denoted as D9, D7, D5, D3. The (2, 0) theory has solutions V5, V3 and $N = 2$ supergravity has a $N = 1$ solution. The *red circles* indicate the basic super-Maxwell theories that we started from in the bottom-up approach and are obtained as maximal p theories

There are 16 ϵ and 16 ζ symmetries, and the remainders of the rigid Poincaré transformations in $D = 10$ lead to shift symmetries for the 6 scalars ϕ^I. We can compare this with the usual formulation of the $\mathcal{N} = 4$, $D = 4$ super-Maxwell theory:

$$S_{\text{Maxw}} = \int \mathrm{d}^4x \Big(-\tfrac{1}{4} F_{\mu\nu} F^{\mu\nu} - \bar{\psi}_i \partial\!\!\!/ \psi^i - \tfrac{1}{8} \partial_\mu \varphi_{ij} \partial^\mu \varphi^{ij} \Big). \tag{1.42}$$

$F_{\mu\nu}$ is the field strength of the vector field, the ψ_i are 4 Majorana spinors, written as Weyl spinors using the notations $\psi^i = \frac{1}{2}(1 + \gamma_*)\psi^i$ and $\psi_i = \frac{1}{2}(1 - \gamma_*)\psi_i$. The 6 scalar fields are here represented as antisymmetric tensors φ_{ij}, with

$$\varphi^{ij} \equiv (\varphi_{ij})^* = -\tfrac{1}{2} \varepsilon^{ijk\ell} \varphi_{k\ell}. \tag{1.43}$$

One can find (1.42) and the transformation laws as the $\alpha = 0$ part of (1.40) and (1.36), by making some identifications. The scalars ϕ^I representing the 6 remaining coordinates in $D = 10$ according to (1.33) are divided in two triplets ϕ_a and ϕ_{a+3} and we identify

$$\alpha \varphi_{ij} = \phi_a \beta^a_{ij} - \mathrm{i} \phi_{a+3} \alpha^a_{ij}\,, \qquad a = 1, 2, 3, \tag{1.44}$$

where α^a_{ij} and β^a_{ij} are the Gliozzi–Scherk–Olive 4×4 matrices [23, 24]. These are also used to identify the $D = 10$ Majorana–Weyl spinor λ introduced in (1.33), with the 4 Majorana spinors ψ^i. This is done with the $D = 10$ gamma matrix representation

$$\Gamma^\mu = \gamma^\mu \otimes \mathbb{1}_8, \quad \Gamma^a = \gamma_* \otimes \begin{pmatrix} 0 & \beta^a \\ -\beta^a & 0 \end{pmatrix}, \quad \Gamma^{a+3} = \gamma_* \otimes \begin{pmatrix} 0 & i\alpha^a \\ i\alpha^a & 0 \end{pmatrix},$$
$$C_{10} = C_4 \otimes \begin{pmatrix} 0 & \mathbb{1}_4 \\ \mathbb{1}_4 & 0 \end{pmatrix}, \qquad \Gamma_* = \gamma_* \otimes \begin{pmatrix} \mathbb{1}_4 & 0 \\ 0 & -\mathbb{1}_4 \end{pmatrix}, \tag{1.45}$$

where C_{10} and C_4 are the charge conjugation matrices (for notation, see [1]) in 10 and 4 dimensions, and γ^μ are the $D = 4$ gamma matrices. In this basis, λ is decomposed as

$$\lambda = \begin{pmatrix} \psi^i \\ \psi_i \end{pmatrix}. \tag{1.46}$$

With these identifications, the $\alpha = 0$ part of (1.40) agrees with (1.42). Since the action (1.40) is invariant to all orders in α under the 16 + 16 supersymmetries, it gives the fully consistent deformation of the $\mathcal{N} = 4,\ D = 4$ gauge multiplet. It has both type of supersymmetries: ordinary SUSY and VA-type supersymmetry. It can be written in the usual 4-dimensional notations using the translations (1.44) and (1.46), but the $D = 10$ formulation is much simpler.

1.5.4 Worldvolume Theory in AdS Background

In order to make progress for $\mathcal{N} = 4,\ D = 4$ supergravity, we would need the deformed gauge multiplet with the superconformal symmetries. The extra VA symmetries are not of the type of S-supersymmetry. Inspiration may come from old work [25–27] where the worldvolume theories of branes were considered in an AdS background, leading to a superconformal theory on the brane. The AdS backgrounds exist only in particular dimensions and extensions, corresponding to the fact that the superconformal theories also only exist for particular cases as explained at the end of Sect. 1.3, see Table 1.7. These actions on the brane are of the form

$$\begin{aligned} S_{cl} &= S_{\rm DBI} + S_{\rm WZ}, \\ S_{\rm DBI} &= -\int d^{p+1}\sigma \sqrt{-\det\left(g^{\rm ind}_{\mu\nu} + F_{\mu\nu}\right)}, \\ g^{\rm ind}_{\mu\nu} &= \partial_\mu X^M \partial_\nu X^N G_{MN}, \end{aligned} \tag{1.47}$$

where G_{MN} denotes the AdS $\times$ sphere metric that is a solution of the embedding theory. The theory has then rigid symmetries inherited from the solution. These are the AdS isometries and the isometries of the sphere and the corresponding supersymmetries. The brane theory has as in Sect. 1.5.2 the worldvolume general coordinate transformations and kappa symmetries as local symmetries. After gauge fixing these, the remaining (global) symmetries appear as conformal symmetries on the brane. The fermionic ones are then ϵ ordinary supersymmetry and η special supersymmetry. Hence this is very similar to the appearance of ordinary and VA type supersymmetries in our new work [19]. This gives us a hope to obtain an all-order deformation of gauge multiplet theories with superconformal symmetries in the cases where the superalgebras exist, which includes the D3 brane with $\mathcal{N} = 4$, $D = 4$ supersymmetry.

1.6 Conclusions

Superconformal symmetry has been used as a tool for constructing classical actions of supergravity. Also higher-derivative terms can be constructed with superconformal tensor calculus [16, 28–32]. Quantum calculations show that there are unknown relevant properties of supergravity theories. We have investigated the possibility that (broken) superconformal symmetry be such an extra quantum symmetry [18]. The non-existence of (broken) superconformal-invariant counterterms and anomalies for $\mathcal{N} = 4$, $D = 4$ supergravity could in that case explain 'miraculous' vanishing results. However, we do not have a systematic knowledge of which higher-derivative supergravity actions can be invariant under supersymmetry at all orders in derivatives.

In order to get more insight, we have been looking to gauge multiplets in global supersymmetry [19]. We first considered a perturbative approach, i.e. constructing actions and transformation laws order by order in a dimensionful parameter α, which can be related to the string coupling constant. Starting from Dp brane actions in $D = 10$ we can construct DBI-type actions that have ordinary supersymmetry plus VA-type supersymmetry with $16 + 16$ components. They are related to IIB supergravity, and thus exist for $p = 9, 7, 5, 3, \ldots$, leading to global supersymmetry actions for gauge multiplets in $p + 1$ dimensions. For $p = 3$ this is the deformation of $\mathcal{N} = 4$, $D = 4$ with higher order derivatives. One can also start from the iib theory in $D = 6$. Also in that case DBI-VA actions (related to objects called vector branes or 'V-branes' [22]) with $8 + 8$ supersymmetries. This leads e.g. to the deformation of $D = 4$, $\mathcal{N} = 2$ vector multiplets. We hope that insight in these new constructions can lead also to supergravity actions using the superconformal methods.

Acknowledgments Most results in this paper are obtained in collaboration with E. Bergshoeff, F. Coomans, R. Kallosh, S. Ferrara and C. S. Shahbazi, and part of the talk has been prepared in collaboration with R. Kallosh. We thank S. Bellucci and other organizers of this topical meeting, which was very stimulating. This work was supported in part by the FWO–Vlaanderen, Project No. G.0651.11, and in part by the Interuniversity Attraction Poles Programme initiated by the Belgian Science Policy (P7/37).

References

1. D.Z. Freedman, A. Van Proeyen, *Supergravity* (Cambridge University Press, Cambridge, 2012)
2. J. Strathdee, Extended Poincaré supersymmetry. Int. J. Mod. Phys. **A2**, 273 (1987)
3. F. Cordaro, P. Fré, L. Gualtieri, P. Termonia, M. Trigiante, N = 8 gaugings revisited: an exhaustive classification. Nucl. Phys. **B532**, 245–279 (1998). arXiv:hep-th/9804056

4. H. Nicolai, H. Samtleben, Compact and noncompact gauged maximal supergravities in three dimensions. JHEP **04**, 022 (2001). arXiv:hep-th/0103032
5. B. de Wit, H. Samtleben, M. Trigiante, Magnetic charges in local field theory. JHEP **09**, 016 (2005). arXiv:hep-th/0507289
6. E. Bergshoeff, M. de Roo, B. de Wit, Extended conformal supergravity. Nucl. Phys. **B182**, 173 (1981)
7. M. de Roo, Matter coupling in N = 4 supergravity. Nucl. Phys. **B255**, 515 (1985)
8. W. Nahm, Supersymmetries and their representations. Nucl. Phys. **B135**, 149 (1978)
9. J.W. van Holten, A. Van Proeyen, N = 1 supersymmetry algebras in d = 2, 3, 4 mod. 8. J. Phys. **A15**, 3763 (1982)
10. R. D'Auria, S. Ferrara, M.A. Lledó, V.S. Varadarajan, Spinor algebras. J. Geom. Phys. **40**, 101–128 (2001). arXiv:hep-th/0010124
11. E. Bergshoeff, M. de Roo, B. de Wit, Conformal supergravity in ten dimensions. Nucl. Phys. **B217**, 489 (1983)
12. Z. Bern, J. Carrasco, L.J. Dixon, H. Johansson, D. Kosower et al., Three-loop superfiniteness of $\mathcal{N} = 8$ supergravity. Phys. Rev. Lett. **98**, 161303 (2007). arXiv:hep-th/0702112
13. Z. Bern, J. Carrasco, L.J. Dixon, H. Johansson, R. Roiban, The ultraviolet behavior of $\mathcal{N} = 8$ supergravity at four loops. Phys. Rev. Lett. **103**, 081301 (2009). arXiv:0905.2326 [hep-th]
14. Z. Bern, S. Davies, T. Dennen, Y.-T. Huang, Absence of three-loop four-point divergences in $\mathcal{N} = 4$ supergravity. Phys. Rev. Lett. **108**, 201301 (2012). arXiv:1202.3423 [hep-th]
15. S. Cecotti, S. Ferrara, Supersymmetric Born-Infeld Lagrangians. Phys. Lett. **B187**, 335 (1987)
16. B. de Wit, S. Katmadas, M. van Zalk, New supersymmetric higher-derivative couplings: full $N = 2$ superspace does not count! JHEP **1101**, 007 (2011). arXiv:1010.2150 [hep-th]
17. W. Chemissany, S. Ferrara, R. Kallosh, C. Shahbazi, N = 2 supergravity counterterms, off and on shell. JHEP **12**, 089 (2012). arXiv:1208.4801 [hep-th]
18. S. Ferrara, R. Kallosh, A. Van Proeyen, Conjecture on hidden superconformal symmetry of N = 4 supergravity. Phys. Rev. **D87**, 025004 (2013). arXiv:1209.0418 [hep-th]
19. E. Bergshoeff, F. Coomans, R. Kallosh, C. Shahbazi, A. Van Proeyen, Dirac-Born-Infeld-Volkov-Akulov and deformation of supersymmetry. JHEP **08**, 100 (2013) arXiv:1303.5662 [hep-th]
20. E. Bergshoeff, M. Rakowski, E. Sezgin, Higher derivative super-Yang-Mills theories. Phys. Lett. **B185**, 371 (1987)
21. M. Aganagic, C. Popescu, J.H. Schwarz, Gauge invariant and gauge fixed D-brane actions. Nucl. Phys. **B495**, 99–126 (1997). arXiv:hep-th/9612080
22. E.A. Bergshoeff, F. Riccioni, Heterotic wrapping rules. JHEP **1301**, 005 (2013). arXiv:1210.1422 [hep-th]
23. L. Brink, J.H. Schwarz, J. Scherk, Supersymmetric Yang-Mills theories. Nucl. Phys. **B121**, 77 (1977)
24. F. Gliozzi, J. Scherk, D.I. Olive, Supersymmetry, supergravity theories and the dual spinor model. Nucl. Phys. **B122**, 253–290 (1977)
25. P. Claus, R. Kallosh, A. Van Proeyen, M 5-brane and superconformal (0, 2) tensor multiplet in 6 dimensions. Nucl. Phys. **B518**, 117–150 (1998). arXiv:hep-th/9711161
26. P. Claus, R. Kallosh, J. Kumar, P.K. Townsend, A. Van Proeyen, Conformal theory of M2, D3, M5 and D1+D5 branes. JHEP **06**, 004 (1998). arXiv:hep-th/9801206
27. P. Claus, M. Derix, R. Kallosh, J. Kumar, P.K. Townsend, A. Van Proeyen, Black holes and superconformal mechanics. Phys. Rev. Lett. **81**, 4553–4556 (1998). arXiv:hep-th/9804177
28. K. Hanaki, K. Ohashi, Y. Tachikawa, Supersymmetric completion of an R^2 term in five-dimensional supergravity. Prog. Theor. Phys. **117**, 533 (2007). arXiv:hep-th/0611329
29. E.A. Bergshoeff, J. Rosseel, E. Sezgin, Off-shell D = 5, N = 2 Riemann squared supergravity. Class. Quant. Grav. **28**, 225016 (2011). arXiv:1107.2825 [hep-th]
30. E. Bergshoeff, F. Coomans, E. Sezgin, A. Van Proeyen, Higher derivative extension of 6D chiral gauged supergravity. JHEP **1207**, 011 (2012). arXiv:1203.2975 [hep-th]
31. M. Ozkan, Y. Pang, Supersymmetric completion of Gauss-Bonnet combination in five dimensions. JHEP **1303**, 158 (2013). arXiv:1301.6622 [hep-th]
32. M. Ozkan, Y. Pang, All off-shell R^2 invariants in five dimensional $\mathcal{N} = 2$ supergravity. JHEP **1308**, 042 (2013) arXiv:1306.1540 [hep-th]

Chapter 2
Constitutive Relations, Off Shell Duality Rotations and the Hypergeometric Form of Born-Infeld Theory

Paolo Aschieri, Sergio Ferrara and Stefan Theisen

We review equivalent formulations of nonlinear and higher derivatives theories of electromagnetism exhibiting electric-magnetic duality rotations symmetry. We study in particular on shell and off shell formulations of this symmetry, at the level of action funcitonals as well as of equations of motion. We prove the conjecture that the action functional leading to Born-Infeld nonlinear electromagnetism, that is duality rotation invariant off shell and that is known to be a root of an algebraic equation of fourth order, is a hypergeometric function.

2.1 Introduction

Electric-magnetic duality is a symmetry of Maxwell electromagnetism and also, as remarked by Schrödinger [1], of the nonlinear theory of electromagnetism proposed by Born and Infeld [2]. This symmetry does not leave the Lagrangians invariant, only

P. Aschieri (✉)
Dipartimento di Scienze e Tecnologie Avanzate, Università del Piemonte Orientale, INFN, Sezione di Torino, Gruppo collegato di Alessandria, Viale T. Michel 11, 15121 Alessandria, Italy
e-mail: aschieri@to.infn.it

S. Ferrara
Physics Department, Theory Unit, CERN, CH 1211, 23 Geneva, Switzerland

S. Ferrara
INFN, Laboratori Nazionali di Frascati, Via Enrico Fermi 40, I-00044 Frascati, Italy

S. Ferrara
Department of Physics and Astronomy, University of California Los Angeles, Los Angeles, CA 90095-1547, USA
e-mail: sergio.ferrara@cern.ch

S. Theisen
Max-Planck-Institut für Gravitationsphysik, Albert-Einstein-Institut, 14476 Golm, Germany
e-mail: stefan.theisen@aei.mpg.de

S. Bellucci (ed.), *Breaking of Supersymmetry and Ultraviolet Divergences in Extended Supergravity*, Springer Proceedings in Physics 153,
DOI: 10.1007/978-3-319-03774-5_2,

the equations of motion, and therefore it is not immediately detectable. This symmetry was subsequently discovered to be present in extended supergravity theories [3–6]. In [4] the first example of a noncompact duality rotation group was considered, it is due to scalar fields transforming nonlinearly under duality rotations. These results triggered further investigations in the general structure of self-dual theories. In particular the symplectic formalism for nonlinear electromagnetism coupled to scalar and fermion fields was initiated in [7], there the duality groups were shown to be subgroups of noncompact symplectic groups (compact groups being recovered in the absence of scalar fields). Also nonlinear theories admit noncompact duality symmetry, a most studied example is Born-Infeld electrodynamics coupled to axion and dilaton fields [8]. A relevant aspect of Born-Infeld theory [9] is that the spontaneous breaking of $N = 2$ rigid supersymmetry to $N = 1$ can lead to a Goldstone vector multiplet whose action is the supersymmetric and self-dual Born-Infeld action [10, 11]. Higher supersymmetric Born-Infeld type actions are also self-dual and related to spontaneous supersymmetry breakings in field theory [12–15] and in string theory [16, 17].

Another recent motivation for the renewed study of duality symmetry is due to its relevance for investigating the structure of possible counterterms in extended supergravity. After the explicit computations that showed the 3-loop UV finiteness of $N = 8$ supergravity [18], an explanation based on $E_{7(7)}$ duality symmetry was provided [19–22]. Furthermore duality symmetry arguments have also been used to suggest all loop finiteness of $N = 8$ supergravity [23, 24]. Related to these developments, in [25] a proposal on how to implement duality rotation invariant counterterms in a corrected action $S[F]$ leading to a self-dual theory was put forward under the name of "deformed twisted self-duality conditions". The proposal (renamed "nonlinear twisted self-duality conditions") was further elaborated in [26] and [27]; see also [28], and [29–31], for the supersymmetric extensions and examples. The proposal encompasses theories that depend nonlinearly on the field strength F and also on the partial derivative terms $\partial F, \partial\partial F, \ldots$. That is why we speak of *nonlinear and higher derivatives theories*.

The proposal is equivalent to a formulation of self-dual theories using auxiliary fields studied in [32] and [33] in case of nonlinear electromagnetism without higher derivatives of the field strength. This coincidence has been brought to light in a recent paper [34]. In [35] two of us presented a systematic and general study of the different formulations of $U(1)$ gauge theories and of self-dual ones. This lead to a closed form expression of the duality invariant action functional describing Born-Infeld theory.

Before outlining the content of the present work let us recall the notion of *constitutive relations*. A nonlinear and higher derivative electromagnetic theory is determined by defining, eventually implicitly, the relation between the electric field strength F (given by the electric field $\vec{E}$ and the magnetic induction $\vec{B}$) and the magnetic field strength G (given by the magnetic field $\vec{H}$ and the electric displacement $\vec{D}$). We call *constitutive relations* the relations defining G in terms of F or vice versa. Different constitutive relations determine different $U(1)$ gauge theories.

In this paper we first review and clarify the relations between constitutive relations and action functionals in nonlinear and higher derivative electromagnetism. Then we provide a pedagogical analysis of the "deformed twisted self duality conditions" and introduce the action functional $\mathcal{I}[T^-, \overline{T^-}]$ obtained via a Legendre transformation from the usual $S[F]$ action functional in the field strength F. All theories defined via an action functional $S[F]$ and having duality symmetry have a formulation via an action functional $\mathcal{I}[T^-, \overline{T^-}]$ that is off shell invariant under duality rotations.

We then further study the different formulations of the constitutive relations of nonlinear and higher derivatives electromagnetism and then of self-dual theories. These different formulations are all equivalent *on shell*. Finally we prove the conjecture formulated in [35] concerning the hypergeometric function expression of the functional $\mathcal{I}$ of Born-Infeld theory. The proof uses Cauchy residue theorem in order to show that the hypergeometric function satisfies the algebraic quartic equation characterizing the functional $\mathcal{I}$.

2.2 U(1) Duality Rotations in Nonlinear and Higher Derivatives Electromagnetism

2.2.1 Action Functionals from Equations of Motion

Nonlinear and higher derivatives electromagnetism is described by the equations of motion

$$\partial_\mu \widetilde{F}^{\mu\nu} = 0, \tag{2.1}$$

$$\partial_\mu \widetilde{G}^{\mu\nu} = 0, \tag{2.2}$$

$$\widetilde{G}^{\mu\nu} = h^{\mu\nu}[F, \lambda]. \tag{2.3}$$

The first two simply state that the 2-forms F and G are closed, $dF = dG = 0$, indeed $\widetilde{F}^{\mu\nu} \equiv \frac{1}{2}\varepsilon^{\mu\nu\rho\sigma}F_{\rho\sigma}$, $\widetilde{G}^{\mu\nu} \equiv \frac{1}{2}\varepsilon^{\mu\nu\rho\sigma}G_{\rho\sigma}$ (with $\varepsilon^{0123} = 1$). The last set $\widetilde{G}^{\mu\nu} = h^{\mu\nu}[F, \lambda]$, where λ is the dimensionful parameter typically present in a nonlinear theory,[1] are the constitutive relations. They specify the dynamics and determine the magnetic field strength G as a functional in terms of the electric field strength F, and, vice versa, determine F in term of G, indeed F and G should be treated on equal footing in (2.1)–(2.3). The square bracket notation $h^{\mu\nu}[F, \lambda]$ stems from the possible dependence of $h^{\mu\nu}$ on derivatives of F.

Since in general we consider curved background metrics $g_{\mu\nu}$, it is convenient to introduce the $*$-Hodge operator; on an arbitrary antisymmetric tensor $F_{\mu\nu}$ it is defined by

$$^*F_{\mu\nu} = \frac{1}{2\sqrt{g}} g_{\mu\alpha} g_{\nu\beta}\, \varepsilon^{\alpha\beta\rho\sigma} F_{\rho\sigma} = \frac{1}{\sqrt{g}} \widetilde{F}_{\mu\nu}, \tag{2.4}$$

[1] Nonlinear and higher derivatives theories of electromagnetism admit one (or more) dimensionful coupling constant(s) λ.

where $g = -\det(g_{\mu\nu})$, and it squares to minus the identity. The constitutive relations (2.3) implicitly include also a dependence on the background metric $g_{\mu\nu}$ and for example in case of usual electromagnetism they read $G_{\mu\nu} = {}^*F_{\mu\nu} = \frac{1}{\sqrt{g}}\widetilde{F}_{\mu\nu}$, while for Born-Infeld theory,

$$S_{BI} = \frac{1}{\lambda}\int d^4x\,\sqrt{g}\Big(1 - \sqrt{1 + \frac{1}{2}\lambda F^2 - \frac{1}{16}\lambda^2(F^*F)^2}\,\Big), \tag{2.5}$$

where $F^2 = FF = F_{\mu\nu}F^{\mu\nu}$ and $F^*F = F_{\mu\nu}{}^*F^{\mu\nu}$, they read

$$G_{\mu\nu} = \frac{{}^*F_{\mu\nu} + \frac{1}{4}\lambda(F^*F)\,F_{\mu\nu}}{\sqrt{1 + \frac{1}{2}\lambda F^2 - \frac{1}{16}\lambda^2(F^*F)^2}}. \tag{2.6}$$

The constitutive relations (2.3) define a nonlinear and higher derivatives extension of electromagnetism because we require that setting $\lambda = 0$ in (2.3) we recover usual electromagnetism: $G_{\mu\nu} = {}^*F_{\mu\nu}$.

We now recall [35] that in the general nonlinear case (where the constitutive relations do not involve derivatives of F) the equations of motion (2.1)–(2.3) can always be obtained from a variational principle provided they satisfy the integrability conditions

$$\frac{\partial h^{\mu\nu}}{\partial F_{\rho\sigma}} = \frac{\partial h^{\rho\sigma}}{\partial F_{\mu\nu}}. \tag{2.7}$$

These conditions are necessary in order to obtain (2.3) from an action $S[F] = \int d^4x\mathcal{L}(F)$. Indeed if[2] $h^{\mu\nu} = 2\frac{\partial\mathcal{L}}{\partial F_{\mu\nu}}$ then (2.7) trivially holds.

In order to show that (2.7) is also a sufficient condition we recall that the field strength $F_{\mu\nu}(x)$ locally is a map from spacetime to $\mathbb{R}^6$ (with coordinates $F_{\mu\nu}$, $\mu < \nu$). We assume $h^{\mu\nu}(F,\lambda)$ to be well defined functions on $\mathbb{R}^6$ or more generally on an open submanifold $M \subset \mathbb{R}^6$ that includes the origin ($F_{\mu\nu} = 0$) and that is a star shaped region w.r.t. the origin (e.g. a 6-dimensional ball or cube centered in the origin).

Then condition (2.7) states that the 1-form $\hbar = h^{\mu\nu}dF_{\mu\nu}$, is closed, and hence, by Poincaré lemma, exact on M; we write $\hbar = d\mathcal{L}$. We have $\mathcal{L}(F) - \mathcal{L}(0) = \int_\gamma \hbar$ for any curve $\gamma(c)$ of coordinates $\gamma_{\mu\nu}(c)$ such that $\gamma_{\mu\nu}(0) = 0$ and $\gamma_{\mu\nu}(1) = F_{\mu\nu}$. In particular, choosing the straight line from the origin to the point with coordinates $F_{\mu\nu}$, and setting $S = \int d^4x\,\mathcal{L}(F)$, we immediately conclude:

Under the integrability conditions (2.7) locally the equations of motion of nonlinear electromagnetism (2.1)–(2.3) can be obtained from the action

$$S = \frac{1}{2}\int d^4x \int_0^1 dc\, cF\widetilde{G}_c, \tag{2.8}$$

[2] The factor 2 is due to the convention $\frac{\partial F_{\rho\sigma}}{\partial F_{\mu\nu}} = \delta^\mu_\rho\delta^\nu_\sigma$ adopted in [7] and in the review [37]. It will be used throughout the paper.

where $\widetilde{G}_c = \frac{1}{c} h(cF, \lambda)$.

One can also consider the more general case of nonlinear and higher derivatives electromagnetism. Here too if the theory is obtained from an action functional $S[F]$ then we have

$$S[F] = \frac{1}{2} \int d^4x \int_0^1 dc\, F h[cF, \lambda], \tag{2.9}$$

that we simply rewrite $S = \frac{1}{2} \int d^4x \int_0^1 dc\, cF\widetilde{G}_c$.

Proof Consider the one parameter family of actions $S_c[F] = \frac{1}{c^2} S[cF]$. Deriving with respect to c we obtain

$$-c\frac{\partial S_c}{\partial c} = 2S_c - \int d^4x\, F \frac{\delta S_c[F]}{\delta F}, \tag{2.10}$$

i.e. $-c\frac{\partial S_c}{\partial c} = 2S_c - \frac{1}{2} \int d^4x\, F\widetilde{G}_c$. It is easy to see that $S_c = \frac{1}{2c^2} \int d^4x \int_0^c dc'\, c' F \widetilde{G}_{c'}$ is the primitive with the correct behaviour under rescaling of c and F. We conclude that $\frac{1}{c^2} S[cF] = \frac{1}{2c^2} \int d^4x \int_0^c dc'\, c' F \widetilde{G}_{c'}$, and setting $c = 1$ we complete the proof.

An equivalent form of the expression $S = \frac{1}{2} \int d^4x \int_0^1 dc\, cF\widetilde{G}_c$ has been considered, for self-dual theories, in [27] and called reconstruction identity. It has been used to reconstruct the action S from equations of motion with duality rotation symmetry in examples with higher derivatives of F.

2.2.2 Conditions for $U(1)$ Duality Rotation Symmetry of the Equations of Motion

Nonlinear and higher derivatives electromagnetism admits $U(1)$ duality rotation symmetry if given a field configuration F, G that satisfies (2.1)–(2.3) then the rotated configuration

$$\begin{pmatrix} F' \\ G' \end{pmatrix} = \begin{pmatrix} \cos\alpha & -\sin\alpha \\ \sin\alpha & \cos\alpha \end{pmatrix} \begin{pmatrix} F \\ G \end{pmatrix}, \tag{2.11}$$

that is trivially a solution of $\partial_\mu \widetilde{F}^{\mu\nu} = 0$, $\partial_\mu \widetilde{G}^{\mu\nu} = 0$, satisfies also $\widetilde{G}'_{\mu\nu} = h_{\mu\nu}[F', \lambda]$, so that F', G' is again a solution of the equations of motion. If we consider an infinitesimal duality rotation, $F \to F + \Delta F$, $G \to G + \Delta G$ then condition $\widetilde{G}'_{\mu\nu} = h_{\mu\nu}[F', \lambda]$ reads $\Delta \widetilde{G}_{\mu\nu} = \int d^4x \frac{\delta h_{\mu\nu}}{\delta F_{\rho\sigma}} \Delta F^{\rho\sigma}$, i.e., $\widetilde{F}_{\mu\nu} = -\int d^4x \frac{\delta h_{\mu\nu}}{\delta F_{\rho\sigma}} G^{\rho\sigma}$, that we simply rewrite

$$\widetilde{F}_{\mu\nu} = -\int d^4x \frac{\delta \widetilde{G}_{\mu\nu}}{\delta F_{\rho\sigma}} G^{\rho\sigma}. \tag{2.12}$$

It is straightforward to check that electromagnetism and Born-Infeld theory satisfy (2.12).

If the theory is obtained from an action functional $S[F]$ (in the field strength F and its derivatives) then (2.3) is given by

$$\widetilde{G}^{\mu\nu} = 2\frac{\delta S[F]}{\delta F_{\mu\nu}}. \tag{2.13}$$

In particular it follows that

$$\frac{\delta \widetilde{G}^{\mu\nu}}{\delta F_{\rho\sigma}} = \frac{\delta \widetilde{G}^{\rho\sigma}}{\delta F_{\mu\nu}}, \tag{2.14}$$

hence the duality symmetry condition (or self-duality condition) (2.12) equivalently reads $\widetilde{F}_{\mu\nu} = -\int d^4x \, \frac{\delta \widetilde{G}_{\rho\sigma}}{\delta F_{\mu\nu}} G^{\rho\sigma}$. Now writing $\widetilde{F}_{\mu\nu} = \frac{\delta}{\delta F_{\mu\nu}} \frac{1}{2}\int d^4x \, F_{\rho\sigma}\widetilde{F}^{\rho\sigma}$ we equivalently have

$$\frac{\delta}{\delta F_{\mu\nu}} \int d^4x \, (F\widetilde{F} + G\widetilde{G}) = 0, \tag{2.15}$$

where $F\widetilde{F} = F_{\rho\sigma}\widetilde{F}^{\rho\sigma}$ and similarly for $G\widetilde{G}$. We require this condition to hold for any field configuration F (i.e. off shell of (2.1), (2.2)) and hence we obtain the Noether-Gaillard-Zumino (NGZ) self-duality condition[3]

$$\int d^4x \, (F\widetilde{F} + G\widetilde{G}) = 0. \tag{2.16}$$

The vanishing of the integration constant is determined for example by the condition $G = {}^*F$ for weak and slowly varying fields, i.e. by the condition that in this regime the theory is approximated by usual electromagnetism.

We also observe that the NGZ self-duality condition (2.16) is equivalent to the invariance of $S^{inv} = S - \frac{1}{4}\int d^4x \, F\widetilde{G}$, indeed under a rotation (2.11) with infinitesimal parameter α we have $S^{inv}[F'] - S^{inv}[F] = -\frac{\alpha}{4}\int d^4x \, (F\widetilde{F} + G\widetilde{G}) = 0$.

From this relation it follows that the action $S[F]$ is not invariant under duality rotations and that under a finite transformation (2.11) we have

$$S[F'] = S[F] + \frac{1}{8}\int d^4x \left(\sin(2\alpha)(F\widetilde{F} - G\widetilde{G}) - 4\sin^2(\alpha) F\widetilde{G} \right). \tag{2.17}$$

Thus the action changes by the integral of the four-forms $F \wedge F - G \wedge G$ and $F \wedge G$, that, on the equations of motion $dF = dG = 0$ ((2.1) and (2.2)), are locally total derivatives. This is a sufficient condition for the transformation (2.11) with $\widetilde{G}^{\mu\nu} = 2\frac{\delta S[F]}{\delta F_{\mu\nu}}$ to be a symmetry.

We summarize the results thus far obtained: The self-duality condition (2.16) is off shell of (2.1) and (2.2) but on shell of (2.3). The action functional $S[F]$ provides

[3] Note that (2.16) (the integrated form of the more restrictive self-duality condition $F\widetilde{F} + G\widetilde{G}$) also follows in a straightforward manner by repeating the passages in [7] but with G the functional derivatives of the action rather than the partial derivatives of the lagrangian [13, 37]. This makes a difference for nonlinear theories which also contain terms with derivatives of F.

a variational principle for the equation (2.3) and under duality rotations changes by a term that on shell of (2.1) and (2.2) is a total derivative.

2.2.3 Off Shell Formulation of Duality Symmetry

We here provide an off shell formulation of duality symmetry by considering a Legendre transformation to new variables. The new action functional, off shell of the equations of motion (2.1)–(2.3), is invariant under duality rotations. This formulation allows for a classification of duality rotation symmetric theories (an ackward task using the action functional $S[F]$).

An example of functional invariant under duality rotations is provided by the Hamiltonian action functional. Indeed the Hamiltonian itself (and more generally the energy-momentum tensor) of duality symmetric theories is invariant under duality rotations [7].[4] The problem with the Hamiltonian formulation is however the lack of explicit Lorentz covariance.

These observations lead to consider a Legendre transformation of $S[F]$ to an action functional in new variables that transform linearly under duality rotations and that are Lorentz tensors.

The action $S[F]$ determines the submanifold of equations $\widetilde{G} = 2\frac{\partial S[F]}{\partial F}$ in the plane of coordinates F and G. Equivalently, defining the complex self-dual combinations

$$F^- = \frac{1}{2}(F - i{}^*F), \tag{2.18}$$

$$G^- = \frac{1}{2}(G - i{}^*G), \tag{2.19}$$

and their complex conjugates $\overline{F^-} = F^+ = \frac{1}{2}(F + i{}^*F)$, $\overline{G^-} = G^+ = \frac{1}{2}(G + i{}^*G)$, the action $S[F^-, \overline{F^-}] = S[F]$ determines the submanifold of equations $G^- = -2i\frac{\partial S}{\partial F^-}$ in the plane of coordinates F^-, G^-.

We want to retrieve this submanifold using the new variables

$$T^- = F^- - iG^-, \tag{2.20}$$

$$\overline{T^+} = F^- + iG^- = 2F^- - T^-, \tag{2.21}$$

and their complex conjugates $\overline{T^-} = F^+ + iG^+$, $T^+ = F^+ - iG^+ = 2F^+ - \overline{T^-}$. These variables transform simply with a phase under duality rotations, $T^{-\,\prime} =$

---

[4] In a general nonlinear theory the Hamiltonian depends on the magnetic field $\vec{B}$ and on the electric displacement $\vec{D} = \frac{\delta S[F]}{\delta \vec{E}}$, that rotate into each other under the duality (2.11), $\begin{pmatrix} \vec{B}' \\ -\vec{D}' \end{pmatrix} = \begin{pmatrix} \cos\alpha & -\sin\alpha \\ \sin\alpha & \cos\alpha \end{pmatrix} \begin{pmatrix} \vec{B} \\ -\vec{D} \end{pmatrix}$. Since the composite fields $\vec{B}^2 + \vec{D}^2$ and $(\vec{B} \times \vec{D})^2$ are duality invariant, Hamiltonians that depend upon these combinations and their derivatives are trivially duality invariant and lead to duality symmetric theories.

$e^{i\alpha}T^-$, $\overline{T^+}' = e^{-i\alpha}\overline{T^+}$; hence the formulation of a theory symmetric under duality rotations should be facilitated in these variables. The change of variables $(F^-, G^-) \rightarrow (T^-, \overline{T^+})$ is achieved by first changing from G^- to T^-, then by a Legendre transformation so that T^- become the independent variables and F^- the dependent ones, and finally changing further the dependent variables from F^- to $\overline{T^+} = 2F^- - iT^-$. Schematically we undergo the following chain of change of variables

$$(F^-, G^-) \longrightarrow (F^-, T^-) \longrightarrow (T^-, F^-) \longrightarrow (T^-, \overline{T^+}). \tag{2.22}$$

More explicitly the equation in the (F^-, G^-)-plane

$$G^- = -2i\frac{\partial S}{\partial F^-} \tag{2.23}$$

is equivalent to the equation in the (F^-, T^-)-plane

$$T^- = \frac{\partial U}{\partial F^-} \tag{2.24}$$

where $U[F^-, F^+] = -2S[F^-, F^+] + \frac{1}{2}\int d^4x\sqrt{g}(F^{-2} + F^{+2})$. Furthermore, via Legendre transform, this last equation is equivalent to the equation in the (T^-, F^-)-plane

$$F^- = \frac{\delta V}{\delta T^-} \tag{2.25}$$

where $V[T^-, \overline{T^-}] = -U[F^-, F^+] + \int d^4x\sqrt{g}\ (T^-F^- + \overline{T^-}F^+)$. Finally we rewite this equation in the $(T^-, \overline{T^+})$-plane as

$$\overline{T^+} = \frac{\delta \mathcal{I}}{\delta T^-} \tag{2.26}$$

where

$$\mathcal{I}[T^-, \overline{T^-}] = 2V[T^-, \overline{T^-}] - \frac{1}{2}\int d^4x\sqrt{g}\ (T^{-2} + \overline{T^-}^2). \tag{2.27}$$

In conclusion, as pioneered in [33] (in the case of no derivatives of F in the action), we have that $\mathcal{I}[T^-, \overline{T^-}]$ and $S[F]$ are related by

$$\begin{aligned}\frac{1}{4}\mathcal{I}[T^-, \overline{T^-}] = S[F] + \int d^4x\sqrt{g}\Big(&\frac{1}{2}T^-F^- - \frac{1}{8}T^{-2} \\ &- \frac{1}{4}F^{-2} + \frac{1}{2}\overline{T^-}F^+ - \frac{1}{8}\overline{T^-}^2 - \frac{1}{4}F^{+2}\Big).\end{aligned} \tag{2.28}$$

The equations of motion (2.26) were studied in [25], where a nontrivial example of a self-dual action with an infinite number of derivatives of the field strength F is considered (see also the generalizations in the appendix of [35]).

Let's now study duality rotations. We consider F to be the elementary fields and let $S[F]$ be the action functional of a self-dual theory. Under infinitesimal duality rotations (2.11), $F \to F + \Delta F = F - \alpha G$, $G \to G + \Delta G = G + \alpha F$ we have (since $T^- = F^- - \frac{2}{\sqrt{g}}\frac{\delta S}{\delta F^-}$) that $T^- \to T^- + \Delta T^- = T^- - i\alpha T^-$. We calculate the variation of (2.28) under duality rotations. After a little algebra we see that

$$\begin{aligned}\Delta\mathcal{I} &= \mathcal{I}[T^- + \Delta T^-, \overline{T^-} + \Delta\overline{T^-}] - \mathcal{I}[T^-, \overline{T^-}]\\ &= S[F + \Delta F] - S[F] + \frac{\alpha}{4}\int d^4x\sqrt{g}\,(G\widetilde{G} - F\widetilde{F})\\ &= -\frac{\alpha}{4}\int d^4x\sqrt{g}\,(G\widetilde{G} + F\widetilde{F}) = 0\end{aligned} \tag{2.29}$$

where we used that $S[F + \Delta F] - S[F] = \int d^4x\, \frac{\delta S}{\delta F}\Delta F = -\frac{\alpha}{2}\int d^4x\, \widetilde{G}G$, and the self-duality conditions (2.16). Hence $\mathcal{I}$ is invariant under duality rotations.

Vice versa, we can consider T^-, $\overline{T^-}$ to be the elementary fields and assume $\mathcal{I}[T^-, \overline{T^-}]$ to be duality invariant. Then from $2F^- - T^- = \frac{1}{\sqrt{g}}\frac{\delta\mathcal{I}[T^-,\overline{T^-}]}{\delta\overline{T}^-_{\mu\nu}}$, and $F^- - iG^- = T^-$, it follows that under the infinitesimal rotation $T^- \to T^- + \Delta T^- = T^- - i\alpha T^-$ we have $F \to F + \Delta F = F - \alpha G$, $G \to G + \Delta G = G + \alpha F$, and from (2.29) we recover the self-duality conditions (2.16) for the action $S[F]$.

This shows the equivalence betweeen the $S[F]$ and the $\mathcal{I}[T^-, \overline{T^-}]$ formulations of self-dual constitutive relations. Hence the deformed twisted self-duality condition proposal originated in the context of supergravity counterterms is actually the general framework needed to discuss self-dual theories starting from a variational principle.

We stress that while we needed to use the equations of motion in order to verify that the action $S[F]$ leads to a duality rotation symmetric theory, we do not need to use the equations of motion in order to verify that the action $\mathcal{I}[T^-, \overline{T^-}]$ is duality invariant. In the formulation with the $\mathcal{I}[T^-, \overline{T^-}]$ action functional duality rotations are an off shell symmetry provided that $\mathcal{I}[T^-, \overline{T^-}]$ is invariant under $T^- \to e^{i\alpha}T^-$ and $\overline{T^-} \to e^{-i\alpha}\overline{T^-}$.

2.3 Constitutive Relations without Self-Duality

2.3.1 The $\mathcal{N}$ and $\mathcal{M}$ Matrices

More insights in the constitutive relations (2.3) can be obtained if we restrict our study to the wide subclass that can be written as

$$^*G_{\mu\nu} = \mathcal{N}_2\, F_{\mu\nu} + \mathcal{N}_1\, {}^*F_{\mu\nu}, \tag{2.30}$$

where $\mathcal{N}_2$ is a real scalar field, while $\mathcal{N}_1$ is a real pseudo-scalar field (i.e., it is not invariant under parity, or, if we are in curved spacetime, it is not invariant under an orientation reversing coordinate transformation). As usual in the literature we set

$$\mathcal{N} = \mathcal{N}_1 + i\mathcal{N}_2. \tag{2.31}$$

In nonlinear theories $\mathcal{N}$ depends on the field strength F, and in higher derivatives theories also on derivatives of F, we have therefore in general a functional dependence $\mathcal{N} = \mathcal{N}[F, \lambda]$. Furthermore $\mathcal{N}$ is required to satisfy $\mathcal{N} \to -i$ in the limit $\lambda \to 0$ so that we recover classical electromagnetism when the coupling constant(s) $\lambda \to 0$, or otherwise stated, in the weak and slowly varying field limit, i.e., when we discard higher powers of F and derivatives of F. Since $\mathcal{N}_2 \to -1$ for $\lambda \to 0$, $\mathcal{N}_2$, at least for sufficiently weak and slowly varying fields, is invertible. It follows that the constitutive relation (2.30) is equivalent to the more duality symmetric one

$$\begin{pmatrix} {}^*F \\ {}^*G \end{pmatrix} = \begin{pmatrix} 0 & -1 \\ 1 & 0 \end{pmatrix} \mathcal{M} \begin{pmatrix} F \\ G \end{pmatrix} \tag{2.32}$$

where the matrix $\mathcal{M}$ is given by

$$\mathcal{M}(\mathcal{N}) = \begin{pmatrix} 1 & -\mathcal{N}_1 \\ 0 & 1 \end{pmatrix} \begin{pmatrix} \mathcal{N}_2 & 0 \\ 0 & \mathcal{N}_2^{-1} \end{pmatrix} \begin{pmatrix} 1 & 0 \\ -\mathcal{N}_1 & 1 \end{pmatrix} = \begin{pmatrix} \mathcal{N}_2 + \mathcal{N}_1 \mathcal{N}_2^{-1} \mathcal{N}_1 & -\mathcal{N}_1 \mathcal{N}_2^{-1} \\ -\mathcal{N}_2^{-1} \mathcal{N}_1 & \mathcal{N}_2^{-1} \end{pmatrix}. \tag{2.33}$$

The matrix $\mathcal{M}$ is symmetric and sympletic and $\mathcal{M} \to -1$ for $\lambda \to 0$. Actually any such matrix is of the kind (2.33) with $\mathcal{N}_1$ real and $\mathcal{N}_2$ real and negative.

Finally, in order to really treat on equal footing the electric and magnetic field strengths F and G, we should consider functionals $N_1[F, G, \lambda]$ and $N_2[F, G, \lambda]$ such that the constitutive relations ${}^*G = N_2[F, G, \lambda]\, F + N_1[F, G, \lambda]\, {}^*F$ are equivalent to (2.30), i.e., such that on shell of these relations, $N_1[F, G, \lambda] = \mathcal{N}_1[F, \lambda]$ and $N_2[F, G, \lambda] = \mathcal{N}_2[F, \lambda]$. Henceforth, with slight abuse of notation, from now on the $\mathcal{N}, \mathcal{N}_1, \mathcal{N}_2$ fields in (2.30)–(2.33) will in general be functionals of both F and G.

We now reverse the argument that led from (2.30) to (2.32). We consider constitutive relations of the form

$$\begin{pmatrix} {}^*F \\ {}^*G \end{pmatrix} = \begin{pmatrix} 0 & -1 \\ 1 & 0 \end{pmatrix} \mathcal{M}[F, G, \lambda] \begin{pmatrix} F \\ G \end{pmatrix} \tag{2.34}$$

that treat on equal footing F and G, and where $\mathcal{M} = \mathcal{M}[F, G, \lambda]$ is now an *arbitrary* real 2×2 matrix (with scalar entries $\mathcal{M}_{ij}$). We require $\mathcal{M} \to -1$ for $\lambda \to 0$, so that we recover classical electromagnetism when the coupling constant $\lambda \to 0$. A priory (2.34) is a set of 12 real equations, twice as much as those present in the constitutive relations (2.30). We want only 6 of these 12 relations to be independent in order to be able to determine G in terms of independent fields F (or equivalently F in terms of independent fields G). Only in this case the constitutive relations are well given. In [35] we show,

Proposition 1 *The constitutive relations (2.34) with* $\mathcal{M}[F, G, \lambda]|_{\lambda=0} = -1$ *are well given if and only if on shell of (2.34) the matrix* $\mathcal{M}[F, G, \lambda]$ *is symmetric and symplectic. They are equivalent to the constitutive relations (2.30) provided that on shell the relation between the* $\mathcal{M}$ *and* $\mathcal{N}$ *matrices is as in (2.33).*

Notice that off shell of (2.34) the matrix $\mathcal{M}$ does not need to be symmetric and symplectic. This is what happens with Schrödinger's formulation of Born-Infeld theory (see (2.50) and comments thereafter).

2.3.2 Schrödinger's Variables

Following Schrödinger [1, 36] it is fruitful to consider the complex variables

$$T = F - iG, \quad \overline{T} = F + iG. \tag{2.35}$$

The transition from the real to the complex variables is given by the symplectic and unitary matrix $\mathcal{A}^t$ where

$$\mathcal{A} = \frac{1}{\sqrt{2}}\begin{pmatrix} 1 & 1 \\ -i & i \end{pmatrix}, \quad \mathcal{A}^{-1} = \mathcal{A}^{\dagger}. \tag{2.36}$$

The equation of motions in these variables read $dT = 0$, with constitutive relations obtained applying the matrix $\mathcal{A}^t$ to (2.34):

$$\begin{pmatrix} {}^*T \\ {}^*\overline{T} \end{pmatrix} = -i\begin{pmatrix} 1 & 0 \\ 0 & -1 \end{pmatrix} \mathcal{A}^t \mathcal{M} \overline{\mathcal{A}} \begin{pmatrix} T \\ \overline{T} \end{pmatrix}, \tag{2.37}$$

where $\mathcal{A}^t \mathcal{M} \overline{\mathcal{A}}$, on shell of (2.37), is complex symplectic and pseudounitary w.r.t the metric $\left(\begin{smallmatrix} 1 & 0 \\ 0 & -1 \end{smallmatrix}\right)$, i.e. it belongs to $Sp(2, \mathbb{C}) \cap U(1, 1) = SU(1, 1)$. It is also Hermitian and negative definite. These properties uniquely characterize the matrices $\mathcal{A}^t \mathcal{M} \overline{\mathcal{A}}$ as the matrices

$$\begin{pmatrix} -\sqrt{1+\tau\overline{\tau}} & -i\tau \\ i\overline{\tau} & -\sqrt{1+\tau\overline{\tau}} \end{pmatrix} \tag{2.38}$$

where $\tau = \tau[T, \overline{T}]$ is a complex field that depends on $T, \overline{T}$ and possibly also their derivatives. We then see that the constitutive relations (2.37) are equivalent to the equations

$${}^*T_{\mu\nu} = i\sqrt{1+\tau\overline{\tau}}\, T_{\mu\nu} - \tau\, \overline{T}_{\mu\nu}. \tag{2.39}$$

In conclusion the most general set of equations in the T variables that is well defined in the sense that it allows to express $G = \frac{i}{2}(T + \overline{T})$ in terms of $F = \frac{1}{2}(T + \overline{T})$ as in (2.30) (equivalently F in terms of G) is equivalent, on shell, to the equations (2.39) for a given $\tau = \tau[T, \overline{T}]$. In this sense equations (2.39) are the most general way of defining constitutive relations of electromagnetism. The constitutive relations (2.30) are determined by the complex function $\mathcal{N}$ (depending on F, G and their derivatives $\mathcal{N} = \mathcal{N}[F, G]$) the equivalent constitutive relations (2.39) are determined by the complex function τ (depending on $T, \overline{T}$ and their derivatives $\tau = \tau[T, \overline{T}]$).

2.4 Schrödinger's Approach to Self-Duality Conditions

In the previous section we have clarified the structure of the constitutive relations for an arbitrary nonlinear theory of electromagnetism. The theory can also be with higher derivatives of the field strength because the complex field $\mathcal{N}$, or equivalently the matrix $\mathcal{M}$ in (2.34) of (pseudo)scalar entries, can depend also on derivatives of the electric and magnetic field strengths F and G.

We now further examine the constitutive relations for theories that satisfy the NGZ self-duality condition

$$F\widetilde{F} + G\widetilde{G} = 0, \tag{2.40}$$

i.e., $\overline{T}\widetilde{T} = 0$, or equivalently,

$$\overline{T}{}^*T = 0. \tag{2.41}$$

We multiply (2.39) by ${}^*T$ and obtain

$$-T^2 = i\sqrt{1 + \tau\overline{\tau}}\, T{}^*T \tag{2.42}$$

It is convenient to consider modulus and argument of these complex scalar expressions. Setting

$$T^2 = |T^2|e^{i\alpha} \tag{2.43}$$

from (2.42) we have

$$T{}^*T = |T{}^*T|ie^{i\varphi} \tag{2.44}$$

We also contract (2.39) with ${}^*\overline{T}^{\mu\nu}$ and obtain $-T\overline{T} = -\tau\overline{T{}^*T}$ that implies

$$|\tau| = \frac{T\overline{T}}{|T{}^*T|}. \tag{2.45}$$

Use of (2.42) then gives the moduli relations

$$|T^2|^2 = |T\,{}^*T|^2 + (T\overline{T})^2. \tag{2.46}$$

The constitutive relations (2.39) can also be rewritten using the chiral variables $T^{\pm} = T \pm i\,{}^*T$, they read

$$T^+_{\mu\nu} = te^{i\varphi}\overline{T^-}_{\mu\nu} \tag{2.47}$$

where $t = \frac{T\overline{T}}{|T^2|+|T\,{}^*T|}$. In order to obtain the explicit relation between the ratio $|\tau| = T\overline{T}/|T\,{}^*T|$ and t we calculate

$$|T^{-2}|(1-t^2) = \frac{1}{2}(|T^2| + |T\,{}^*T|)(1-t^2) = |T\,{}^*T|, \tag{2.48}$$

multiply this last equality by $|\tau|$ and obtain

$$(1-t^2)|\tau| = 2t. \tag{2.49}$$

Example 1 Linear electromagnetism ($G = {}^*F$) corresponds to $|\tau| = 0$. *Born-Infeld nonlinear theory* satisfies the relations

$${}^*T_{\mu\nu} = -\frac{T^2}{T\,{}^*T}T_{\mu\nu} - \frac{\lambda}{8}(T\,{}^*T)\,\overline{T}_{\mu\nu} \tag{2.50}$$

as remarked by Schrödinger [1], see [36] for a clear account in nowadays notations. Comparison with (2.39) shows that, on shell of (2.50) and (2.41), i.e. using (2.42) and (2.45), $\frac{T^2}{T\,{}^*T} = i\sqrt{1+\tau\overline{\tau}}$ and $\tau = \frac{\lambda}{8}T\,{}^*T$. Hence Born-Infeld theory is determined by

$$|\tau| = \frac{\lambda}{8}|T\,{}^*T|. \tag{2.51}$$

Schrödinger's formulation of Born-Infeld theory uses the freedom, dicussed in Proposition 1, of considering a matrix $\mathcal{M}$ that off shell of (2.34) is not symmetric and symplectic. Indeed the term $\frac{T^2}{T\,{}^*T}$ is not pure imaginary off shell. Schrödinger's elegant variational principle formulation of Born-Infeld constitutive relations is also due to this freedom. Defining the "Lagrangian" $\Upsilon(T) = \frac{4\,T^2}{T\,{}^*T}$ we have that (2.50) is equivalent to

$$\lambda^*\overline{T}^{\mu\nu} = \frac{\partial}{\partial T_{\mu\nu}}\Upsilon(T). \tag{2.52}$$

2.5 Nonlinear Theories without Higher Derivatives

We now consider theories (possibly in curved spacetime) that depend only on the (pseudo) scalars F^2 and F^*F, or T^{-2} and $\overline{T^{-}}^2$. Since the action functional $\mathcal{I}[T^-, \overline{T^-}]$ studied in Sect. 2.2.3 and the scalar field t defined in (2.47) are duality invariant, and under a duality of angle α we have the phase rotation $T^{-2} \to e^{2i\alpha}T^{-2}$, we conclude that $\mathcal{I}$ and t depend only on the modulus of T^{-2}, hence $\mathcal{I} = \mathcal{I}[T^-, \overline{T^-}]$ and $t = t[T^-, \overline{T^-}]$ simplify to

$$\mathcal{I} = \frac{1}{\lambda}\int d^4x\sqrt{g}\, I(u), \quad t = t(u), \tag{2.53}$$

where $I(u)$ is an adimensional scalar function, and the variable u is defined by

$$u \equiv 2\lambda|T^{-2}| = \lambda(|T^2| + |T^*T|). \tag{2.54}$$

Similarly, the constitutive relations (2.26) simplify to

$$T^{+\mu\nu} = \frac{1}{\lambda}\frac{\partial I}{\partial \overline{T^-}_{\mu\nu}} = \frac{1}{\lambda}\frac{dI}{du}\frac{\partial u}{\partial \overline{T^-}_{\mu\nu}}, \tag{2.55}$$

and comparison with (2.47) leads to

$$t = 2\frac{dI}{du}. \tag{2.56}$$

(Hint: calculate $\frac{\partial u^2}{\partial \overline{T^-}_{\mu\nu}}$ and use $T^{-2} = |T^{-2}|e^{i\varphi}$).

2.5.1 *Born-Infeld Nonlinear Theory*

We determine the scalar field $t = t(u) = 2\frac{dI}{du}$ in case of Born-Infeld theory. This is doable thanks to Schrödinger's formulation (2.50) of Born-Infeld theory, that explicitly gives $|\tau| = \frac{\lambda}{8}|T^*T|$, see (2.51). Then from (2.48) we have

$$|\tau| = \frac{1}{16}u(1 - t^2), \tag{2.57}$$

and recalling (2.49) we obtain [34, 35]

$$(1 - t^2)^2 u = 32t. \tag{2.58}$$

Now in the limit $u \to 0$, i.e., $\lambda \to 0$, we see from the definition of t that $t \to 0$. The function $t = t(u)$ defining Born-Infeld theory is then given by the unique positive root of the fourth order polynomial equation (2.58) that has the limit $t \to 0$ for $\lambda \to 0$. Explicitly,

$$t = \frac{1}{\sqrt{3}}\Big(\sqrt{1+s+s^{-1}} - \sqrt{2-s-s^{-1}+\frac{24\sqrt{3}}{u\sqrt{1+s+s^{-1}}}}\,\Big), \tag{2.59}$$

where

$$s = \frac{1}{u}\Big(216\,u + 12\sqrt{3}\sqrt{108+u^2}\,u + u^3\Big)^{\frac{1}{3}}. \tag{2.60}$$

2.5.2 The Hypergeometric Function and its Hidden Identity

In [26] the action functional $\mathcal{I}$ and the function $t(u)$ corresponding to the Born-Infeld action were found via an iterative procedure order by order in λ (or equivalently in u). The first coefficients of the power series expansion of $t(u)$ were recognized to be those of a generalized hypergeometric function, leading to the conclusion

$$\begin{aligned} t(u) &= \frac{u}{32}\,{}_3F_2\Big(\frac{1}{2},\frac{3}{4},\frac{5}{4};\frac{4}{3},\frac{5}{3};-\frac{u^2}{3^3\cdot 2^2}\Big),\\ &= \frac{2u}{32}\sum_{k=0}^{\infty}\frac{(4k+1)!}{(3k+2)!k!}\Big(-\frac{u^2}{4^5}\Big)^k \end{aligned} \tag{2.61}$$

and, integrating (2.56),

$$I(u) = 6\bigg(1 - {}_3F_2\Big(-\frac{1}{2},-\frac{1}{4},\frac{1}{4};\frac{1}{3},\frac{2}{3};-\frac{u^2}{3^3\cdot 2^2}\Big)\bigg). \tag{2.62}$$

In [35] we conjectured, and checked up to order $O(u^{1000})$, that the expansion in power series of u of the closed form expression of $t(u)$ derived in (2.59), (2.60) coincides with the power series expansion in (2.61).

We here present a proof by showing that the power series in (2.61) satisfies the quartic equation (2.58). We consider the generic power series

$$t = \sum_{m=1}^{\infty} a_m u^m \tag{2.63}$$

with the initial condition $t = \mathcal{O}(u)$ for $u \to 0$, and determine the coefficients a_m so as to satisfy the quartic equation (2.58). The initial condition $t = \mathcal{O}(u)$ for $u \to 0$ is compatible with (2.58), indeed from (2.58) we see that for $u \to 0$ we have $t = \frac{u}{32}$.

We extend the variables t and u to the complex plane so that use of Cauchy's residue theorem gives

$$a_m = \frac{1}{2\pi i} \oint_{C_0} t u^{-m-1} du \tag{2.64}$$

We next calculate from (2.58) the differential

$$du = 32 d \frac{t}{(1-t^2)^2} = 32 \frac{1+3t^2}{(1-t^2)^3} dt, \tag{2.65}$$

and observe that, since for $u \to 0$, $t = \mathcal{O}(u)$, infinitesimal closed paths surrounding the origin of the complex u-plane are mapped to infinitesimal ones surrounding the origin of the complex t-plane (that we still denote C_0). We hence obtain

$$\begin{aligned} a_m &= \frac{32}{2\pi i} \oint_{C_0} \frac{t+3t^3}{(1-t^2)^3} \frac{(1-t^2)^{2m+2}}{(32t)^{m+1}} dt \\ &= \frac{1}{32^m\, 2\pi i} \oint_{C_0} (t^{-m} + 3t^{2-m})(1-t^2)^{2m-1} dt \\ &= \frac{1}{32^m\, 2\pi i} \oint_{C_0} (t^{-m} + 3t^{2-m}) \sum_{n=0}^{2m-1} (-1)^n t^{2n} \binom{2m-1}{n} dt \\ &= \frac{1}{32^m} \sum_{n=0}^{2m-1} (-1)^n \binom{2m-1}{n} (\delta_{2n-m+1,0} + 3\delta_{2n-m+3,0}). \end{aligned} \tag{2.66}$$

We see that only the coefficients a_m with m odd are nonvanishing, setting $m = 2k+1$ we have

$$\begin{aligned} a_{2k+1} &= \frac{(-1)^k}{32^{2k+1}} \left[\binom{4k+1}{k} - 3 \binom{4k+1}{k-1} \right] \\ &= (-1)^k \frac{2}{32^{2k+1}} \frac{(4k+1)!}{(3k+2)!k!} \end{aligned} \tag{2.67}$$

that proves the conjecture.

As a corollary we have that the hypergeometric function in (2.61)

$$\mathfrak{F}(u^2) \equiv {}_3F_2\left(\frac{1}{2}, \frac{3}{4}, \frac{5}{4}; \frac{4}{3}, \frac{5}{3}; -\frac{u^2}{3^3 \cdot 2^2}\right) = 2 \sum_{k=0}^{\infty} \frac{(4k+1)!}{(3k+2)!k!} \left(-\frac{u^2}{4^5}\right)^k \tag{2.68}$$

has the closed form expression $\mathfrak{F}(u^2) = \frac{32}{u}t(u)$ where $t(u)$ is given in (2.59), (2.60), and, because of (2.58), that it satisfies the "hidden" identity

$$\mathfrak{F}(u^2) = \left(1 - \frac{u^2}{4^5}\mathfrak{F}(u^2)^2\right)^2. \tag{2.69}$$

2.5.3 General Nonlinear Theory

Since Born-Infeld theory is singled out by setting $|\tau| = \frac{\lambda}{8}|T\,{}^*T|$, and Maxwell theory by setting $|\tau| = 0$ (Example 1), it is convenient to describe a general nonlinear theory without higher derivatives by setting

$$|\tau| = \frac{\lambda}{8}|T\,{}^*T| f(u)/u \tag{2.70}$$

where $f(u)$ is a positive function of u. We require the theory to reduce to electromagnetism in the weak field limit, i.e., ${}^*G_{\mu\nu} = -F + o(F)$ for $F \to 0$. Then we have $T^- = \mathcal{O}(F)$, $T^+ = o(F)$, $u = \mathcal{O}(F^2)$. Hence from (2.47) we obtain $\lim_{u\to 0} t = 0$. Moreover from (2.49), $r = \mathcal{O}(t)$ and from $r = \frac{1}{16}f(u)(1-t^2)$ (that follows from (2.70) and (2.48)) $f = \mathcal{O}(t)$. Hence the theory reduces to electromagnetism in the weak field limit if and only if $\lim_{u\to 0} f(u) = 0$.

From $r = \frac{1}{16}f(u)(1-t^2)$ (that follows from (2.70) and (2.48)) and (2.49) we obtain that the composite function $t(f(u))$ satisfies the fourth order polynomial equation

$$(1-t^2)^2 f(u) = 32t, \tag{2.71}$$

so that $t(f(u))$ is obtained with the substitution $u \to f(u)$ in (2.59) and (2.60), or in (2.61).

More explicitly, generalizing the results of Example 1, we conclude, as in [35], that the constitutive relations à la Schrödinger

$${}^*T_{\mu\nu} = -\frac{T^2}{T\,{}^*T}T_{\mu\nu} - \frac{\lambda}{8}\frac{f(u)}{u}\,(T\,{}^*T)\,\overline{T}_{\mu\nu}, \tag{2.72}$$

are (on shell) equivalent to the constitutive relations (deformed twisted self-duality conditions)

$$T^{+\mu\nu} = \frac{1}{2\lambda}t(f(u))\,\frac{\partial u}{\partial \overline{T^-}_{\mu\nu}}, \tag{2.73}$$

where $t(f(u))$ satisfies the quartic equation (2.71), and we recall that $u = 2\lambda|T^{-2}| = \lambda(|T^2| + |T\,{}^*T|)$.

In other words the appearence of the quartic equation (2.71) is a general feature of the relation between the constitutive relations (2.72) and (2.73), it appears for any self-dual theory and it is not only a feature of the Born-Infeld theory.

Acknowledgments P.A. acknowledges the hospitality of Max-Planck-Institut für Gravitationsphysik Albert-Einstein-Institut during commencement of the present work, and the hospitality of Galileo Galilei Institute during its completion. This work is supported by the ERC Advanced Grant no. 226455, Supersymmetry, Quantum Gravity and Gauge Fields (SUPERFIELDS).

References

1. E. Schrödinger, Contributions to Born's new theory of the electromagnetic field. Proc. Roy. Soc. (London) **A150**, 465 (1935)
2. M. Born, L. Infeld, Foundations of the new field theory. Proc. Roy. Soc. (London) **A144**, 425 (1934)
3. S. Ferrara, J. Scherk, B. Zumino, Algebraic properties of extended supergravity theories. Nucl. Phys. **B121**, 393 (1977)
4. E. Cremmer, J. Scherk, S. Ferrara, SU(4) invariant supergravity theory. Phys. Lett. B **74**, 61 (1978)
5. E. Cremmer, B. Julia, The SO(8) supergravity. Nucl. Phys. B **159**, 141 (1979)
6. E. Cremmer, B. Julia, The N=8 supergravity theory. 1. The lagrangian. Phys. Lett. B **80**, 48 (1978)
7. M.K. Gaillard, B. Zumino, Duality rotations for interacting fields. Nucl. Phys. B **193**, 221 (1981)
8. G.W. Gibbons, D.A. Rasheed, SL(2, R) invariance of nonlinear electrodynamics coupled to an axion and a dilaton. Phys. Lett. B **365**, 46 (1996). arXiv:hep-th/9509141
9. J. Bagger, A. Galperin, A new goldstone multiplet for partially broken supersymmetry. Phys. Rev. D **55**, 1091 (1997). arXiv:hep-th/9608177
10. S. Deser, R. Puzalowski, Supersymmetric nonpolynomial vector multiplets and causal propagation. J. Phys. **A13**, 2501 (1980)
11. S. Cecotti, S. Ferrara, Supersymmetric Born-Infeld lagrangians. Phys. Lett. B **187**, 335 (1987)
12. S.V. Ketov, A manifestly N=2 supersymmetric Born-Infeld action. Mod. Phys. Lett. A **14**, 501 (1999)
13. S.M. Kuzenko, S. Theisen, Supersymmetric duality rotations. JHEP **0003**, 034 (2000). arXiv: hep-th/0001068
14. S.M. Kuzenko, S. Theisen, Nonlinear selfduality and supersymmetry. Fortsch. Phys **49**, 273 (2001). arXiv:hep-th/0007231
15. S. Bellucci, E. Ivanov, S. Krivonos, Towards the complete N=2 superfield Born-Infeld action with partially broken N=4 supersymmetry. Phys. Rev. **D64**, 025014 (2001). arXiv:hep-th/0101195
16. M. Rocek, A.A. Tseytlin, Partial breaking of global D = 4 supersymmetry, constrained superfields, and three-brane actions. Phys. Rev. D **59**, 106001 (1999). arXiv:hep-th/9811232
17. S.V. Ketov, Born-Infeld-goldstone superfield actions for gauge-fixed D-5 and D-3 branes in 6d. Nucl. Phys. B **553**, 250 (1999). arXiv:hep-th/9812051
18. Z. Bern, J.J.M. Carrasco, L.J. Dixon, H. Johansson, R. Roiban, Manifest ultraviolet behavior for the three-loop four-point amplitude of N=8 supergravity. Phys. Rev. D **78**, 105019 (2008). arXiv:0808.4112 [hep-th]
19. J. Broedel, L.J. Dixon, R**4 counterterm and E(7)(7) symmetry in maximal supergravity. JHEP **1005**, 003 (2010). arXiv:0911.5704 [hep-th]
20. H. Elvang, M. Kiermaier, Stringy KLT relations, global symmetries, and $E_{7(7)}$ violation. JHEP **1010**, 108 (2010). arXiv:1007.4813 [hep-th]

21. N. Beisert, H. Elvang, D.Z. Freedman, M. Kiermaier, A. Morales, S. Stieberger, E7(7) constraints on counterterms in N=8 supergravity. Phys. Lett. B **694**, 265 (2010). arXiv:1009.1643 [hep-th]
22. G. Bossard, P.S. Howe, K.S. Stelle, JHEP **1101**, 020 (2011). arXiv:1009.0743 [hep-th]
23. R. Kallosh, $E_{7(7)}$ symmetry and finiteness of N=8 supergravity. JHEP **1203**, 083 (2012). arXiv: 1103.4115 [hep-th]
24. R. Kallosh, $\mathcal{N} = 8$ counterterms and $E_{7(7)}$ current conservation. JHEP **1106**, 073 (2011). arXiv:1104.5480 [hep-th]
25. G. Bossard, H. Nicolai, Counterterms vs. dualities. JHEP **1108**, 074 (2011). arXiv:1105.1273 [hep-th]
26. J.J.M. Carrasco, R. Kallosh, R. Roiban, Covariant procedures for perturbative non-linear deformation of duality-invariant theories. Phys. Rev. D **85**, 025007 (2012). arXiv:1108.4390 [hep-th]
27. W. Chemissany, R. Kallosh, T. Ortin, Born-Infeld with higher derivatives. Phys. Rev. D **85**, 046002 (2012). arXiv:1112.0332 [hep-th]
28. J. Broedel, J.J.M. Carrasco, S. Ferrara, R. Kallosh, R. Roiban, N=2 supersymmetry and U(1)-duality. Phys. Rev. D **85**, 125036 (2012). arXiv:1202.0014 [hep-th]
29. S.M. Kuzenko, Nonlinear self-duality in N = 2 supergravity. JHEP **1206**, 012 (2012). arXiv: 1202.0126 [hep-th]
30. S.M. Kuzenko, Duality rotations in supersymmetric nonlinear electrodynamics revisited. J. High Energy Phys. **2013**(3), 1–21 (2013)
31. E. Ivanov, O. Lechtenfeld, B. Zupnik, Auxiliary superfields in N=1 supersymmetric self-dual electrodynamics. JHEP **1305**, 133 (2013). arXiv:1303.5962 [hep-th]
32. E.A. Ivanov, B.M. Zupnik, N=3 supersymmetric Born-Infeld theory. Nucl. Phys. B **618**, 3 (2001). arXiv:hep-th/0110074
33. E.A. Ivanov, B.M. Zupnik, New approach to nonlinear electrodynamics: dualities as symmetries of interaction. Phys. Atom. Nucl. **67**, 2188 (2004) [Yad. Fiz. 67 (2004) 2212]. arXiv:hep-th/0303192
34. E.A. Ivanov, B.M. Zupnik, Bispinor auxiliary fields in duality-invariant electrodynamics revisited. Phys. Rev. D **87**(3), 065023 (2013). arXiv:1212.6637 [hep-th]
35. P. Aschieri, S. Ferrara, Constitutive relations and Schroedinger's formulation of nonlinear electromagnetic theories. JHEP **1305**, 087 (2013). arXiv:1302.4737 [hep-th]
36. M. K. Gaillard, B. Zumino, Nonlinear electromagnetic self-duality and Legendre transformations. arXiv:hep-th/9712103
37. P. Aschieri, S. Ferrara, B. Zumino, Duality rotations in nonlinear electrodynamics and in extended supergravity. Riv. Nuovo Cim. **31**, 625 (2008). arXiv:0807.4039 [hep-th]
38. M. Hatsuda, K. Kamimura, S. Sekiya, Electric magnetic duality invariant lagrangians. Nucl. Phys. B **561**, 341 (1999). arXiv:hep-th/9906103

Chapter 3
Vector Branes

Eric A. Bergshoeff and Fabio Riccioni

We show how the recent classification of half-supersymmetric branes of maximal supergravity has a simple group-theoretical characterization in terms of the longest weights of the T-duality representation to which the potentials that couple to these branes belong. We identify the branes of half-maximal supergravity that have Dirac-Born-Infeld-Volkov-Akulov worldvolume dynamics. We show that reducing the branes of ten-dimensional string theory leads to the half-supersymmetric branes in lower dimensions provided we impose simple wrapping rules for these branes. The origin and interpretation of these wrapping rules is discussed.

3.1 Introduction

"Branes", i.e. massive objects with a number of worldvolume and transverse directions, play a crucial role in string theory and M-theory. Historically, the first example of a brane other than a string was the eleven-dimensional supermembrane [1]. An important class of branes are the Dirichlet branes or, shortly, D-branes of ten-dimensional superstring theory [2]. These branes are non-perturbative in the sense that their brane tension scales with the inverse of the string coupling constant. D-branes played a decisive role in the calculation of the entropy of a certain class of black holes [3]. Branes also play a central role in the AdS/CFT correspondence [4] and the brane-world scenario [5, 6].

E. A. Bergshoeff (✉)
Centre for Theoretical Physics, University of Groningen, Nijenborg 4,
9747 AG, Groningen, The Netherlands
e-mail: E.A.Bergshoeff@rug.nl

F. Riccioni
INFN Sezione di Roma, Dipartimento di Fisica, Università di Roma "La Sapienza",
Piazzale Aldo Moro 2, 00185 Roma, Italy
e-mail: Fabio.Riccioni@roma1.infn.it

S. Bellucci (ed.), *Breaking of Supersymmetry and Ultraviolet Divergences in Extended Supergravity*, Springer Proceedings in Physics 153,
DOI: 10.1007/978-3-319-03774-5_3, © Springer International Publishing Switzerland 2014

Much information about branes can be obtained by studying the low-energy approximation of string theory and/or M-theory which is a supergravity theory. For instance, the mere fact that eleven-dimensional supergravity contains a 3-form potential is already indicative of the fact that M-theory contains a membrane since 3-forms naturally couple to membranes. The fact that this membrane is actually a supermembrane which breaks half of the supersymmetry follows from the construction of a kappa-symmetric supermembrane action [1]. Kappa symmetry requires that the worldvolume action describing the dynamics of the brane contains a Nambu-Goto and a Wess-Zumino (WZ) term. The latter describes the coupling of the brane to the potentials of supergravity. A classification of branes therefore necessarily involves a classification of the supergravity potentials.

Due to their different nature it is important to distinguish between branes with more than 2 transverse directions and branes with 2 or less transverse directions. The half-supersymmetric branes with more than 2 transverse directions have been classified a long time ago. We will refer to them collectively as the "standard" branes. The classification of the remaining branes is more subtle and has only recently been obtained [7–9]. We will refer to them as the "non-standard" branes. We call the ones with 2, 1 and 0 transverse directions "defect-branes", "domain-walls" and "space-filling branes", respectively. To summarize:

standard branes: more than 2 transverse directions
non-standard branes: 2,1 or 0 transverse directions

One difference between the standard and non-standard branes is that the non-standard ones are not asymptotically flat. Furthermore, they are only well-defined if one considers multiple brane configurations together with an orientifold. For our purposes it will be enough to consider single brane configurations. Another difference is that the standard branes couple, via the WZ term, to potentials that describe continuous degrees of freedom. For the non-standard branes this is only the case for the defect branes which couple to the dual of the supergravity scalars. Even this case is different from the standard brane case in the sense that the number of dual potentials that fit into a U-duality representation is not equal to the number of physical scalars. From the higher-dimensional point of view the origin of this mismatch is the fact that some of the scalars originate from the higher-dimensional metric for which no dual metric can be defined. The potentials that couple to domain walls can be viewed as dual to a discrete degree of freedom such as a mass parameter or a gauge coupling constant. The space-filling branes are a bit special since they couple to potentials that do not describe any degree of freedom at all.

To verify whether a given potential couples to a half-supersymmetric brane or not we require that a gauge-invariant WZ term can be constructed. This often requires that, besides the embedding scalars, more world-volume potentials are introduced that transform under (some of) the gauge transformations of the supergravity potentials with non-trivial shifts. In this way gauge-invariance of the WZ term can always be achieved but it is not clear whether the newly introduced worldvolume potentials together with the embedding scalars fit into a worldvolume supermultiplet. This so-called "WZ-term requirement" imposes restrictions on the number of half-supersymmetric branes. Using the WZ-term requirement we have found that there is another difference between the standard and non-standard branes. Whereas for standard branes every supergravity potential (and its dual) couples to a half-supersymmetric brane, for the non-standard ones we find that there are less half-supersymmetric branes than there are potentials:

$$\text{\# half-susy standard branes} = \text{\# potentials},$$
$$\text{\# half-susy non-standard branes} < \text{\# potentials}.$$

In the next section we will discuss in more detail the relation between branes and the WZ terms. We will review a so-called "light-cone rule" which provides a simple way, by using a light-cone basis for the T-duality indices, to specify which potentials couple to a half-supersymmetric brane and which do not. In Sect. 3.3 we will show that the light-cone rule has a simple group-theoretical interpretation in terms of a "longest-weight rule" which states that the number of half-supersymmetric branes is equal to the number of longest weights of the T-duality representation to which the potentials in question belong. In Sect. 3.4 we will resolve a puzzle that arises when one investigates the half-supersymmetric branes of half-maximal supergravity. In particular, we will discuss the so-called 'vector branes', i.e. branes whose worldvolume dynamics is described by a single vector super multiplet, and show how these vector branes arise in half-maximal supergravity. Having classified the half-supersymmetric branes of maximal supergravity it is natural to ask how the branes in different dimensions are related to each other via dimensional reduction. In Sect. 3.5 we will show that reducing the branes of ten-dimensional string theory one obtains the half-supersymmetric branes in lower dimensions we just classified provided we impose simple wrapping rules for these branes. In the conclusions, see Sect. 3.6, we will discuss the origin and interpretation of these wrapping rules.

3.2 Branes and Wess-Zumino Terms

It is instructive to first consider the branes of Type IIB string theory. These branes can be analysed by looking at the field content of the low-energy IIB supergravity effective action. This includes not only the propagating fields and their magnetic duals—an $SL(2,\mathbb{R})$ doublet of 2-forms, corresponding to the F1 fundamental string and the D1-brane, the magnetic dual 6-forms, corresponding to the D5-brane and the NS5-brane, and a selfdual 4-form, corresponding to the D3-branes—but also forms of higher rank. These forms can be obtained by imposing the closure of the supersymmetry algebra, and they are a triplet of 8-forms and a quadruplet of 10-forms [10, 11]. In [12] the half-supersymmetric branes associated to these latter fields were derived by looking at the corresponding brane effective action. The result of this analysis is that only two components of the triplet of 8-forms and only two components of the quadruplet of 10-forms are associated to half-supersymmetric branes. A simple explanation of this result can be given by looking at the WZ term in the effective action. Denoting with $A_{8,\alpha\beta}$ the triplet of 8-forms ($\alpha, \beta = 1, 2$ are $SL(2, \mathbb{R})$ doublet indices), gauge invariance implies that such a term must be of the form

$$T^{\alpha\beta}[A_{8,\alpha\beta} + A_{6,(\alpha}\mathcal{F}_{2,\beta)} + \cdots], \tag{3.1}$$

where $T^{\alpha\beta}$ is the 7-brane charge, $A_{6,\alpha}$ are the doublet of 6-forms and $\mathcal{F}_{2,\alpha} = da_{1,\alpha} + A_{2,\alpha}$ are a doublet of world-volume 2-form field-strengths ($a_{1,\alpha}$ are a doublet of world-volume vectors and $A_{2,\alpha}$ are the 2-forms). In order for the effective action to preserve one-half of the supersymmetries, we must impose that the world-volume fields fit in an 8-dimensional 16-supercharge multiplet, that is a vector multiplet (one vector and two scalars). The two scalars are the transverse scalars, while the request that (3.1) contains only one world-volume vector imposes that the charge must be either T^{11} or T^{22} (the third component T^{12} would result in a WZ term containing both components of the doublet of world-volume vectors). The same analysis leads to two 9-branes in the quadruplet of 10-forms. The main lesson of this analysis is that in the IIB theory the number of standard branes is the same as the number of corresponding potentials, while the number of non-standard branes is less than the number of components of the corresponding potentials.

We now move to consider maximally supersymmetric theories in any dimension. A full classification of the potentials of these theories for all dimensions was given in [13, 14] making use of the properties of the very extended Kac-Moody algebra E_{11} [15]. Starting from this result, the study of the half-supersymmetric branes as components of the U-duality representations of the corresponding potentials, based on the analysis of the WZ terms was initiated in [16]. This analysis, completed in [7, 9], shows that as in ten dimensions the number of half-supersymmetric non-standard branes is less than the dimension of the corresponding U-duality representations. Here we are interested in the analogous analysis in terms of representations of the T-duality group. Denoting with E_{11-D} the U-duality group in D dimensions, one has

$$E_{11-D} \supset SO(10-D, 10-D) \times \mathbb{R}^+, \tag{3.2}$$

where $SO(10-D, 10-D)$ is the T-duality group, that will be denoted from now on as $SO(d, d)$, with $d = 10 - D$. Decomposing the U-duality representations under T-duality allows to classify the branes according to the way their tension T scales with respect to the D-dimensional string coupling,

$$\mathrm{T} \sim g_s^{\alpha}, \tag{3.3}$$

where α is related to the $\mathbb{R}^+$ weight. The value of α is always non-positive, and $\alpha = 0$ corresponds to the fundamental branes, while the other branes, with $\alpha < 0$, are non-perturbative objects in string theory.

It turns out that the classification of the potentials associated to branes as representations of SO(d, d) is universal for $\alpha > -4$. For our present purposes, it is enough to consider only fundamental branes ($\alpha = 0$), Dirichlet branes ($\alpha = -1$) and solitonic branes ($\alpha = -2$). The fields with $\alpha = 0$ are a 1-form $B_{1,A}$ in the vector representation of SO(d, d) and a 2-form singlet B_2. The RR fields, with $\alpha = -1$, belong to spinor representations with alternating chirality, and we denote them with $C_{2n+1,a}$ and $C_{2n,\dot{a}}$. The fields with $\alpha = -2$ are D_{D-4}, $D_{D-3,A}$, D_{D-2,A_1A_2}, $D_{D-1,A_1A_3A_4}$ and $D_{D,A_1A_2A_3A_4}$, where sets of indices $A_1 \dots A_n$ are always meant to be antisymmetrised. It turns out that the fields with $\alpha = -3$ are in irreducible tensor-spinor representations.

In [17] the $\alpha = -2$, i.e. solitonic branes were classified by looking at the world-volume field content of the WZ terms. The outcome of that analysis is that the components of the T-duality representations of the $\alpha = -2$ potentials that correspond to branes are obtained from the following "light-cone rule":

We introduce light-like indices $i\pm$, $i = 1, \dots, d$ for SO(d, d). The $\alpha = -2$ fields are then denoted as $D_{D-4+n,i_1\pm \dots i_n\pm}$, with $n = 0, 1, \dots, 4$. The components associated to half-supersymmetric branes are those for which the i's are all different. The number of $(D - 5 + n)$-branes is therefore

$$\binom{d}{n} \times 2^n, \tag{3.4}$$

which is smaller than the dimension of the representation, which is $\binom{2d}{n}$.

As can be deduced from (3.4), there are no solitonic branes with world-volume dimension higher than 6, because they correspond to fields with $n > d$, for which (3.4) clearly gives a vanishing result. The case $n = d$, which can only occur in $D \geq 6$ and always corresponds to a 5-brane, is special because the T-duality representation with d antisymmetric indices of SO(d, d) splits into a selfdual and an anti-selfdual part. Correspondingly, the 2^d branes that come from (3.4) split into 2^{d-1} branes supporting a vector multiplet and 2^{d-1} branes supporting a tensor multiplet. In all the other cases the branes support a world-volume vector multiplet.

The peculiarity of the branes with $\alpha > -4$ is that for each T-duality representation there is always at least one brane that comes from torus dimensional reduction from the 10-dimensional branes (either wrapped or unwrapped along some of the internal directions). There is also an $\alpha = -4$ brane in Type IIB string theory, namely the S-dual of the D9-brane. In D dimensions, this brane wraps along the T^d internal torus to give a spacefilling $(D - 1)$-brane. The potential associated to this brane is the field $F^+_{D,A_1 \dots A_d}$ in the selfdual representation of T-duality with d antisymmetric indices. According to the light-cone rule, the number of branes associated to this potential is given by (3.4) with $n = d$, divided by two because of the selfduality condition. This gives 2^{d-1} branes, which is clearly less than the dimension of the representation, $\frac{1}{2}\binom{2d}{d}$.

Starting from $D = 7$, there are also half-supersymmetric branes with $\alpha \leq -4$ that belong to T-duality representations that do not contain any brane coming from 10 dimensions. These branes will not be discussed here, but their number was obtained by the WZ term analysis in [7, 9].

Before we proceed, it is important to mention that all the results discussed so far have been also obtained in [8] using a different method, namely by counting the real roots of the E_{11} Kac-Moody algebra. In the next section we will show that the light-cone rule discussed above can be replaced by an alternative group-theoretical rule, which we will denominate the "longest-weight rule". This new rule also reveals why the WZ requirement and the E_{11} method give the same result.

3.3 Branes and Weights

The counting of branes that results from the analysis of the WZ terms in [7, 9] has a simple group-theoretical explanation: the components of the U-duality representations of the potentials that correspond to half-supersymmetric branes are those associated to the longest weights [18]. The potentials corresponding to standard branes belong to representations whose weights have all the same length, and this explains why in that case the number of half-supersymmetric branes coincides with the dimension of the representation. On the other hand, the potentials corresponding to non-standard branes belong to representations whose weights have different lengths, and therefore in this case the number of branes is less than the dimension of the representation. As an example one can consider the defect branes, associated to the $(D-2)$-form potentials, which belong to the adjoint representation. The number of such branes is equal to the dimension of the group minus the rank [19], which is the number of roots. Given that the symmetry groups of maximal supergravities are always simply laced, which means that all the roots have the same length, this implies that the roots are the longest weights of the adjoint (the other weights being the Cartan, which have zero length). The longest weights of the U-duality representation precisely correspond to the real roots of E_{11}, and therefore the observation that the branes correspond to the longest weights explains why the WZ analysis of [7, 9] and the E_{11} analysis of [8] give the same result.

We now want to give a characterisation of the length of the various weights within a representation in terms of the so-called "dominant weights". An irreducible representation is denoted in terms of the Dynkin labels of the highest weight. We recall that a weight is defined as the eigenvalue of the Cartan generators in a given representation, and the corresponding eigenvector is called a weight vector. A highest-weight vector is a weight vector annihilated by all the positive-root generators, and the non-zero (i.e. positive) Dynkin labels identify the negative-root generators that do not annihilate the highest-weight vector. As an example we can consider the group

SL(3, $\mathbb{R}$), with simple roots α_1 and α_2. We first consider the fundamental representation, which is the **3**, whose Dynkin labels of the highest weight W^3 are $\boxed{1\,0}$. From this we read the weight $W^3 - \alpha_1$, with Dynkin labels $\boxed{-1\,1}$. The lowest weight of the representation is $W^3 - \alpha_1 - \alpha_2$, with Dynkin labels $\boxed{0\,-1}$. The reader can see that only the highest weight of the **3** has Dynkin labels that are all non-negative. In general one defines a dominant weight as a weight whose Dynkin labels are all non-negative. Clearly a highest weight is a dominant weight, but the opposite is not necessarily true. As we have seen, the **3** of SL(3, $\mathbb{R}$) has only one dominant weight, which is the highest weight. However, this is no longer the case if we consider instead the symmetric product $\mathbf{3} \otimes_S \mathbf{3}$, which is the **6**. This representation has highest weight $\boxed{2\,0}$, but it also contains the weight $\boxed{0\,1}$, which is a dominant weight.

In general, each dominant weight in a representation identifies a set of weights which all have the same length as that dominant weight. We can consider again SL(3, $\mathbb{R}$) as an example. In the case of the **3**, all the weights have the same length, which is the length of the highest weight. In the case of the **6** instead, there are three long weights, one of them being the highest weight $\boxed{2\,0}$, and three short weights, one of them being the second dominant weight $\boxed{0\,-1}$. This implies that the standard branes are associated to potentials belonging to representations that have only one dominant weight (which is the highest weight) while the non-standard branes are associated to potentials that are in representations with more than one dominant weight, and one can count for each representation the number of weights with the same length as each dominant weight. This was done in [18] for the U-duality representations associated to all the non-standard branes in any dimension. Here we want to apply the same analysis of [18] to the representations of the T-duality group SO(d, d) that are associated to branes. We will show that the longest-weight rule of [18] is the same as the light-cone rule reviewed in the previous section.

We first review what are the Dynkin labels of the highest weights of the representations of SO(d, d) that are relevant for our discussion. We are assuming that we are labelling the nodes of the Dynkin diagram of SO(d, d) in the standard way, with the last two nodes (node $d-1$ and node d) corresponding to the two spinor representations. The highest weight of the vector representation is $\boxed{1\,0\,0\ldots0\,0}$, while more generally the highest weight of the representation with n antisymmetric indices ($n < d-1$) has all zero Dynkin labels apart from the nth label, whose value is 1. The highest weight of the representation with $d-1$ antisymmetric indices is $\boxed{0\,0\,0\ldots0\,1\,1}$, and the ones with d antisymmetric indices are $\boxed{0\,0\,0\ldots0\,2\,0}$ (selfdual) and $\boxed{0\,0\,0\ldots0\,0\,2}$ (anti-selfdual). The spinor representations are $\boxed{0\,0\,0\ldots0\,0\,1}$ (chiral, denoted with the index a) and $\boxed{0\,0\,0\ldots0\,1\,0}$ (anti-chiral, denoted with the index $\dot{a}$).

We now discuss the various dominant weights of the T-duality representations associated to the half-supersymmetric branes discussed in the previous section for the different values of α, see (3.3). The $\alpha = 0$, i.e. fundamental, branes correspond to the potentials $B_{1,A}$ (F0-branes) and B_2 (F1-brane). The vector representation clearly has only one dominant weight, which is the highest weight $\boxed{1\,0\,0\ldots0\,0}$. The

F0-branes correspond to the lightlike directions $B_{1,i\pm}$, and their number is $2d$, which is equal to the dimension of the representation. There always is a single F1-brane (i.e. the fundamental string) associated to the T-duality singlet B_2. The $\alpha = -1$ branes, i.e. the Dp-branes, belong to the chiral (p even) and anti-chiral (p odd) representations. These representations have only one dominant weight (i.e. the highest weight) and therefore the number of branes is equal to the dimension of the representation, which is 2^{d-1}.

We then consider the $\alpha = -2$, i.e. solitonic, branes. The discussion for the $(D-5)$-branes and the $(D-4)$-branes is the same as for the fundamental branes, of which they are the magnetic dual. The $(D-3)$-branes correspond to the potentials D_{D-2,A_1A_2}. The dimension of the representation (which is the adjoint of SO(d,d)) is $\binom{2d}{2}$. There are $\binom{d}{2} \times 4$ long weights (i.e. the roots) associated to the dominant weight $\boxed{0\,1\,0\,0\ldots0\,0\,0}$, and d weights of zero length (the Cartan) which means that the dominant weight $\boxed{0\,0\,0\ldots0\,0\,0}$ has multiplicity d. In components, the long weights correspond to $D_{D-2,i_1\pm i_2\pm}$ with $i_1 \neq i_2$, and the short weights to $D_{D-2,i+i-}$, and given that i takes d values, this explains the degeneracy d of the short dominant weight.

The $(D-2)$-branes are associated to the potential $D_{D-1,A_1A_2A_3}$. In the representation of SO(d,d) with three antisymmetric indices, there are $\binom{d}{3} \times 8$ long weights, one of which being the highest weight $\boxed{0\,0\,1\,0\ldots0\,0\,0}$, corresponding to the components $D_{D-1,i_1\pm i_2\pm i_3\pm}$ with i_1, i_2 and i_3 all different. The remaining components are $D_{D-1,i\pm j+j-}$, with $i \neq j$. These are associated to the dominant weight $\boxed{1\,0\,0\,0\ldots0\,0\,0}$, which has multiplicity $(d-1)$ because there are $d-1$ possible values for j once i is fixed. The total number of short weights is $2d(d-1)$. Clearly, the sum of the long weights and the short weights is $\binom{2d}{3}$, which is the dimension of the representation.

The last type of solitonic branes are the $(D-1)$-branes, corresponding to the potentials $D_{D,A_1\ldots A_4}$. The representation has $\binom{d}{4} \times 16$ longest weights, which are of the same length as the highest weight $\boxed{0\,0\,0\,1\ldots0\,0\,0}$. They correspond to the components $D_{D,i_1\pm i_2\pm i_3\pm i_4\pm}$, with the i's all different. The next-to-longest weights correspond to the components of the form $D_{D,i_1\pm i_2\pm j+j-}$, with i_1, i_2 and j all different. The corresponding dominant weight is $\boxed{0\,1\,0\,0\ldots0\,0\,0}$, and its multiplicity is $(d-2)$ because these are the possible choices that can be made for j once i_1 and i_2 are fixed. The total number of next-to-longest weights is $\binom{d}{2} \times 4 \times (d-2)$. Finally, the shortest weights correspond to the components $D_{D,i+i-j+j-}$, with $i \neq j$. They all correspond to the dominant weight $\boxed{0\,0\,0\,0\ldots0\,0\,0}$, with multiplicity $\binom{d}{2}$ because these are all the possible choices for i and j. This gives a total of $\binom{d}{2}$ shortest weights. The sum of the longest, next-to-longest and shortest weights gives the dimension of the representation, which is $\binom{2d}{4}$.

The discussion above is valid if $D \leq 5$. In $D = 6$ the spacefilling branes split into tensor and vector branes, corresponding to the representation with 4 antisymmetric indices of SO$(4,4)$ splitting into selfdual and anti-selfdual. For each of the two irreducible representations the number of longest, next-to-longest and shortest

Table 3.1 This table gives the dominant weights and the number of weights of same length of each dominant weight for the solitonic ($\alpha = -2$) non-standard branes

Branes	Field	Dim. repr.	Dominant weights	Weights
Defect branes	D_{D-2,A_1A_2}	$\binom{2d}{2}$	$\boxed{0100\ldots000}$	$\binom{d}{2} \times 4$
			$d \times \boxed{0000\ldots000}$	d
Domain walls	$D_{D-1,A_1A_2A_3}$	$\binom{2d}{3}$	$\boxed{0010\ldots000}$	$\binom{d}{3} \times 8$
			$(d-1) \times \boxed{1000\ldots000}$	$2d(d-1)$
Space-filling branes	$D_{D,A_1A_2A_3A_4}$	$\binom{2d}{4}$	$\boxed{0001\ldots000}$	$\binom{d}{4} \times 16$
			$(d-2) \times \boxed{0100\ldots000}$	$(d-2)\binom{d}{2} \times 4$
			$\binom{d}{2} \times \boxed{0000\ldots000}$	$\binom{d}{2}$

weights is simply half of what one would get putting $d = 4$ in the analysis above. In $D = 7$ the representation of SO(3, 3) with 4 antisymmetric indices is dualised to the one with two antisymmetric indices. This is consistent with the fact that if one puts $d = 3$ in the formulas above for the longest weights of the representation with four antisymmetric indices one obtains zero, which implies that there are no solitonic spacefilling branes in $D = 7$. Similar considerations apply in $D = 8$ and $D = 9$. The complete result of this analysis for the non-standard solitonic ($\alpha = -2$) is summarised in Table 3.1.

3.4 Vector Branes and Half-Maximal Supergravity

It is well-known that the worldvolume dynamics of the 10D Dirichlet branes of maximal supergravity is governed by a supersymmetric Born-Infeld system with 16 supercharges. For a recent discussion of supersymmetric Born-Infeld systems in the context of finiteness properties of quantum supergravity, see [20]. We will from now on denominate branes whose worldvolume dynamics is governed by a vector multiplet as 'vector branes'. Such branes can be associated with supersymmetric Born-Infeld systems. Examples of $10D$ vector branes of maximal supergravity are the Dirichlet branes and the NS5B brane.

When considering the vector branes of half-maximal supergravity the following puzzle arises. It is well-known that supersymmetric Born-Infeld systems with 16 supercharges can be truncated to Born-Infeld systems with 8 supercharges. One would expect that these truncated worldvolume theories describe the dynamics of some vector branes of half-maximal supergravity. However, when considering the Dirichlet ($\alpha = -1$) branes of half-maximal supergravity one finds that their worldvolume dynamics is not described by a vector multiplet but a hyper multiplet. For

Table 3.2 This table indicates the potentials with $\alpha = 0$ and $\alpha = -2$ of heterotic supergravity

α	Fields
0	$B_{1,A}$ B_2
-2	D_{D-4} $D_{D-3,A}$ D_{D-2,A_1A_2} $D_{D-1,A_1A_2A_3}$ $D_{D,A_1A_2A_3A_4}$

The D-potentials are antisymmetric in the A indices

instance, when truncating IIB supergravity to heterotic supergravity one sets the RR potentials equal to zero which eliminates all Dirichlet branes. On the other hand, if one truncates the IIB theory to Type I supergravity, one sets the NS-NS 2-form field equal to zero. This means no (fundamental) string can end on the brane and the worldvolume dynamics of the relevant $\alpha = -1$ branes is described by a hyper multiplet instead of a vector multiplet.

The following question therefore arises:

where are the half-supersymmetric vector branes of half-maximal supergravity?

To answer this question, it is sufficient to consider heterotic supergravity and to apply the brane classification techniques described in the previous section. The field content of the $D = 10 - d$ dimensional $U(1)^{16}$ heterotic supergravity theory is given by

$$\{e_\mu{}^a, B_2, d \times B_1, \phi\} \text{ plus } (16 + d) \times \{B_1, d \times \phi\}. \tag{3.5}$$

The vectors transform as the fundamental representation of $SO(d, 16+d)$ while the scalars parametrize the coset

$$SO(d, 16 + d)/[SO(d) \times SO(16 + d)]. \tag{3.6}$$

It turns out that the heterotic branes only occur with tensions

$$\mathrm{T_H} \sim (g_s)^\alpha \quad \text{with } \alpha = 0, -2, -4, \ldots$$

In particular, we find the potentials with $\alpha = 0$ and $\alpha = -2$ given in Table 3.2.

It turns out that at this point we find only branes whose worldvolume dynamics is described by hypermultiplets with 8 supercharges. We do not find any vector branes sofar. However, a new feature that arises in $6D$ is the occurrence of a 2-form potential D_2 with $\alpha = -2$, i.e. one that couples to a solitonic string. This gives us the new possibility to have branes on which not fundamental strings but *solitonic* strings may end. Of course, such objects are highly non-perturbative and less under control than the Dirichlet branes since now we are dealing with imposing Dirichlet boundary conditions on the solitonic instead of the fundamental string.

To be concrete, as an example we find that heterotic supergravity allows F-potentials $F_{5,A}$, with $\alpha = -4$, that transform in the fundamental representation of $SO(d, 16 + d)$ [18]. These potentials are dual to a set of gauge coupling constants

g^A or Romans mass parameters m^A. The heterotic supergravity theory allows both deformations [21]. These potentials couple to half-supersymmetric domain walls whose worldvolume dynamics is governed by a vector multiplet with 8 supercharges. The schematic form of these so-called 'V4-branes' is given by

$$\mathcal{L}_{\text{WZ}}(\text{V4}) \sim \text{T}^A\big[\underset{-4}{F_{5,A}} + \underset{-2}{\mathcal{H}_2}\,\underset{-2}{D_{3,A}}\big], \quad \mathcal{H}_2 = dd_1 - D_2, \tag{3.7}$$

where T^A are the tensions and d_1 is the Born-Infeld vector that corresponds to the solitonic string. The numbers indicate the α values of the different fields. Note that both terms in the above Wess-Zumino term have the same total α-weight. This V4 Wess-Zumino term is very similar to the Wess-Zumino term used for Dirichlet-branes. For instance, the Wess-Zumino term for a D4-bran with tension T has the form:

$$\mathcal{L}_{\text{WZ}}(\text{D4}) \sim \text{T}\big[\underset{-1}{C_5} + \underset{0}{\mathcal{F}_2}\,\underset{-1}{C_3} + \cdots\big], \quad \mathcal{F}_2 = db_1 - B_2, \tag{3.8}$$

where b_1 is the Born-Infeld vector corresponding to the fundamental string. The main difference is that the V4-brane has a more negative α-weight than the D4-brane and is therefore more non-perturbative, and also more difficult to study, than the D4-brane.

The V4-brane is just one example of a vector brane in half-maximal supergravity. We find that, similarly, Vp-branes are present in both heterotic, Type I as well as chiral (2,0) supergravity [18]. The world-volume dynamics of all these vector branes is described by a Born-Infeld theory with 8 supercharges. Similar vector branes are expected to arise in supergravity theories with less supersymmetry but none of these vector branes will have $\alpha = -1$.

3.5 Wrapping Rules

Now that we know the numbers of half-supersymmetric branes, resulting from either the light-cone rule or the longest-weight rule, it is natural to ask oneself how all the branes in different dimensions are related to each other via dimensional reduction, if that can be done at all. Since the scaling of the brane tension T with respect to the D-dimensional dilaton g_s does not change under dimensional reduction it is natural to consider the reduction of branes whose tension has a given scaling. We are interested in studying the dimensional reduction of the ten-dimensional branes, whose tension scales like g_s^α with $\alpha = 0, -1, -2, -3, -4$. This means that in any dimension we are interested in the branes with these values of α. Explicitly, we refer to these branes as:

$$\begin{aligned}
&\text{T}_\text{F} \sim 1: && \text{Fundamental branes,}\\
&\text{T}_\text{D} \sim 1/g_s: && \text{Dirichlet branes (D-branes),}\\
&\text{T}_\text{S} \sim 1/g_s^2: && \text{Solitonic branes,}\\
&\text{T}_\text{E} \sim 1/g_s^3: && \text{Exotic branes,}\\
&\text{T}_\text{SF} \sim 1/g_s^4: && \text{Space-filling } \alpha = -4 \text{ branes.}
\end{aligned} \tag{3.9}$$

Note that this distinction of branes with different dilaton scaling is different from the distinction between standard and non-standard branes. For instance, one has both standard as well as non-standard D-branes.

If a brane saw a standard geometry we would expect that upon dimensional reduction it would always lead to two different branes. Either one reduces along a transverse direction or along a worldvolume direction. The latter case corresponds to the wrapping of the brane, which leads to a brane with a reduced world-volume direction. We summarize this by saying that the "wrapping rules" corresponding to standard geometry are given by

$$\text{any brane} \quad \begin{cases} \text{wrapped} & \to \text{undoubled}, \\ \text{unwrapped} & \to \text{undoubled}. \end{cases} \tag{3.10}$$

The use of the word 'undoubled' stresses the fact that in both cases, wrapped or un-wrapped, only a single brane is obtained. Giving these wrapping rules and given the branes of ten-dimensional string theory with a given scaling α it is non-trivial that we precisely obtain the number of half-supersymmetric brane we obtained in our earlier brane classification. Indeed, it turns out that this only happens in the case of D-branes. Given the D-branes of Type IIA or IIB string theory and applying the wrapping rules (3.10) one precisely obtains the lower-dimensional D-branes which organize themselves into spinor representations of the T-duality group.

The same strategy does not work for the fundamental branes. As we saw earlier, in each dimension we have a singlet fundamental string and fundamental 0-branes that form the components of a vector representation of the T-duality group $\mathrm{SO}(d, d)$. This means that we need for each compactified direction *two* fundamental 0-branes. Clearly only one of these two branes can come from a wrapped fundamental string. We need another source to explain the occurrence of the second 0-brane. This is provided by the T-dual of the fundamental string, which is a pp-wave which upon reduction gives rise to the second 0-brane. The extra contribution due to the pp-waves gives rise to the following effective wrapping rules for the fundamental branes [22]:

$$\mathrm{T_F} \sim 1: \quad \begin{cases} \text{wrapped} & \to \text{doubled}, \\ \text{unwrapped} & \to \text{undoubled}. \end{cases} \tag{3.11}$$

These wrapping rules remind the 'doubled geometry' proposal of [23] where each compactified direction is doubled with a T-dual direction. Note that the doubled geometry proposal is based on a perturbative symmetry and therefore only applies to the fundamental branes and not necessarily to the other type of branes. Indeed, as we saw above, the D-branes have their own wrapping rules (3.10) corresponding to standard geometry.

Things get more interesting when we consider the solitonic branes. Again we find that the wrapping rules (3.10) corresponding to standard geometry do not lead to the right number of half-supersymmetric solitonic branes in lower dimensions. In this case, however, the extra input comes from the Kaluza-Klein (KK) monopoles. In each dimension $D \geq 5$ there is a KK monopole which can be considered as the dual of the pp-wave. The KK monopole divides spacetime into *three* different directions:

$$\text{KK monopole:} \quad \begin{cases} p+1 \text{ worldvolume directions,} \\ 1 \text{ isometry direction,} \\ 3 \text{ transverse directions.} \end{cases} \tag{3.12}$$

A brane in lower dimensions is obtained by reducing over the single isometry direction. In each dimension we have a singlet solitonic $(D-5)$-brane which is dual to the fundamental string. This singlet follows from the worldvolume reduction of the ten-dimensional NS5-brane. We also have solitonic $(D-4)$-branes which transform as a vector of the T-duality group. This implies that for each compactified direction we need *two* $(D-4)$-branes. Consider, for instance, the doublet of solitonic 5-branes in 9D. The first 5-brane follows from a transverse reduction of the ten-dimensional NS5-brane. To obtain the second 5-brane we need the help from the 10D KK-monopole. Indeed, the ten-dimensional KK monopole has 6 worldvolume, 1 isometry and 3 transverse directions. Reducing over the isometry direction leads to the second 5-brane. We thus obtain the following effective wrapping rules for solitonic branes [24]:

$$\mathrm{T_S} \sim 1/g_s^2: \quad \begin{cases} \text{wrapped} & \rightarrow \text{undoubled,} \\ \text{unwrapped} & \rightarrow \text{doubled.} \end{cases} \tag{3.13}$$

These rules can be viewed as dual to the fundamental wrapping rules (3.11).

An issue arises if we now also consider the non-standard solitonic $(D-3)$-, $(D-2)$- and $(D-1)$-branes which transform according to anti-symmetric tensor representations of the T-duality group, see Table 3.1. The precise number of such branes, which is given by the red entries in the last column, first three rows of Table 3.1, is reproduced if we apply the solitonic wrapping rules (3.13) also for these cases [24]. However, the KK monopole, upon reduction over the isometry directions, only gives rise to a standard brane with three transverse directions. We need to introduce something new to explain the numbers of the non-standard solitonic branes. One possibility is that one introduces 'generalized' KK monopoles which have less than three transverse directions. Such monopoles have already been considered a long time ago using T-duality arguments [25]. At the moment it is not clear how rigorously such generalized objects can be defined within string theory. Note that such objects, if they exist at all, seem to couple to mixed-symmetry tensors instead of p-form potentials. The possibility of including such mixed-symmetry tensors into a supergravity multiplet is as yet unknown. Another attitude is to say that the solitonic branes 'see' a different so-called 'dual doubled geometry' which is different from the 'doubled' geometry sensed by the fundamental branes or the standard geometry as viewed by the D-branes. In some sense, the generalized KK-monopoles represent information about this dual doubled geometry.

The pattern that arises is that each type of brane, depending on the scaling of the brane tension with the string coupling constant, sees a different geometry. For instance, the ten-dimensional Type IIB string theory has only one brane with a tension that scales with $T \sim 1/g_s^3$. This is the S-dual of the D7-brane. Clearly, this type of brane is highly non-perturbative and therefore difficult to study with the usual string theory techniques that one can use for the Dirichlet branes. Nevertheless, IIB supergravity suggests that this type of 'exotic' branes do exist. We find that, in order to explain the number of 'exotic' $\alpha = -3$ branes in lower dimensions that follows from our classification (see the red entries in the last column, forth and fifth row of Table 3.1) we need to impose the following new wrapping rule:

$$T_S \sim 1/g_s^3: \quad \begin{cases} \text{wrapped} & \rightarrow \text{doubled,} \\ \text{unwrapped} & \rightarrow \text{doubled.} \end{cases} \tag{3.14}$$

We call the new geometry defined by these wrapping rules 'exotic geometry'. Like in the previous cases the realization of this wrapping rule requires the input of new objects. How to precisely define these new objects within string theory is not clear but one could think about them as 'generalized' KK monopoles with less than three transverse and/or more than one isometry direction.

The only other type of brane that exists within ten-dimensional string theory is a space-filling brane whose tension scales as $T \sim 1/g_s^4$. It is the S-dual of the D9-brane. Space-filling branes are a bit special in the sense that they can only wrap to give a space-filling brane in lower dimensions. As we reviewed in the previous sections, the field that contains the $(D-1)$-brane that comes from the wrapping of this brane is the D-form $F^+_{D,A_1\ldots A_d}$, and from the light-cone rule (or the longest-weight rule) one obtains 2^{d-1} branes in D dimensions. To explain this number from the wrapping of the S-dual D9-brane we need to impose the following wrapping rule:

$$T_{SF} \sim 1/g_s^4: \quad \text{wrapped} \rightarrow \text{doubled.} \tag{3.15}$$

Since these branes can only wrap, one cannot tell whether they see a doubled geometry or an exotic geometry.

Based upon the above wrapping rules we conclude that the different branes of ten-dimensional string theory see the following kind of geometries:

fundamental branes:	doubled geometry
Dirichlet branes:	standard geometry
solitonic branes:	dual doubled geometry
exotic branes:	exotic geometry.

This is not yet the end of the story. Starting from $D = 7$, there are additional $\alpha = -4$ branes apart from those associated to the potential $F^+_{D,A_1\ldots A_d}$. More generally, maximal supergravity in lower dimensions suggests the existence of branes with $\alpha < -4$. Clearly, all such branes can never result from the reduction of any brane in ten dimensions. They should either follow from the reduction of new objects within string theory or result as the effect of a new kind of geometry. Clearly, the last word has not been said on this issue.

3.6 Conclusions

We have classified the half-supersymmetric branes of maximal supergravity by investigating the worldvolume WZ term that describes the coupling of the supergravity potentials to these branes. By requiring that a gauge-invariant WZ term could be constructed involving only worldvolume fields that fit into a half-maximally supersymmetric matter multiplet we were able to classify the branes. The worldvolume content of these branes is either a vector multiplet or a 6-dimensional (self-dual) tensor multiplet. We call such branes vector branes and tensor 5-branes, respectively. Note that for branes with a low-dimensional world-volume, such as membranes and strings, the vector multiplet becomes equivalent to a scalar multiplet.[1] The dynamics of vector branes is governed by a Dirac-Born-Infeld/Volkov-Akulov (DBI-VA) action. Such vector branes have recently been considered in discussions on the quantum properties of 4D supergravity theories [20].

The investigation of the WZ term led to two simple, equivalent, rules that specify the number of half-supersymmetric branes. The first, so-called 'light-cone rule', is based on decomposing the SO(d, d) indices into its light-cone directions. The second, so-called 'longest-weight rule', states that the light-cone rule is equivalent to the group-theoretical rule that the half-supersymmetric branes correspond to the longest weights of the T-duality representation in which the supergravity potentials transform. We have not commented on the role of the next-to-longest weights etc. They are related to bound states of half-supersymmetric branes. These states, unlike bound states of standard branes, can be 1/2-supersymmetric threshold bound states [18, 26].

We showed that vector branes not only occur in maximal supergravity but also in supergravity theories with less supersymmetry. In particular, we investigated the case of half-maximal supergravity. The difference with the maximal case is that the vector branes of half-maximal supergravity are more non-perturbative in the sense that they have a lower value of α. We gave an example of a $6D$ domain wall, i.e. a V4-brane, with a solitonic string ending on it. This solves a puzzle which we formulated as a question in Sect. 3.4.

Having classified the branes we went on to investigate the way in which the branes in different dimensions are related to each other by dimensional reduction. This led us to consider the introduction of several wrapping rules, one set of rules for each brane with a given brane tension scaling α for $\alpha = 0, -1, -2, -3, -4$. These wrapping rules can be found in (3.11), (3.10), (3.13), (3.14) and (3.15), respectively. In some cases, the origin of the doubling in the wrapping rules is understood. They come from pp-waves and KK monopoles that upon reduction give rise to additional branes. But this is not enough. In order to explain the doubling in all cases something new is needed. Here there are two different points of view. Either one introduces new objects in string theory. We called them 'generalized KK monopoles' but the precise status of these monopoles in string theory is not clear. They seem to be related to the issue whether mixed-symmetry tensors can be introduced in supergravity. Another point

[1] In 3D a vector is dual to a scalar, whereas in 2D a vector is equivalent to an integration constant.

of view is to say that the extra branes result from a new geometry that is described by the brane wrapping rules. This is more in line with the doubled geometry proposal that can be used to explain the wrapping rules of the fundamental branes.

We did not discuss several other interesting brane properties that follow from our methods. For instance, our techniques allow to determine the BPS conditions of the branes and their relation to the central charges in the supersymmetry algebra.[2] Again, we find here an important distinction between the standard and non-standard branes. Whereas for the standard branes each brane has its own BPS condition, in the case of non-standard branes the same BPS condition can be satisfied by several branes. We have calculated the degeneracies of each BPS condition [18]. Apart from this, one may also study brane orbits and multi-charge configurations [9, 18].[3]

It remains to be seen what the precise role is of the non-standard branes we discussed. Recently, is has been argued that in particular the defect branes play a role in describing the microscopic degrees of freedom of black holes [30] and that they are related to non-geometric Q-fluxes [31].

Our methods may be generalized and applied to study the half-supersymmetric branes of supergravities with less supersymmetry. The branes of half-supersymmetric supergravity have already been studied [32]. We hope to report on the half-supersymmetric branes of a quarter-supersymmetric supergravity shortly [33].

Finally, we hope that all this new information on branes will lead to a better understanding of their role in string theory and, most importantly, of the geometry underlying string theory.

Acknowledgments E. A. Bergshoeff wishes to thank the organizers of the BUDS2013 meeting in Frascati for giving the opportunity to give a talk and for providing a stimulating atmosphere.

References

1. E. Bergshoeff, E. Sezgin, P.K. Townsend, Supermembranes and eleven-dimensional supergravity. Phys. Lett. B **189**, 75 (1987)
2. J. Polchinski, Dirichlet branes and Ramond-Ramond charges. Phys. Rev. Lett. **75**, 4724 (1995). arXiv:hep-th/9510017
3. A. Strominger, C. Vafa, Microscopic origin of the Bekenstein-Hawking entropy. Phys. Lett. B **379**, 99 (1996). arXiv:hep-th/9601029
4. J.M. Maldacena, The Large N limit of superconformal field theories and supergravity. Adv. Theor. Math. Phys. **2**, 231 (1998). arXiv:hep-th/9711200
5. L. Randall, R. Sundrum, A Large mass hierarchy from a small extra dimension. Phys. Rev. Lett. **83**, 3370 (1999). arXiv:hep-ph/9905221
6. L. Randall, R. Sundrum, An alternative to compactification. Phys. Rev. Lett. **83**, 4690 (1999). arXiv:hep-th/9906064
7. E. A. Bergshoeff, F. Riccioni, The D-brane U-scan. arXiv:1109.1725 [hep-th]
8. A. Kleinschmidt, Counting supersymmetric branes. JHEP **1110**, 144 (2011). arXiv:1109.2025 [hep-th]
9. E.A. Bergshoeff, A. Marrani, F. Riccioni, Brane orbits. Nucl. Phys. B **861**, 104 (2012). arXiv: 1201.5819 [hep-th]

[2] For an early discussion of these properties, for the standard branes, see [27].

[3] For earlier work, see [28, 29].

10. E.A. Bergshoeff, M. de Roo, S.F. Kerstan, F. Riccioni, IIB supergravity revisited. JHEP **0508**, 098 (2005). arXiv:hep-th/0506013
11. E.A. Bergshoeff, J. Hartong, P.S. Howe, T. Ortín, F. Riccioni, IIA/IIB Supergravity and Ten-forms. JHEP **1005**, 061 (2010). arXiv:1004.1348 [hep-th]
12. E.A. Bergshoeff, M. de Roo, S.F. Kerstan, T. Ortín, F. Riccioni, SL(2, R)-invariant IIB brane actions. JHEP **0702**, 007 (2007). arXiv:hep-th/0611036
13. F. Riccioni, P.C. West, The E(11) origin of all maximal supergravities. JHEP **0707**, 063 (2007). arXiv:0705.0752 [hep-th]
14. E.A. Bergshoeff, I. De Baetselier, T.A. Nutma, E(11) and the embedding tensor. JHEP **0709**, 047 (2007). arXiv:0705.1304 [hep-th]
15. P.C. West, E(11) and M theory. Class. Quant. Grav. **18**, 4443 (2001). arXiv:hep-th/0104081
16. E.A. Bergshoeff, F. Riccioni, D-brane Wess-Zumino terms and U-duality. JHEP **1011**, 139 (2010). arXiv:1009.4657 [hep-th]
17. E.A. Bergshoeff, F. Riccioni, String solitons and T-duality. JHEP **1105**, 131 (2011). arXiv: 1102.0934 [hep-th]
18. E.A. Bergshoeff, F. Riccioni, L. Romano, Branes, weights and central charges. JHEP **1306**, 019 (2013). arXiv:1303.0221 [hep-th]
19. E.A. Bergshoeff, T. Ortín, F. Riccioni, Defect branes. Nucl. Phys. B **856**, 210 (2012). arXiv: 1109.4484 [hep-th]
20. E. Bergshoeff, F. Coomans, R. Kallosh, C. S. Shahbazi, A. Van Proeyen, Dirac-Born-Infeld-Volkov-Akulov and deformation of supersymmetry. arXiv:1303.5662 [hep-th]
21. E.A. Bergshoeff, J. Gomis, T.A. Nutma, D. Roest, Kac-Moody spectrum of (half-)maximal supergravities. JHEP **0802**, 069 (2008). arXiv:0711.2035 [hep-th]
22. E.A. Bergshoeff, F. Riccioni, Branes and wrapping rules. Phys. Lett. B **704**, 367 (2011). arXiv: 1108.5067 [hep-th]
23. C.M. Hull, R.A. Reid-Edwards, Gauge symmetry T-duality and doubled geometry. JHEP **0808**, 043 (2008). arXiv:0711.4818 [hep-th]
24. E.A. Bergshoeff, F. Riccioni, Dual doubled geometry. Phys. Lett. B **702**, 281 (2011). arXiv: 1106.0212 [hep-th]
25. E. Lozano-Tellechea, T. Ortín, 7-branes and higher Kaluza-Klein branes. Nucl. Phys. B **607**, 213 (2001). arXiv:hep-th/0012051
26. E.A. Bergshoeff, A. Kleinschmidt, F. Riccioni, Supersymmetric domain walls. Phys. Rev. D **86**, 085043 (2012). arXiv:1206.5697 [hep-th]
27. S. Ferrara, J.M. Maldacena, Branes, central charges and U-duality invariant BPS conditions. Class. Quant. Grav. **15**, 749 (1998). arXiv:hep-th/9706097
28. S. Ferrara, M. Gunaydin, Orbits of exceptional groups, duality and BPS states in string theory. Int. J. Mod. Phys. A **13**, 2075 (1998). arXiv:hep-th/9708025
29. H. Lu, C.N. Pope, K.S. Stelle, Multiplet structures of BPS solitons. Class. Quant. Grav. **15**, 537 (1998). arXiv:hep-th/9708109
30. J. de Boer, M. Shigemori, Exotic Branes in string theory. JHEP **532**, 65–118 (2013). arXiv: 1209.6056 [hep-th]
31. F. Hassler, D. Lust, Non-commutative/non-associative IIA (IIB) Q- and R-branes and their intersections. JHEP **1307**, 048 (2013). arXiv:1303.1413 [hep-th]
32. E.A. Bergshoeff, F. Riccioni, Heterotic wrapping rules. JHEP **1301**, 005 (2013). arXiv:1210. 1422 [hep-th]
33. E.A. Bergshoeff, F. Riccioni, L. Romano, Towards a classification of branes in theories with eight supercharges. arXiv:1402.2557 [hep-th]

Chapter 4
Pure Spinor Superfields: An Overview

Martin Cederwall

Maximally supersymmetric theories do not allow off-shell superspace formulations with traditional superfields containing a finite set of auxiliary fields. It has become clear that off-shell supersymmetric action formulations of such models can be achieved by the introduction of pure spinors. In this talk, an overview of this formalism is given, with emphasis on $D = 10$ super-Yang–Mills theory and $D = 11$ supergravity. This a somewhat expanded version of a talk presented at the workshop "Breaking of supersymmetry and ultraviolet divergences in extended supergravity" (BUDS), Laboratori Nazionali di Frascati, March 25–28, 2013.

4.1 Introduction

The search for formalisms treating maximally supersymmetric models in a "covariant" way—covariance here taken in the sense of manifestly exhibiting Lorentz symmetry as well as the full supersymmetry—has a long history. To a large extent it has been pursued in terms of first-quantised particle (or string) theories, with the purpose of then applying second quantisation to obtain a covariant field theory. Let us remind how the problem arises, first in a particle or string theory, and then in field theory.

The Brink–Schwarz superparticle [1, 2], where the fermions are Lorentz spinors, exhibits a problematic mixture of first and second class constraints, as does the Green–Schwarz superstring [3]. That this must be the case is realised already from a counting of the fermionic degrees of freedom describing massless supermultiplets, i.e., from the 1/2-BPS property of a massless (short) supermultiplet. There is half a spinor of first class constraint and half a spinor of second class constraints [4–6], and these can not be separated in a Lorentz-covariant manner. The first class constraints generate

M. Cederwall (✉)
Department of Fundamental Physics, Chalmers University of Technology, Gothenburg, Sweden
e-mail: martin.cederwall@chalmers.se

S. Bellucci (ed.), *Breaking of Supersymmetry and Ultraviolet Divergences in Extended Supergravity*, Springer Proceedings in Physics 153,
DOI: 10.1007/978-3-319-03774-5_4,

the so called κ-symmetry [4]. Some attempts to a direct covariant treatment of the κ-symmetry have appeared (see e.g. [7, 8]), but most of the proposed solutions to the problem have involved drastic changes of variables, such as twistor [9] methods.

Supertwistors solve the problem of covariant quantisation of superparticles in 3, 4, 6 and 10 dimensions [10–15], and make manifest not only super-Poincaré but the whole superconformal symmetry (except, of course in $D = 10$). We mention the supertwistor track here partly since it has similarities with our main focus of attention, pure spinors, in that both twistors and pure spinors are bosonic spinors (i.e., of "wrong" statistics), and partly since twistor methods (of a different flavor) have been of revived interest later and used for amplitude calculations [16–22]. Some works seems to point towards a deeper relation between pure spinors and twistors [23]. It should be mentioned that, although some attempts have been made [24, 25], twistor transform methods seem less powerful in string theory than in particle theory, due to the massive spectrum.

The corresponding problem is of course seen also in field theory. There, the natural way of manifesting supersymmetry is to use superfields, that depend not only on the bosonic coordinates x^m, but also on some fermions θ^μ, that together form a (Wess–Zumino) superspace [26]. If the field theory in question is a gauge theory [27], the superfield formulation will be a gauge theory on superspace [28–32], and if it contains gravity [33–37], it will be described as superspace geometry [26, 37–44]. In both cases, the maximally supersymmetric models (which means 16 supercharges for super-Yang–Mills theory (SYM) and 32 for supergravity (SG)) only have on-shell formulations in superspace. This can be stated in a couple of equivalent ways. The supersymmetry transformations close only modulo the equations of motion. In a component formalism, there is no set of auxiliary (non-dynamical) fields, that can be added so that the bosonic and fermionic numbers of fields agree off-shell and fill a representation of supersymmetry. We will come back to the superspace formulations of some maximally supersymmetric models later, and examine them in more detail, because it is precisely the traditional superspace theories that form the basis of the pure spinor superfield formalism.

Pure spinors are interesting objects from a mathematical point of view. The original definition by E. Cartan [45, 46] is valid in even dimensions. A Cartan pure spinor is a spinor annihilated by half-dimensional isotropic (light-like) subspace. If the dimension is $D = 2n$, then this can be expressed as $\gamma^{+i}\lambda = 0$, $i = 1, \ldots, n$, for a suitable choice of basis (depending on the pure spinor λ). Here, we think of the signature of space-time as split. For euclidean signature, take the γ-matrices with holomorphic indices. Modulo a complex scale, the pure spinor space is isomorphic to the space of isotropic n-planes, which is $SO(2n)/U(n)$. This condition can be translated into certain bilinear conditions on the spinor. The first case where the pure spinor condition is non-trivial is $n = 4$. Up to $n = 6$, the pure spinors form the only non-trivial orbit of the rotation group in between the full orbit of unconstrained spinors and the trivial orbit of 0, but for higher n there are more orbits [47–49], of which the pure spinor is the most constrained.

The "pure spinors" we will use sometimes coincide with Cartan pure spinors, sometimes not. The canonical example of $D = 10$ SYM is an example where they are identical. The important and defining property, that we will give a geometric interpretation, is a bilinear identity $(\lambda\gamma^a\lambda) = 0$, which in $D = 10$ coincides with the constraint on a Cartan pure spinor. Even if Cartan pure spinors are uninteresting in $D < 8$, we will encounter non-trivial "pure spinor" constraints e.g. in $D = 6$ and $D = 3$, essentially due to the presence of R-symmetry. We will also use the bilinear constraint in odd dimensions, notably $D = 11$.

We are mainly concerned with field theories, including supergravity, and will not say much about the use of pure spinors in superstring theory. From investigations of the superspace formulation of maximally supersymmetric theories, it was early recognised that pure spinors might have a rôle to play in an off-shell formulation [50–52]. The discovery of the precise rôle of pure spinors came from two independent (but in retrospect clearly related) lines of research. One, the covariant quantisation of the superstring, provided a valid set of ghost variables for a covariant superstring, and thereby also for its massless sector [53, 54]. The other was the systematic search for higher-derivative terms in maximally supersymmetric theories, where revisiting the structure of the superspace constraints revealed a cohomological structure of the deformations [55–57], which later was realised to be equivalent to that of the pure spinor BRST operator. The latter formalism led to results on deformations of SYM [55, 58, 59] (e.g. the full form of the terms related to F^4) as well as SG [60–65] models.

Pure spinor superfield models have been given for SYM [54, 55, 58, 59, 66–68] for $D = 11$ supergravity [69, 70] and for $D = 3$ superconformal models [71–73]. It is quite clear that the method applies to any maximally supersymmetric model that does not contain selfdual fields.

The wide breakthrough of the use of pure spinors in connection with supersymmetry came with the realisation of Berkovits that they provide a good set of variables for covariant quantisation of the superstring [53, 74, 75]. The formalism has been extensively used in superstring theory, see e.g. [76–99]. Applications to supermembrane theory have also been attempted, but with less clear results [100–102].

This presentation takes its starting point in the traditional superspace formulation of supersymmetric field theories. In Sect. 4.2 we explain why the basics of the pure spinor superfield formalism is (almost) inherent in the superspace formalism. We derive the BRST operator of the linearised models. Section 4.3 deals with the calculation of the field content, i.e., the BRST cohomology, which is illustrated with some examples. In order to formulate actions, a measure is needed, which is developed in Sect. 4.4, based on the "non-minimal" variables of Berkovits. Section 4.5 gives the field–antifield machinery needed in order to formulate consistent interactions. The following sections deal with gauge fixing, necessary for quantum calculations, and with an application: to find higher-derivative terms. Finally, in Sect. 4.8 some (hopefully) interesting open questions and possible developments are mentioned.

4.2 Pure Spinors from Superspace

We denote bosonic and fermionic indices in coordinate basis ("curved indices") by $M, N \ldots = (m, n, \ldots; \mu, \nu, \ldots)$ and in Lorentz basis ("flat indices") by $A, B, \ldots = (a, b, \ldots; \alpha, \beta, \ldots)$. Wess–Zumino superspace has a torsion

$$T_{\alpha\beta}{}^{a} = 2\gamma^{a}_{\alpha\beta} \tag{4.1}$$

(there might be slight formal variations on this expression, e.g. when there is some R-symmetry in case of extended supersymmetry, but with a liberal interpretation (4.1) is always true). Note that we always express components in Lorentz indices, since fermionic directions otherwise can not be seen as spinors. This is typically the only non-vanishing torsion component at dimension zero (in on-shell theories), dimension here being defined so that a bosonic derivative has dimension 1 and a fermionic $\frac{1}{2}$. In flat superspace, this statement amounts to the anticommutator between fermionic covariant derivatives being

$$\{D_\alpha, D_\beta\} = -T_{\alpha\beta}{}^{a}\partial_a = -2\gamma^{a}_{\alpha\beta}\partial_a \ . \tag{4.2}$$

In flat space, these are the ordinary derivatives

$$D_\alpha = \frac{\partial}{\partial\theta^\alpha} - \left(\gamma^a\theta\right)_\alpha \partial_a \ , \tag{4.3}$$

which anticommute with the global supersymmetry generators (superspace Killing vectors)

$$Q_\alpha = \frac{\partial}{\partial\theta^\alpha} + (\gamma^a\theta)_\alpha\partial_a \ . \tag{4.4}$$

Some special possible rôle of pure spinors can be seen already here. Suppose that λ is pure (in the sense mentioned in the introduction), i.e., that

$$(\lambda\gamma^a\lambda) = 0 \ . \tag{4.5}$$

If one forms the scalar fermionic operator

$$Q = \lambda^\alpha D_\alpha \ , \tag{4.6}$$

it becomes immediately clear from (4.1) and (4.5) that

$$Q^2 = 0 \ . \tag{4.7}$$

It is possible to think of Q as a BRST operator, and examine its cohomology. This cohomology will be non-trivial due to the pure spinor constraint. This will actually be the BRST operator used in the (minimal) pure spinor formalism, and its cohomology will consist of the physical states.

In order to understand why this happens, and why it indeed is little more than a reformulation of the traditional superspace formalism, it is suitable to reexamine the canonical example, $D = 10$ SYM (the procedure describes equally well its dimensional reductions) [31, 50]. For simplicity, we will use an abelian field.

Note that we aim at going directly to the field theory, without passing via a first-quantised superparticle model. The BRST operator (4.6) is not obtained as the BRST operator for some local symmetry on the world-line of a superparticle, but postulated more or less *ad hoc*. It will soon be motivated from superspace arguments, though. Some work has been done on showing the equivalence of the first-quantised superparticle or string with the formulation based on Q [103–105]. We take a more pragmatic point of view—if the correct field theories are produced we are happy with that.

4.2.1 SYM

We work in $D = 10$, where a chiral spinor has 16 components. The theory starts from a gauge theory on superspace [31, 50]. This means that the connection 1-form *a priori* is completely general,

$$A = E^A A_A = E^a A_a(x,\theta) + E^\alpha A_\alpha(x,\theta) \tag{4.8}$$

(where $E^A = dZ^M E_M{}^A$ is the superspace vielbein). In order to reduce the very large number of component fields, some constraints must be imposed. One such constraint, which goes under the name of conventional constraint, completely expresses the superfield A_a in terms of A_α. This is desirable, since there is another component 1-form at level θ in A_α, and only one in the physical theory. The conventional constraint is formulated in terms of the field strength, in order not to destroy gauge symmetry, and reads (in the abelian case)

$$\gamma_a^{\alpha\beta} F_{\alpha\beta} = 0 \ . \tag{4.9}$$

Since this part of F is expressed as

$$F_{\alpha\beta} = 2D_{(\alpha}A_{\beta)} + T_{\alpha\beta}{}^a A_a \ , \tag{4.10}$$

the conventional constraint does exactly what it is supposed to. Then, one is left with A_α, the lowest-dimensional superfield.

In order to take the fields on-shell the remaining part of $F_{\alpha\beta}$ is also set to zero. This is a selfdual 5-form. We will not exhibit the detailed calculation here, but contend ourselves with the well known statement the setting the dimension-0 field strength to zero gives the equations of motion for the component fields. These sit in the superfield at order θ (the gauge connection) and θ^2 (the fermion) (and of course also at higher orders if they contain non-zero modes). Traditionally, to keep gauge invariance manifest, the superfield A_a is not actually eliminated. Instead one uses the Bianchi identities for the superspace field strength F, which will give the equations of motions once $F_{\alpha\beta} = 0$. This is not the path taken here. Instead we leave A_a completely aside and focus on A_α.

We can then observe that the conditions imposed are exactly those implied by demanding that a field $\Psi = \lambda^\alpha A_\alpha(x,\theta)$ is annihilated by the BRST operator $Q = \lambda^\alpha D_\alpha$. The fermionic covariant derivative acts on the superfield A_β, and the bilinear in λ contains only the 5-form part, due to the pure spinor condition. In addition, gauge invariance is implemented as $\delta_\Lambda \Psi = Q\Lambda$ (that this is true for the the bosonic connection at level θ of course requires a small calculation), which makes clear that the cohomology of Q describes precisely the on-shell physical fields. The cohomology will be examined to greater generality in the following section.

Expanding out the λ-dependence of the field Ψ, we thus have an infinite set of superfields,

$$\Psi(x,\theta,\lambda) = \sum_{n=0}^{\infty} \lambda^{\alpha_1} \ldots \lambda^{\alpha_n} A_{\alpha_1 \ldots \alpha_n}(x,\theta) . \qquad (4.11)$$

In order for $Q = \lambda D$ to behave as a BRST operator, it is natural to assign a ghost number 1 to λ. We have already mentioned that the cohomology of Q at order λ reproduces the gauge connection and the fermion, subject to their linearised equations of motion (the remaining cohomology will be left for Sect. 4.3). The field Ψ then also carries ghost number 1, so that the physical fields have ghost number 0.

Already at this point we see that relaxing the equations of motion is equivalent to relaxing the condition $Q\Psi = 0$. If a suitable integration measure is found, a true off-shell formulation could be provided by an action of the type $S \sim \int \Psi Q \Psi + \cdots$, which will be the objective of Sects. 4.4 and 4.5.

4.2.2 SG

What will be said in this section will apply to $D = 11$ supergravity, and its dimensional reductions.

A spinor in $D = 11$ has 32 components. The symmetric spinor bilinears are a 1-form, a 2-form and a 5-form. In addition to the metric field, $D = 11$ SG also contains a 3-form potential C with 4-form field strength $H = dC$ and a gravitino. The component action for the bosonic fields,

$$S = \frac{1}{2\kappa^2} \left(\int d^{11}x \left(R - \frac{1}{48} H^2 \right) + \frac{1}{6} \int C \wedge H \wedge H \right) , \qquad (4.12)$$

contains a Chern–Simons term for C.

There are two ways of approaching the superspace construction of the supergravity. The first one is via the actual supergeometry, examined in [41, 42, 60–62, 106]. Here one starts with the vielbein on superspace $E_M{}^A$ together with a Lorentz algebra-valued connection Ω_M. Just like in the case of gauge theory, all the superfields except the one of lowest dimension, $E_\mu{}^a$, are effectively eliminated as independent degrees of freedom via conventional constraints [62, 107, 108]. This is slightly more involved than in the SYM case, and we refer to [62] for a complete treatment. Essentially, by formulating constraints on the superspace torsion,

$$T^A = dE^A + E^B \wedge \Omega_B{}^A \ , \tag{4.13}$$

all connection superfields and all of the vielbein become expressible in $E_\mu{}^a$. The conventional constraints reduce the possible dimension-0 torsion $T^a_{\alpha\beta}$ (apart from the standard part $2\gamma^a_{\alpha\beta}$) to the irreducible modules

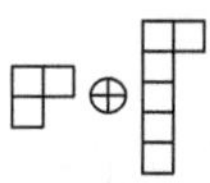 , (4.14)

where the 2 or 5 antisymmetrised indices come from the contraction of the two spinor indices with γ^{ab} or γ^{abcde}.

Like in SYM, the standard procedure for deriving the full equations of motion is not to actually solve for the vielbein and spin connection superfields, but to use torsion Bianchi identities [39],

$$DT^A = E^B \wedge R_B{}^A \ , \tag{4.15}$$

to obtain the equations of motion without giving up any manifest gauge invariance.

Suppose we now want to interpret this, at the linearised level, in terms of pure spinors. Then we again leave all the superfield except the lowest-dimensional one out. After converting the form index on $E_\mu{}^a$ to a flat spinor index, we have a field $\phi_\alpha{}^a$. It is actually only its γ-traceless part that is not eliminated by conventional constraints. Note that the spinor bilinears appearing above in the torsion $T_{\alpha\beta}{}^a$ after conventional constraints have been used, the 2-form and and 5-form, are exactly those which are non-vanishing for a pure spinor. It looks reasonable to think of the linearised superfield $\phi_\alpha{}^a$ as appearing at order λ in a pure spinor superfield $\Phi^a(x, \theta, \lambda)$. The linearised equations of motion then come from $Q\Phi^a = 0$. There is only a small ingredient missing here, namely that $\phi_\alpha{}^a$ is γ-traceless, as are the two torsion modules. This is achieved by declaring an equivalence relation

$$\Phi^a \approx \Phi^a + (\lambda\gamma^a\varrho) \ . \tag{4.16}$$

We call this type of equivalence relation a "shift symmetry" [69–72, 109], and we will come back to its rôle in the following sections.

The other way of obtaining the linearised equations of motion is from the 3-form C, which extends to a 3-form on superspace. This method has not traditionally been used alone as a formulation of supergravity, since the geometry (via the torsion) will enter its Bianchi identities. Nevertheless, at the linearised level this produces all the supergravity fields, without involving superspace geometry; this will be made clear in Sect. 4.3. Without going into details about conventional constraints, it is again the lowest-dimensional superfield that is relevant. This is $C_{\alpha\beta\gamma}$, of dimension $-\frac{3}{2}$, and actually only the irreducible modules consisting of γ-traceless 2-form- and 5-form-spinors. These modules fit perfectly in the expansion of a scalar pure spinor superfield $\Psi(x,\theta,\lambda)$ to third order in λ,

$$\Psi = \cdots + \frac{1}{6}\lambda^{\alpha}\lambda^{\beta}\lambda^{\gamma}C_{\alpha\beta\gamma} + \cdots \tag{4.17}$$

The linearised supergravity equations of motion come from demanding that

$$H_{\alpha\beta\gamma\delta}\Big|_{[2,2]\oplus[2,2,1,1,1]\oplus[2,2,2,2,2]} = 0\ , \tag{4.18}$$

which is equivalent to the condition

$$Q\Psi = 0\ , \tag{4.19}$$

since these three irreducible modules are precisely the ones occurring in a quadrilinear of a pure spinor.

4.2.3 Summary

We have seen, in the two main examples of $D = 10$ SYM and $D = 11$ SG, that the linearised equations of motion (and gauge symmetries) are reproduced precisely by considering the physical fields as part of a pure spinor superfield with appropriate properties annihilated by the pure spinor BRST operator $Q = \lambda D$. The price paid for this is that interactions are (for the moment) ignored, and that only some lowest-dimensional superfield is considered. This also means that gauge symmetry (including diffeomorphisms and local supersymmetry in the SG case) are not kept "manifest" or "geometrical". We will comment more on this issue when interactions are introduced, in Sect. 4.5.2.

4.3 Cohomology

In this section, we will take a closer look at the cohomology of the BRST operator in the two examples of Sect. 4.2 and some other models. The statements about it reproducing the fields of the models in question will be made more precise, and some interesting structure pointing forward to a field–antifield formalism will be pointed out.

Notice that if λ had been unconstrained (and there was no shift symmetry, for the case of non-scalar fields), the cohomology had been trivial. It is the pure spinor property of λ that gives room for some interesting cohomology. Consider, for example, a scalar pure spinor superfield $\Psi(x,\theta,\lambda)$, and let us for the moment forget about the x-dependence. A field $\Psi=(\lambda\gamma^a\theta)A_a$ represents cohomology: acting with Q gives

$$Q\cdot(\lambda\gamma^a\theta)A_a=(\lambda\frac{\partial}{\partial\theta})\cdot(\lambda\gamma^a\theta)A_a=(\lambda\gamma^a\lambda)A_a=0\ , \tag{4.20}$$

and it is also obvious that such a field can not be written as a Q-exact expression. In the SYM case, this cohomology is precisely the zero mode of the gauge connection. Obviously, Ψ should be taken to be fermionic.

It is clear that the algebraic properties of the pure spinor λ play a decisive rôle for determining the cohomology. Indeed, as we will see in the following sections, a partition function for the pure spinor contains essentially all information needed to determine the full cohomology.

We have seen one example above of an element of the cohomology of a scalar superfield, the zero mode of the gauge connection. We also argued in Sect. 4.2.1 that the cohomology at order λ precisely reproduces the fields of $D=10$ SYM, subject to the linearised equations of motion. What is the general cohomology? One more example is the constant field, $\Psi=c$. This is a cohomology of ghost number 1 (given the ghost number assignment of Sect. 4.2.1), and given the gauge transformation of Ψ it is natural to identify it as the ghost for the gauge symmetry.

Both these examples concern zero mode cohomology, i.e., elements of cohomology independent of the coordinates x. It turns out to be very instructive to first consider general zero mode cohomology. Not only is it much easier to calculate, since it is a purely algebraic problem (the operator Q reduces to $\lambda^\alpha\frac{\partial}{\partial\theta^\alpha}$), it will also give all essential information concerning the full cohomology. Namely, consider a zero mode cohomology of Ψ at order $\lambda^p\theta^q$. Such a cohomology will have ghost number gh#$(\Psi)-p$ and dimension $\dim(\Psi)+\frac{1}{2}(p+q)$. If then x-dependence is introduced, how will the corresponding cohomology behave? The only possibility is to have some field in the same module as the zero mode, but subject to some differential equation, an equation of motion. This equation of motion must in turn have support in the zero mode cohomology. This means that the zero mode cohomology can be used to read off the possible full cohomology. If there is also a zero mode cohomology at $\lambda^{p+1}\theta^{q+2n-1}$ (i.e., at ghost number gh#$(\Psi)-p-1$ and dimension $\dim(\Psi)+\frac{1}{2}(p+q)+n$), a field $\phi(x)$ in some module determined by the zero mode cohomology at $\lambda^p\theta^q$ can be subject to a (linearised) equation of motion of the form $\partial^n\phi=0$, given that the modules of the two zero mode cohomologies match. The corresponding x-dependent cohomology will of course take the generic form

$$\Psi\sim\lambda^p(\theta^q\phi+\theta^{q+2}\partial\phi+\theta^{q+4}\partial^2\phi+\cdots)\ . \tag{4.21}$$

4.3.1 SYM

As mentioned, the algebraic problem of calculating the zero mode cohomology can be used to gain information about the full cohomology [54, 57, 110]. The problem can be solved by computer methods [57] or algebraically [111]. For the field Ψ of ghost number 1 and dimension 0, the result may be summarised in Table 4.1, where the horizontal direction is the expansion in λ (i.e., decreasing ghost number of the component fields) and the vertical is the expansion in θ (i.e., increasing dimension within each superfield). The expansion of the superfields in θ has been shifted, so that components on the same horizontal level have the same dimension. The modules have been labeled by the Dynkin labels of the Lorentz group $Spin(1,9)$. As already discussed we see the gauge ghost at λ^0 and the physical fields (gauge connection A_a and spinor χ^α) at λ^1. In addition there are cohomologies at λ^2 and λ^3. The ones at λ^2 indicate, according to the discussion above, that the physical fields are subject to equations of motion. Their interpretation as components of the field Ψ is as *antifields* A^{*a} and χ^*_α, fields of ghost number -1 with the same dimensions as the equations of motion. The singlet at $\lambda^3\theta^5$ is the ghost antifield c^*. Its presence in cohomology in turn implies the divergencelessness of the on-shell antifield, corresponding to conservation of the gauge current. This is then strong evidence that using a pure spinor to go off shell implies introducing a Batalin–Vilkovisky field–antifield structure. This will be formalised in detail in Sect. 4.5.

As argued in the beginning of the present section, there is a more direct way of deducing the zero mode cohomology (and thereby the full cohomology) from the partition function for a pure spinor. Consider the expansion of a function $f(\lambda)$ in a power series expansion in λ, just as we have done for the pure spinor superfield. The pure spinor λ itself is in the module (00001), and the pure spinor constraint ensures that only the module $(0000n)$ occurs at λ^n. Therefore, the component fields in the expansion will come in the conjugate module $R_n = (000n0)$. A formal partition function [111–113] containing all information about the expansion is

$$\mathscr{P}(t) = \bigoplus_{n=0}^{\infty} R_n t^n = \bigoplus_{n=0}^{\infty} (000n0) t^n \ . \tag{4.22}$$

A less refined partition function is one that only counts the dimensions of the modules, i.e.,

$$\begin{aligned} P(t) &= \sum_{n=0}^{\infty} \dim(000n0) t^n = \sum_{n=0}^{\infty} \tfrac{1}{10} \binom{n+7}{7}\binom{n+5}{3} t^n \\ &= (1-t)^{-11}(1+t)(1+4t+t^2) \\ &= (1-t)^{-16}(1-10t^2+16t^3-16t^5+10t^6-t^8) \ . \end{aligned} \tag{4.23}$$

Various information can be collected here. The next to last line indicates that the number of degrees of a pure spinor in $D = 10$ is 11 (more on this in Sect. 4.4). The last line (where the factor $(1-t)^{-16}$ represents the partition function of an unconstrained spinor) is where the zero mode cohomology can be read off: note

Table 4.1 The zero mode cohomology in Ψ for $D = 10$ super-Yang–Mills theory

dim	gh# 1	0	−1	−2	−3
0	(00000)				
$\frac{1}{2}$	•	•			
1	•	(10000)	•		
$\frac{3}{2}$	•	(00001)	•	•	
2	•	•	•	•	•
$\frac{5}{2}$	•	•	(00010)	•	•
3	•	•	(10000)	•	•
$\frac{7}{2}$	•	•	•	•	•
4	•	•	•	(00000)	•
$\frac{9}{2}$	•	•	•	•	•

The horizontal direction represents the expansion of the superfield in terms of λ whereas the corresponding for the vertical (in each row) is θ (downward). The irreducible representations of the component fields are listed at the positions which describe their ghost numbers and dimensions

the agreement between the numbers in the polynomial $1 - 10t^2 + 16t^3 - 16t^5 + 10t^6 - t^8$ and the dimensions of the modules in Table 4.1. In addition, the signs of the monomials indicate the bosonic (plus) or fermionic (minus) character of the cohomologies (remember that Ψ is fermionic, so all signs change). This property is of course expressible also in the more refined partition $\mathscr{P}$, which can be shown to be

$$\begin{aligned}\mathscr{P}(t) = &\left(\bigoplus_{k=0}^{\infty} \vee^k (00010) t^k\right)\\ &\otimes \big((00000) \oplus (10000)(-t^2) \oplus (00001)t^3\\ &\oplus (00010)(-t^5) \oplus (10000)t^6 \oplus (00000)(-t^8)\big)\ , \qquad (4.24)\end{aligned}$$

where $\vee$ denotes the symmetric product, and the first line is the refined partition function for an unconstrained spinor. This unconstrained factor can formally be written as $(1-t)^{-(00010)}$, see [114], where the pure spinor partition function is related to a certain Borcherds algebra.

4.3.2 Supergravity

The analogous procedure can be performed for $D = 11$ supergravity, and the resulting zero mode cohomologies [57] are listed in Table 4.2. This list is based on the cohomologies in a *scalar* superfield of ghost number 3 and dimension -3, i.e., the field Ψ of Sect. 4.2.2, based on the superspace 3-form. This field must indeed be taken as the basic field of $D = 11$ supergravity, since the "geometric field" Φ^a does not exhibit the gauge invariance of the C-field—only the field strength H appears in the torsion—so one can not hope to reproduce the Chern–Simons term of the action of (4.12) from Φ^a alone (although the equations of motion are reproducible, one of them being the Bianchi identity for H). We will not bother to write down the detailed partition function for the $D = 11$ pure spinor [111]; the relation to the cohomology is completely analogous to the case of SYM.

The reason for Ψ having ghost number -3 is now obvious; the lowest cohomology represents the ghost for ghost for ghost of the the twice reducible gauge transformations of the 3-form field. Consequently, the "highest" cohomology, the corresponding antifield, is a scalar at $\lambda^7\theta^9$. The content of Table 4.2 verifies that indeed all degrees of freedom of the supergravity are present at λ^3, also the gravitational ones (and even some without local degrees of freedom, related to the Weyl invariance of [106]). We also note the presence of ghosts for diffeomorphisms and local supersymmetry, appearing alongside the ghost for tensor gauge transformations at λ^2. As in the SYM case, the zero mode cohomology (and the partition function) is completely symmetric with respect to exchange of fields and antifields.

4.3.3 Other Models

The method may be extended to other models. Specifically, it has been used [71–73] for superconformal models in $D = 3$: the $N = 8$ Bagger–Lambert–Gustavsson (BLG) [115–117] and $N = 6$ Aharony–Bergman–Jafferis–Maldacena (ABJM) [118] models. Here the Chern–Simons connection comes in one (scalar) pure spinor superfield, and the matter multiplets in another, which, in the absence of ghosts, comes in the same module as the scalar fields, subject to a shift symmetry. We refer to the papers [71–73] for details.

We can also note that models containing selfdual fields follow part of the pattern. Take for example the $N = (2, 0)$ tensor multiplet in $D = 6$. Without exhibiting the details [57] here, we note that the correct cohomologies for fields and ghosts are produced. When it comes to "antifields", however, the pattern is broken. The equation of motion for the tensor field is the selfduality of its field strength, and there is no symmetry between fields and antifields in the cohomology. Therefore, equations of motion $Q\Psi = 0$ are meaningful, but the construction of an action along the lines of Sect. 4.5 becomes obstructed.

Table 4.2 The zero mode cohomology in Ψ for $D = 11$ supergravity

dim	gh# 3	2	1	0	−1	−2	−3	−4	−5
-3	(00000)								
$-\frac{5}{2}$	•	•							
-2	•	(10000)	•						
$-\frac{3}{2}$	•	•	•	•					
-1	•	•	(01000) (10000)	•	•				
$-\frac{1}{2}$	•	•	(00001)	•	•	•			
0	•	•	•	(00000)	•	•	•		
$\frac{1}{2}$	•	•	•	(00100) (20000) (00001) (10001)	•	•	•	•	
1	•	•	•	•	•	•	•	•	•
$\frac{3}{2}$	•	•	•	•	(00001)	•	•	•	•
2	•	•	•	•	(10001) (00000) (00100) (20000)	•	•	•	•
$\frac{5}{2}$	•	•	•	•	•	(00001)	•	•	•
3	•	•	•	•	•	(01000) (10000)	•	•	•
$\frac{7}{2}$	•	•	•	•	•	•	•	•	•
4	•	•	•	•	•	•	(10000)	•	•
$\frac{9}{2}$	•	•	•	•	•	•	•	•	•
5	•	•	•	•	•	•	•	(00000)	•
$\frac{11}{2}$	•	•	•	•	•	•	•	•	•

Table 4.3 The zero mode cohomology in Ψ for $D = 6$ $N = (1, 0)$ super-Yang–Mills theory

dim	gh#		
	1	0	−1
0	(000)(0)		
$\frac{1}{2}$	•	•	
1	•	(100)(0)	•
$\frac{3}{2}$	•	(001)(1)	•
2	•	(000)(2)	•
$\frac{5}{2}$	•	•	•

4.3.4 Less than Maximal Supersymmetry

The procedure sketched here is not unique for maximally supersymmetric models, although it is there that it seems to have its highest potential. What happens if the method is attempted for a theory with less than maximal supersymmetry? If the pure spinors are appropriately chosen, the traditional superspace formulation should be reproduced also here. This is indeed the case. If such a superspace formulation results in an *off-shell* supermultiplet including auxiliary fields, this also happens in the pure spinor formulation. The result, then, will be a cohomology without the antifields, since we have argued that the presence of antifield cohomology is what puts the physical fields on shell.

This can be illustrated by $N = (1, 0)$ SYM in $D = 6$ [119]. There is an $SU(2)$ R-symmetry, and with standard assignment of Dynkin labels for $Spin(1, 5) \times SU(2)$ we let λ^{α} transform in the module $(001)(1)$. With the pure spinor constraint $(\lambda\gamma^{a}\lambda) = 0$, the only remaining spinor bilinear is the $SU(2)$ triplet selfdual 3-form $(002)(2)$. Note that such a pure spinor is non-trivially constrained, unlike a Cartan pure spinor in $D = 6$, which has no R-symmetry. The superfields in the λ expansion of a scalar pure spinor superfield Ψ are fields $A_{\alpha_1 \ldots \alpha_n}$ in $(00n)(n)$. A direct calculation of the zero mode cohomology, or equivalently, of the pure spinor partition function, gives at hand that cohomology only occurs at λ^0 (the ghost) and λ^1 (the physical fields). No higher cohomologies exist, and there is no room for equations of motion for the physical fields. The cohomology is listed in Table 4.3, where it is clear that in addition to the gauge connection and fermion field, the triplet of auxiliary fields also appears.

Since all equations of motion follow from setting the auxiliary fields to zero, it is natural that the antifields should occur as cohomology of a separate pure spinor superfield of dimension 2 and ghost number −1 transforming as a triplet. This is indeed the case. The antifields (or, the current multiplet) is described by a pure

Table 4.4 The zero mode cohomology in Ψ^{*I} for the antifields of $D = 6$ $N = (1, 0)$ super-Yang–Mills theory

dim	gh# -1	-2	-3
2	(000)(2)		
$\frac{5}{2}$	(010)(1)	•	
3	(100)(0)	•	•
$\frac{7}{2}$	•	•	•
4	•	(000)(0)	•
$\frac{9}{2}$	•	•	•

spinor superfield Ψ^{*I}, which has a shift symmetry of the form

$$\Psi^{*I} \approx \Psi^{*I} + (\lambda\sigma^I\rho) . \tag{4.25}$$

The cohomology in Ψ^* is the mirror of the one in Ψ, and listed in Table 4.4.

The condition for Ψ being on-must be separately formulated as another condition $s^I\Psi = 0$, where s^I is an operator with ghost number -1 and dimension 2, such that $s^I\Psi$ effectively starts out with the auxiliary field [119].

Similar considerations could be applied to other non-maximally supersymmetric models. It has been used to check the multiplet structure of $D = 3$, $N = 8$ supergravity [120]. The cohomology (Cederwall, unpublished) of $D = 10$, $N = 1$ SG has also been verified to agree with known results [122, 123].

4.4 Pure Spinor Space and Integration

As noted in Sect. 4.2.1, if a reasonable (non-degenerate) integration measure $[dZ]$ (Z denoting the ordinary superspace coordinates together with the pure spinor variables) can be found, an action of the form

$$S = \frac{1}{2}\int [dZ]\Psi Q\Psi + \text{interactions} \tag{4.26}$$

will provide an off-shell formulation of the model in question, and a solution to the problem of finding an action for maximally supersymmetric models. In view of the discussion on cohomology of the previous section, such an action would be a classical Batalin–Vilkovisky (field–antifield) action (see Sect. 4.5).

A measure on the pure spinor space has to fulfil a number of requirements. First, as already noted, it has to be non-degenerate in order that the variation of the action actually implies the equations of motion $Q\Psi = 0$. In addition, and depending on the model at hand, there are restrictions on the dimension and ghost number of the integration.

For the case of $D = 10$ SYM, Ψ has ghost number 1 and dimension 0. Therefore $\int[dZ]$ needs to have have ghost number -3, and since $\frac{1}{g^2}\int d^{10}x\, d^{16}\theta$ has dimension $-4 + \frac{1}{2} \times 16 = 4$, "$\int[d\lambda]$" must have dimension 4. Correspondingly, in $D = 11$ SG, the pure spinor integration measure must contribute ghost number -7 and, since the dimension of $\frac{1}{\kappa^2}\int d^{11}x\, d^{32}\theta$ is $-2 + \frac{1}{2} \times 32 = 14$ and that of Ψ is -3, it also must give dimension -8. In addition the measures should have the property that $\int[dZ]Q\Lambda = 0$, so that BRST-trivial states have zero integral and partial integration with respect to Q is possible.

The second thing to note is there are natural operations with precisely these quantum numbers. If we check the highest ghost antifield cohomology, they come at $\lambda^3\theta^5$ and $\lambda^7\theta^9$, respectively. So, an "integration" that picks out the corresponding term in the expansion of a pure spinor superfield would have (gh# , dim) $= (-3, 4)$ and $(-7, 8)$ respectively, as desired. This is correct in spirit, but is still a degenerate measure, since the expansion in λ only contains positive powers. Some adjustment is needed.

The solution to this problem was provided, for $D = 10$ pure spinors, by Berkovits [74] with the introduction of so called non-minimal variables. By the introduction of another set of pure spinors called $\bar{\lambda}_\alpha$ and a spinor of fermionic variables r_α which is pure relative $\bar{\lambda}$, i.e., fulfilling $(\bar{\lambda}\gamma^a r) = 0$, the measure could be made non-degenerate. Non-minimal sets of variables are quite standard when it comes to field-antifield quantisation, but the present ones are even more natural, even from a purely geometric point of view. Namely, although solutions to the pure spinor constraints are complex (unless one is in split signature), we have so far assumed that the fields depend on λ and not on $\bar{\lambda}$. Unless we have some kind of residue measure, it seems more natural to integrate over the full complex variable $(\lambda, \bar{\lambda})$. The interpretation of the fermion r_α is as the differential $d\bar{\lambda}_\alpha$ (with the fermionic statistics coming from the wedge product), which obviously satisfies $(\bar{\lambda}\gamma^a d\bar{\lambda}) = 0$ [80]. When more variables are introduced, the BRST operator must be changed accordingly in order to keep the cohomology intact. This is done by adding a term to Q:

$$Q = (\lambda D) + \left(r\frac{\partial}{\partial\bar{\lambda}}\right) = Q_0 + \left(d\bar{\lambda}\frac{\partial}{\partial\bar{\lambda}}\right) = Q_0 + \bar{\partial}\,, \tag{4.27}$$

where $\bar{\partial}$ is the antiholomorphic exterior derivative, the Dolbeault operator. The cohomology is unchanged, and any cohomology will have a representative that is independent of $\bar{\lambda}$ and $d\bar{\lambda}$.

A field $\Psi\left(x, \theta; \lambda, \bar{\lambda}, d\bar{\lambda}\right)$ is then seen as an antiholomorphic form on pure spinor space (meaning, it can depend on both λ and $\bar{\lambda}$, but has only antiholomorphic indices, seen as a tensor). A suitable assignment of quantum numbers for $\bar{\lambda}$ and $d\bar{\lambda}$ is that

$\bar{\lambda}$ has ghost number -1 and dimension $\frac{1}{2}$ (the opposite to λ), while $d\bar{\lambda}$ has ghost number 0 and dimension $\frac{1}{2}$ (there is some irrelevant arbitrariness in the assignment, as long as it comes out right for the BRST operator).

Suppose that the integration can be written as an integral of a form over the pure spinor space. Since no fields contain $d\lambda$, the integration measure needs to contain a top form Ω with the maximum number of holomorphic indices. In $D = 10$, this number is 11 (see below). In order for partial integration of $\bar{\partial}$ to be allowed, this form should in addition depend on λ only, so that $\bar{\partial}\Omega = 0$. We now try an expression for the full integral over the non-minimal pure spinor variables,

$$\int [d\lambda]X = \int \Omega \wedge X \,. \tag{4.28}$$

Again counting quantum numbers (for the $D = 10$ case), the λ and $\bar{\lambda}$ integrals cancel, while the r integration ("removal of $d^{11}\bar{\lambda}$") provides ghost number 0 and dimension $-\frac{11}{2}$. In order to land at the desired quantum numbers for the integration, ghost number -3 and dimension -4, the components of Ω must have ghost number -3 and dimension $\frac{3}{2}$, which is accomplished by precisely three negative powers of λ,

$$\Omega \sim \lambda^{-3} d^{11}\lambda \tag{4.29}$$

(we leave it as a trivial exercise to show that the same applies to any assignment of quantum numbers to $\bar{\lambda}$ and $d\bar{\lambda}$ that respects the ones of Q, and that the assignments for $d\lambda$ are irrelevant).

The requirement that the holomorphic top form with $\bar{\partial}\Omega$ exists is equivalent to the existence of a Calabi–Yau structure on the pure spinor space, defined by Ω. There is indeed a unique $Spin(10)$-invariant Calabi–Yau metric (up to a scale) on the pure spinor space, following from the Kähler potential [124]

$$K(\lambda, \bar{\lambda}) = \left(\lambda\bar{\lambda}\right)^{8/11} \,. \tag{4.30}$$

The pure spinor constraint may be solved in a basis where manifest $Spin(10)$ is broken to $SU(5) \times U(1)$. Then, $\mathbf{16} \to \mathbf{1}_{-5/2} \oplus \mathbf{10}_{-1/2} \oplus \bar{\mathbf{5}}_{3/2}$, and a spinor is represented by a 0-form ℓ, a 2-form Λ and a 4-form M. The pure spinor constraint reads $\ell M - \frac{1}{2}\Lambda \wedge \Lambda = 0$, so the 11 coordinates can be taken as ℓ and Λ in a patch where $\ell \neq 0$. It is obvious that

$$\Omega = \ell^{-3} d\ell d^{10}\Lambda \tag{4.31}$$

has vanishing $U(1)$ charge, and it can be checked that it is fully $Spin(10)$-invariant. In [124], it was checked by explicit calculation that this is the Calabi–Yau top form corresponding to the Kähler potential (4.30). It can of course also be given a covariant form. The expression

$$\Omega \sim (\lambda\bar{\lambda})^{-3}\bar{\lambda}_{\alpha_1}\bar{\lambda}_{\alpha_2}\bar{\lambda}_{\alpha_3}\star T^{\alpha_1\alpha_2\alpha_3}{}_{\beta_1\ldots\beta_{11}} d\lambda^{\beta_1} \wedge \ldots \wedge d\lambda^{\beta_{11}}$$

is indeed independent of $\bar{\lambda}$ [125] (which thus can be replaced by any constant spinor), where the the tensor T is precisely what, after dualisation of the 11 antisymmetric lower indices to 5 upper ones, defines the ghost antifield cohomology,

$$\begin{aligned}\Psi &\sim T_{\alpha_1\alpha_2\alpha_3,\beta_1\beta_2\beta_3\beta_4\beta_5}\lambda^{\alpha_1}\lambda^{\alpha_2}\lambda^{\alpha_3}\theta^{\beta_1}\theta^{\beta_2}\theta^{\beta_3}\theta^{\beta_4}\theta^{\beta_5}\\ &\sim (\lambda\gamma^a\theta)(\lambda\gamma^b\theta)(\lambda\gamma^c\theta)(\theta\gamma_{abc}\theta)\ .\end{aligned} \tag{4.32}$$

This whole procedure may be repeated for the $D = 11$ pure spinors. The introduction of non-minimal variables is completely analogous, as is the formulation of the integration in terms of a Calabi–Yau top form. The dimension of the pure spinor space is 23, which can be deduces from an explicit solution similar to the one for $D = 10$. When $Spin(11) \to SU(5) \times U(1)$,

$$\mathbf{32} \to \mathbf{1}_{-5/2} \oplus \mathbf{5}_{-3/2} \oplus \mathbf{10}_{-1/2} \oplus \bar{\mathbf{10}}_{1/2} \oplus \bar{\mathbf{5}}_{3/2} \oplus \mathbf{1}_{5/2}\ . \tag{4.33}$$

A spinor is thus parametrised by an arbitrary form. If we write it as

$$\lambda = \ell \oplus \bigoplus_{p=1}^{5} \Lambda_p \tag{4.34}$$

(ℓ being the 0-form, and the subscript p denoting form degree), the solution to the pure spinor constraint is

$$\begin{aligned}\Lambda_3 &= \ell^{-1}\Lambda_1 \wedge \Lambda_2 + \Sigma,\\ \Lambda_4 &= \ell^{-1}(-\Lambda_1 \wedge \Lambda_3 + \tfrac{1}{2}\Lambda_2 \wedge \Lambda_2),\\ \Lambda_5 &= \ell^{-2}\Lambda_2 \wedge \Lambda_3 - \tfrac{1}{2}\Lambda_1 \wedge \Lambda_2 \wedge \Lambda_2,\end{aligned} \tag{4.35}$$

where Σ is a 3-from satisfying

$$\imath_v \Sigma \wedge \Sigma = 0 \tag{4.36}$$

for all vectors v, i.e., $\epsilon^{jklmn}\Sigma_{ijk}\Sigma_{lmn} = 0$ [111, 126].

An important difference compared to the $D = 10$ pure spinors is that there is a singular locus away from the origin, where the 3-form Σ vanishes. It is straightforward to see that then $(\lambda\gamma^{ab}\lambda) = 0$. This is the space of $D = 12$ Cartan pure spinors, a 16-dimensional space. The degrees of freedom contained in Σ consists, modulo a scale, of the Grassmannian $Gr(2,5) = \frac{SU(5)}{S(U(3)\times U(2))}$ of 2-planes in 5-dimensions. So the appearance of Σ provides 14 more real, or 7 complex dimensions, to make a total of 23. A similar parametrisation of the solution of the constraint on Σ in terms of modules of $su(3) \oplus su(2) \oplus u(1)$, with s being the singlet, gives at hand that the the measure, i.e., the holomorphic top form carries the factor $\ell^{-5}s^{-2}$ [126], and here is the ghost number -7 as announced. Again, the measure can be cast in a Lorentz-covariant form, but we will not go into the details (see [69, 100, 127]). The above reflects the fact that the top cohomology at $\lambda^7\theta^9$ contains 2 powers of

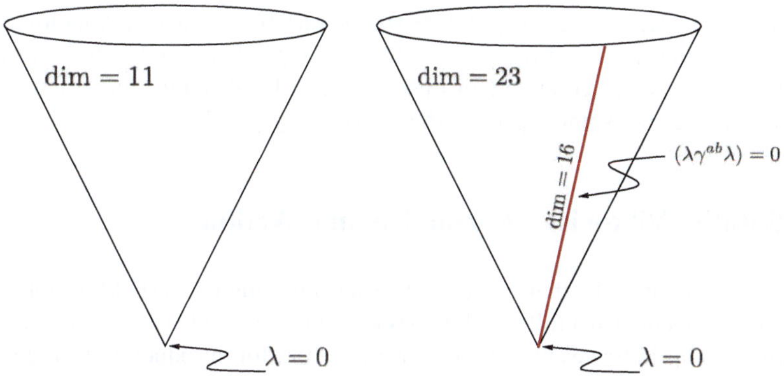

Fig. 4.1 A rough sketch of the $D = 10$ and $D = 11$ pure spinor spaces, with their respective singular subspaces marked

$(\lambda\gamma^{(2)}\lambda)$. The corresponding Kähler potential and metric have not been explicitly constructed, but this should be straightforward.

We finally want to say a few words about integration and regularisation [74]. It was mentioned that the cohomology, also after the introduction of $(\bar{\lambda}, d\bar{\lambda})$, has representatives that are independent of these variables. In other words, they are holomorphic functions (0-forms). How can integrals of (products of) such functions give a non-vanishing result? One will always obtain 0, due the undersaturation of the form degree (the fermionic variables). On the other hand, the polynomial behaviour of the cohomologies at infinity gives ∞, if radial integration is performed first. The integrals are ill-defined, of the form $0 \times \infty$. This can been remedied in two (equivalent) ways. Either we note that the representatives in the minimal variables are a bad choice, and change them into some BRST-equivalent representatives that give well-defined integrals, or we use a BRST-invariant regularisation of the measure. The same type of regulator, an expression of the form $e^{-t\{Q,\chi\}}$, works in both cases. A standard choice for χ is $\chi = \theta^\alpha \bar{\lambda}_\alpha$, giving a regulator

$$e^{-t((\lambda\bar{\lambda})+(\theta d\bar{\lambda}))} . \tag{4.37}$$

If such a regulated measure (with $t > 0$) is used with the minimal representatives, we see that it regulates the bosonic integrals at infinity. At the same time 11 ($D = 10$) or 23 ($D = 11$) $d\bar{\lambda}$'s are needed to saturate the form degree (fermionic integral), and the corresponding term in the expansion of the exponential carries 11 (23) θ's. In order to saturate the θ integration, another 5 (9) are needed, and we see that this agrees with picking out the top cohomology, as was the first, too naïve, candidate for integration. It is thus no coincidence that the number of θ's in the top cohomology agrees with the number of independent constraints on a pure spinor (Fig. 4.1).

The regulated integrals will of course be independent of the parameter t. This looks much like localisation—taking t to be very big localises the integral close to the origin. The dependence on the pure spinor variables is indeed "topological", in the sense that they do not provide new functional dependence, only a finite spectrum. We have not seen any good way of making use of localisation. The origin is not a regular point in pure spinor space, rather a boundary [128].

4.5 Batalin–Vilkovisky Formalism and Actions

We have seen in Sect. 4.3 that the content of the pure spinor superfields is not only the physical fields, but also a full set of ghosts and antifields (at least for maximal supersymmetry). This indicates that the proper framework for introducing interactions (so far, everything has been at a linearised level) is the Batalin–Vilkovisky formalism [129–131].

4.5.1 Field-Antifield Structure

The Batalin–Vilkovisky (BV) formalism can be thought of in several ways. It seems to have originated as an attempt to find something similar to a Hamiltonian formalism, without breaking manifest Lorentz symmetry, in that sense uniting the advantages of the Lagrange and Hamilton methods. Another way of viewing it is that it naturally lifts the BRST method to possible include nonlinear terms and transformations, i.e., interactions. It should be noted that some textbooks (e.g. [132]) introduce the BV formalism in connection with gauge fixing, which tends to somewhat obscure the simplicity. What we will do here is classical BV field theory, although we will discuss gauge fixing in Sect. 4.7.

In the BV framework, a ghost field is introduced for each gauge symmetry (and reducibility) and each of the fields ϕ^I (by which is meant physical fields as well as ghosts) is supplemented by its antifield ϕ^*_I with opposite statistics and a ghost number assignment fulfilling gh# (ϕ)+gh# ($\phi*$) $= -1$. A fermionic bracket, the so called antibracket, between functions of fields and antifields is introduced as

$$(A, B) = \int d^D x \left(A \frac{\overleftarrow{\delta}}{\delta\phi^I(x)} \frac{\overrightarrow{\delta}}{\delta\phi^*_I(x)} B - A \frac{\overleftarrow{\delta}}{\delta\phi^*_I(x)} \frac{\overrightarrow{\delta}}{\delta\phi^I(x)} B \right) . \tag{4.38}$$

The (classical) BV action is defined as a solution to the master equation

$$(S, S) = 0 , \tag{4.39}$$

which reduces to the action for the physical fields when ghosts and antifields are removed. The action itself generates gauge transformations via the antibracket (in a generalised sense, where e.g. antifields are transformed by the equations of motion for the physical fields), so the master equation (4.39) can be seen as the invariance of the action itself.

In the situation at hand, with the pure spinor superfields for maximally supersymmetric theories, we have seen that the cohomology describes both fields and antifields, so a split in the two sets looks problematic. In addition, it is of course necessary to define the antibracket off shell, so that also components outside cohomology takes part. The field–antifield symmetry of the cohomology makes it natural to think of a field Ψ as self-conjugate with respect to the antibracket, and define it as [69]

$$(A, B) = \int A \frac{\overleftarrow{\delta}}{\delta\Psi(Z)} [dZ] \frac{\overrightarrow{\delta}}{\delta\Psi(Z)} B \, . \tag{4.40}$$

It is straightforward to show that this antibracket (in all cases we have considered) carries the correct quantum numbers, and that a free action of the form

$$S_2 = \frac{1}{2} \int [dZ] \Psi Q \Psi \tag{4.41}$$

indeed generates gauge transformations. At this non-interacting level, the master equation is equivalent to the nilpotency of the BRST operator. Actions of this form thus describes both SYM and SG at linearised order.

4.5.2 Interactions from the Master Equation

We now have at our disposal all ingredients necessary to introduce interactions in a consistent way. The guiding principle is the master equation (4.39).

4.5.2.1 SYM

The SYM case is easy. The linearised action has the form of an abelian Chern–Simons action, and since Ψ and Q carry the same quantum numbers a Ψ^3 term can be added, turning the full action into Chern–Simons form,

$$S = \int [dZ] \operatorname{tr} \left(\frac{1}{2} \Psi Q \Psi + \frac{1}{3} \Psi^3 \right) \, . \tag{4.42}$$

This leads to equations of motion

$$Q\Psi + \Psi^2 = 0 \, , \tag{4.43}$$

which could of course equally well be directly deduced from the superspace formalism, where its restriction to the ghost number zero fields reads $\lambda^\alpha \lambda^\beta F_{\alpha\beta} = 0$.

A notable feature is that although the component action contains 4-point couplings, such terms are not present in the manifestly supersymmetric pure spinor superfield action. Instead they are reproduced when the equations of motion are solved sequentially in the θ expansion of the superfields A_α. Such simplifications are typical. We mentioned them in passing for the 3-dimensional conformal models of Sect. 4.3.3, and similar simplifications turn out to happen also for supergravity.

4.5.2.2 SG

The interactions of $D = 11$ supergravity [69, 70] are more subtle. Remember that Q has ghost number 1 and dimension 0, while Ψ has ghost number 3 and dimension -3. The first step will be to construct a 3-point coupling. How can it be formed, given that the integrand in the action must have ghost number 7 and dimension -6?

Here, the geometric field Φ^a comes into play. We remind that it has ghost number 1 and dimension -1. It contains the field strength H but not the potential C. Guided by the form of the Chern–Simons term $C \wedge H \wedge H$, is it possible that something like $\Psi \Phi^2$ may work? Such a combination has ghost number 5 and dimension -5. If it is supplemented by two powers of λ, the quantum numbers are the correct ones. A hypothetical 3-point coupling is then

$$S_3 \sim \int [dZ](\lambda\gamma_{ab}\lambda)\Psi\Phi^a\Phi^b \,. \tag{4.44}$$

Apart from the matching of quantum numbers, the factor $(\lambda\gamma_{ab}\lambda)$ has two other rôles: the antisymmetry in $[ab]$ makes it possible to contract the indices on the (fermionic) Φ fields; and it ensures the invariance under the shift symmetry of (4.16), thanks to the Fierz identity $(\gamma^b\lambda)_\alpha(\lambda\gamma_{ab}\lambda) = 0$, satisfied by a pure spinor λ (but not by an unconstrained one).

This is of course not the final answer for the 3-point coupling. We have argued that Ψ is the fundamental field, but (4.44) is meaningless until we declare how Φ^a is formed from Ψ. Let us assume that there is some operator R^a of ghost number -2 and dimension 2 (defined modulo shift symmetry) such that

$$\Phi^a = R^a\Psi \,. \tag{4.45}$$

Then the master equation, stating the consistency of the tentative 3-point coupling, demands that $[Q, R^a] = 0$ (again modulo shift symmetry). Such an operator was constructed in [69], and it takes the form

$$R^a = \eta^{-1}\left(\bar\lambda\gamma^{ab}\bar\lambda\right)\partial_b + \cdots \,, \tag{4.46}$$

where $\eta = (\lambda\gamma^{ab}\lambda)(\bar{\lambda}\gamma_{ab}\bar{\lambda})$ is the scalar invariant vanishing on the the codimension-7 subspace of 12-dimensional pure spinors, and where the ellpsis denotes terms with $d\bar{\lambda}$ and $d\bar{\lambda}^2$.

This means that we have a consistent 3-point interaction. It is clearly also non-trivial, and since already the 3-point coupling for gravity is cohomologically unique [133], it must be the full 3-point coupling of $D = 11$ SG in Minkowski space. A concrete check on component field couplings would nevertheless be encouraging. In [69, 70], it has been verified that the Chern–Simons term is correctly reproduced, and that the ghost couplings corresponding to the diffeomorphism algebra are the right ones.

Surprisingly, the 3-point interactions provide almost the full answer. When checking the master equation to higher order in Ψ, a very simple 4-point coupling arises, containing a simple nilpotent operator T. The properties of this operator ensures that the master equation is satisfied to all orders, and the full action for $D = 11$ SG is

$$S = \int [dZ] \left[\frac{1}{2}\Psi Q\Psi + \frac{1}{6}(\lambda\gamma_{ab}\lambda)\left(1 - \frac{3}{2}T\Psi\right)\Psi R^a\Psi R^b\Psi \right] . \qquad (4.47)$$

We refer to [70] for the details.

Strikingly enough, the full action for $D = 11$ supergravity becomes polynomial. The 4-point coupling may even be removed by a field redefinition (at the price of having a redefined field which is not canonical with respect to the antibracket, and has a less standard kinetic term). However, it should be said that geometry is somewhat obscured. By basing the formulation on the lowest-dimensional part of the superspace fields, and treating the fields as deformation of the flat background, geometry is not manifest. Still, the appearance of all ghosts, including the ones for diffeomorphisms and local supersymmetry, in the cohomology, together with the master equation, ensures full gauge invariance, although in a form that is not easily recognisable as geometric. Therefore it may be interesting to try to "rebuild" a geometric picture based on the present formalism. We do not have any concrete ideas about how this may be done, but it might involve further variables, reintroducing the superfields that were discarded (the higher-dimensional parts of the super-vielbein). Formally, an analogue statement is true for the SYM action, but the simple Chern–Simons form there makes gauge invariance (almost) manifest. In close connection with this, it is not clear how to best find solutions to the equations of motion. It is not known even how to embed simple, purely gravitational, solutions like the Schwarzschild geometry into the superfield Ψ. For perturbation theory around flat space, on the other hand, the formulation is ideal, both for keeping control over the symmetries and for having a very limited number of couplings, and it has been used for amplitude calculations [126, 134].

4.5.2.3 Other Models

Actions, along the lines drawn up here, can also be constructed for the BLG and ABJM models described briefly in Sect. 4.3.3. Since the fields describing the scalar multiplets are non-scalar, their kinetic terms contain extra λ's ensuring shift symmetry. The interactions consist essentially of a minimal coupling to the Chern–Simons field, replacing and reproducing the higher order interactions among the component fields (e.g. a sixth order potential in the scalars). We again refer to [71–73] for details.

In principle, actions could be formed also for models with less supersymmetry. Then we know from the discussion in Sect. 4.3.4 that separate pure spinor superfields must be introduced for the fields and the antifields. The full formalism for lower supersymmetry has not been developed. In [119] minimal $D = 6$ SYM was treated, but only at the level of equations of motion, and in a minimal pure spinor formalism. Especially issues concerning gauge fixing may turn out to be easier in such models (see Sect. 4.7). In particular, $D = 10$, $N = 1$ supergravity and its dimensional reductions may be interesting, e.g. concerning the investigation of possible counterterms.

4.6 Higher Derivative Terms and Born–Infeld Theory

As an example of an application of our formalism, we will briefly describe the construction of a higher-derivative term. Even though the example is specific—the F^4 deformation of $D = 10$, $N = 1$ SYM, it may be applied to any supersymmetric deformation of a maximally supersymmetric model with a pure spinor action. As we will see, the drastic simplifications of interaction terms persist also here, and although an F^4 deformation in component language will come together with an infinite number of terms of arbitrarily high order in derivatives, a single quartic term turns out to contain the full deformation in the pure spinor superfield language for the abelian model. We conjecture that it describes Born–Infeld theory.

The question addressed here was actually one starting point for the development of the present formalism [55–59]. The work described in this section is based on [109].

Precisely as for any interaction term, the guide to consistent deformation is the master equation. What is needed is some Ansatz for the form of the interactions. In [55, 58], it was observed that the 5-form part of $F_{\alpha\beta} = 0$ must be changed in order to deform the theory. It was also noted that the appropriate $\alpha'^2 F^4$ terms for SYM were generated by

$$F^A_{\alpha\beta} \sim \alpha'^2 t^A{}_{BCD} (\gamma^a \chi^B)_\alpha (\gamma^b \chi^C)_\beta F^D_{ab} \, , \tag{4.48}$$

where t is a symmetric invariant tensor, and χ and F denote the superfields with the corresponding component fields as lowest components. We will from now on drop the explicit factor α'^2. This was then used in [58] in order to derive for the first time the complete deformation at this order, including all fermion couplings.

We need some systematics for lifting expressions like (4.48) to full pure spinor superfield expressions, containing not only fields of definite ghost number. The method introduced in [109] was to form "physical operators", solving this problem. Take for example the physical fermion. We would like to find an operator $\hat{\chi}^\alpha$ that, roughly speaking, strips the pure spinor superfield Ψ of one power of λ and two powers of θ and forms a pure spinor superfield that "starts" with χ^α, and similarly for other component fields. These operators were systematically constructed in the non-minimal formalism. For example, the operator $\hat{\chi}^\alpha$ takes the form

$$\hat{\chi}^\alpha = \frac{1}{2}(\lambda\bar{\lambda})^{-1}(\gamma^a\bar{\lambda})^\alpha\partial_a + \cdots , \tag{4.49}$$

with the ellipsis denoting terms with more singular behaviour in $(\lambda\bar{\lambda})$ and with one or two powers of $d\bar{\lambda}$. The physical operators turn out to satisfy a number of interesting algebraic and differential relations (among them, a somewhat surprising relation to the b operator of Sect. 4.7).

We found that a quartic term in the action

$$S_4 = \frac{1}{4}\int [dZ]\Psi(\lambda\gamma^a\hat{\chi})\Psi(\lambda\gamma^b\hat{\chi})\Psi\hat{F}_{ab}\Psi \tag{4.50}$$

solves the master equation in the Maxwell case, not only to this order but to all orders, and conjectured that it describe supersymmetric Born–Infeld theory. In the non-abelian case, the same term, dressed up with a four-index tensor, describes the full totally symmetric part of the interaction to all orders. We found various ways of rewriting this 4-point coupling in more symmetric ways, and refer to [109] for the details.

The generalisation to supergravity has not been performed, but should not present any other difficulties than purely technical, and may be useful in the search for supersymmetric counterterms. Note that, while in a component language one must make separate Ansätze for the deformed action and the deformed supersymmetry, here everything is uniformly encoded in the master equation.

4.7 Gauge Fixing

We will finally briefly mention gauge fixing, which is an important issue when it comes to quantum calculations and path integrals.

There is a well developed theory of gauge fixing in the BV framework. One must of course eliminate the antifields as independent propagating degrees of freedom, and this is achieved by the introduction of a gauge fermion χ. One then demands that

$$\phi_I^* = \frac{\delta\chi}{\delta\phi^I} \,. \tag{4.51}$$

This makes physical quantities independent of gauge choice. Normally, in a gauge theory, this procedure involves extra non-minimal fields, the "antighost" and Nakanishi-Lautrup fields.

In the pure spinor superfield framework (for maximally supersymmetric models), we have fields Ψ which effectively contain both fields and antifields and are self-conjugate under the antibracket. We can not form a condition like (4.51) without a contrived and unnatural splitting of the field Ψ. Therefore it is necessary to fix the gauge in some other way.

A standard way to fix gauge in string theory is Siegel gauge [135]. The gauge fixing condition is

$$b\Psi = 0 \,, \tag{4.52}$$

where b is a ghost field corresponding to the Virasoro constraint. However, in the pure spinor formalism, no world-sheet or world-line reparametrisation is *a priori* present—as we have seen the equations of motion of the massless fields is an "indirect" consequence of cohomology, and do not follow from "$p^2 = 0$" of some particle model with reparametrisation symmetry. Such a b operator has to be constructed as a composite operator if it exists. This was done for string theory in [74]. The field theory version of this b operator, relevant for SYM, is

$$\begin{aligned} b = &-\frac{1}{2}(\lambda\bar\lambda)^{-1}(\bar\lambda\gamma^a D)\partial_a + \frac{1}{16}(\lambda\bar\lambda)^{-2}(\bar\lambda\gamma^{abc}d\bar\lambda)\left(N_{ab}\partial_c + \frac{1}{24}(D\gamma_{abc}D)\right) \\ &-\frac{1}{64}(\lambda\bar\lambda)^{-3}(d\bar\lambda\gamma^{abc}d\bar\lambda)(\bar\lambda\gamma_a D)N_{bc} \\ &-\frac{1}{1024}(\lambda\bar\lambda)^{-4}(\bar\lambda\gamma^{abe}d\bar\lambda)(d\bar\lambda\gamma^{cd}{}_e d\bar\lambda)N_{ab}N_{cd} \,, \end{aligned} \tag{4.53}$$

where $N_{ab} = (\lambda\gamma_{ab}\frac{\partial}{\partial\lambda})$. The defining property of the b operator is

$$\{Q, b\} = \Box \,. \tag{4.54}$$

The whole purpose of gauge fixing is of course to make the kinetic operator (in this case Q) invertible. With this gauge choice, the propagator G ("Q^{-1}") is formally

$$G = \frac{b}{\Box} \,. \tag{4.55}$$

So, even if b is a complicated operator, it does precisely what is needed for gauge fixing: it eliminates almost all the antifields and implies Lorenz gauge for the gauge connection. By "almost all" we mean that there is a small remainder of the antifield A^{*a}, connected to its on-shell divergencelessness, that gives place for the antighost, which otherwise is normally introduced by hand. That this happens follows from the deliberations in [87].

The consistency of the gauge fixing also relies on the property $b^2 = 0$. This identity is quite cumbersome to show—in string theory so much so that the full calculation was performed only recently [136, 137].

In $D = 11$ the b operator is quite complicated,

$$b = \frac{1}{2}\eta^{-1}(\bar{\lambda}\gamma_{ab}\bar{\lambda})(\lambda\gamma^{ab}\gamma^c D)\partial_c + \cdots \tag{4.56}$$

We will not display it in full detail here, and refer to [126].

The fact that the b operators, and also other operators carrying negative ghost number such as the R^a operator of the supergravity and the physical operators of Sect. 4.6, have quite complicated expression has been the source of some activity searching for simpler versions. See e.g. [78, 81, 126, 138].

Once gauge fixing has thus been performed, it is possible to use the pure spinor superfield formalism for calculation of amplitudes. There will be further (resolvable) questions about regularisation that we will completely forgo here, see [80, 126, 139, 140]. In [126], amplitudes derived from the supergravity action were shown to be finite up to six loops, in agreement with [139, 140] (see the talk presented by Anna Karlsson [134]).

It might be expected that gauge fixing in models with less than maximal supersymmetry can be performed in a way which is more along the standard lines of the BV formalism, i.e., with a gauge fixing fermion, since then fields and antifields are naturally separated in different pure spinor superfields. This remains to be investigated.

4.8 Discussion

We have given a brief overview of the pure spinor superfield formalism, and how it leads to off-shell superfield actions for maximally supersymmetric models. The main focus has been on $D = 10$ SYM and $D = 11$ SG, but also other models have been mentioned. Some of the more technically intricate parts of the formalism have been left out, but we hope that the general message is clear: this is a solution to the problem of going off-shell with maximal supersymmetry.

We have repeatedly pointed out the simplicity of the resulting actions. Indeed, the many terms in a supersymmetric component action generically reduce to some quite simple expression, which is of lower order in fields than the component interactions. In a couple of cases, we even get polynomial expressions where the component ones are non-polynomial. This is of course an advantage when it comes to quantum calculations: the number of vertices is very limited. The other advantage for amplitude calculations is that the presence of an action (as opposed to a first-quantised formalism) directly yields the form of the vertices consistent with all symmetries.

The formulation of supergravity has some drawbacks, though. Since only part of the supervielbein is used, the geometric structure of the theory is obscured. Background invariance is not manifest, since some background is needed in order even to define the BRST operator. In this sense, the behaviour is similar to closed string field theory [135]. It is not clear whether geometry, or some aspects of it can be regained without losing the obvious advantages of the pure spinor formalism. This means also that solutions beyond the linearised level around some background are difficult to find, as is e.g. the dynamics of extended abjects and their coupling to supergravity.

We believe that there is something to learn from the application of pure spinor techniques to theories with less supersymmetry. This is however a largely unexplorsed subject.

Finally, we would be very interested in extending the formalism to other structure groups. The type of models we primarily have in mind are models with "manifest U-duality", formulated as gauge theories within the framework of generalised geometry. Some supermultiplets are already known in connection with U-duality [141–144], and it would be very interesting to continue to a superfield formalism and maybe a (generalisation of the) pure spinor version. A manifest control over both supersymmetry and U-duality would be the ideal situation for examining the ultraviolet properties of maximal supergravity.

References

1. L. Brink, J.H. Schwarz, Quantum superspace. Phys. Lett. B **100**, 310 (1981)
2. R. Casalbuoni, The classical mechanics for Bose-Fermi systems. Nuovo Cim. A **33**, 389 (1976)
3. M.B. Green, J.H. Schwarz, Covariant description of superstrings. Phys. Lett. B **136**, 367 (1984)
4. W. Siegel, Hidden local supersymmetry in the supersymmetric particle action. Phys. Lett. B **128**, 397 (1983)
5. I. Bengtsson, M. Cederwall, Covariant superstrings do not admit covariant gauge fixing. Göteborg-ITP-84-21
6. T. Hori, K. Kamimura, Canonical formulation of superstring. Prog. Theor. Phys. **73**, 476 (1985)
7. E. Bergshoeff, R. Kallosh, Unconstrained BRST for superparticles. Phys. Lett. B **240**, 105 (1990)
8. P.A. Grassi, G. Policastro, M. Porrati, Covariant quantization of the Brink-Schwarz superparticle. Nucl. Phys. B **606**, 380 (2001). arXiv:hep-th/0009239
9. R. Penrose, M.A.H. MacCallum, Twistor theory: an approach to the quantization of fields and space-time. Phys. Rept. **6**, 241 (1972)
10. T. Shirafuji, Lagrangian mechanics of massless particles with spin. Prog. Theor. Phys. **70**, 18 (1983)
11. A.K.H. Bengtsson, I. Bengtsson, M. Cederwall, N. Linden, Particles, superparticles and twistors. Phys. Rev. D **36**, 1766 (1987)
12. I. Bengtsson, M. Cederwall, Particles, twistors and the division algebras. Nucl. Phys. B **302**, 81 (1988)

13. N. Berkovits, A supertwistor description of the massless superparticle in ten-dimensional superspace. Phys. Lett. B **247**, 45 (1990). Nucl. Phys. B **350**, 193 (1991)
14. M. Cederwall, Octonionic particles and the S^7 symmetry. J. Math. Phys. **33**, 388 (1992)
15. E. Witten, An interpretation of classical Yang-Mills theory. Phys. Lett. B **77**, 394 (1978)
16. E. Witten, Twistor-like transform in ten-dimensions. Nucl. Phys. B **266**, 245 (1986)
17. E. Witten, Perturbative gauge theory as a string theory in twistor space. Commun. Math. Phys. **252**, 189 (2004). arXiv:hep-th/0312171
18. N. Berkovits, An alternative string theory in twistor space for $N = 4$ super-Yang-Mills. Phys. Rev. Lett. **93**, 011601 (2004). arXiv:hep-th/0402045
19. N. Berkovits, E. Witten, Conformal supergravity in twistor-string theory. JHEP **0408**, 009 (2004). arXiv:hep-th/0406051
20. F. Cachazo, P. Svrček, E. Witten, Twistor space structure of one-loop amplitudes in gauge theory. JHEP **0410**, 074 (2004). arXiv:hep-th/0406177
21. F. Cachazo, P. Svrček, E. Witten, MHV vertices and tree amplitudes in gauge theory. JHEP **0409**, 006 (2004). arXiv:hep-th/0403047
22. R. Boels, L.J. Mason, D. Skinner, Supersymmetric gauge theories in twistor space. JHEP **0702**, 014 (2007). arXiv:hep-th/0604040
23. T. Adamo, M. Bullimore, L. Mason, D. Skinner, Scattering amplitudes and Wilson loops in twistor space. J. Phys. A **44**, 454008 (2011). arXiv:1104.2890 [hep-th]
24. N. Berkovits, Pure spinors, twistors, and emergent supersymmetry. JHEP **1212**, 006 (2012). arXiv:1105.1147 [hep-th]
25. M. Cederwall, An extension of the twistor concept to string theory. Phys. Lett. B **226**, 45 (1989)
26. N. Berkovits, Twistors and the Green-Schwarz superstring. Conf. Proc. C **9115201**, 310 (1991)
27. J. Wess, B. Zumino, Superspace formulation of supergravity. Phys. Lett. B **66**, 361 (1977)
28. L. Brink, J.H. Schwarz, J. Scherk, Supersymmetric Yang-Mills theories. Nucl. Phys. B **121**, 77 (1977)
29. J. Wess, B. Zumino, Supergauge invariant extension of quantum electrodynamics. Nucl. Phys. B **78**, 1 (1974)
30. S. Ferrara, B. Zumino, Supergauge invariant Yang-Mills theories. Nucl. Phys. B **79**, 413 (1974)
31. R. Grimm, M. Sohnius, J. Wess, Extended supersymmetry and gauge theories. Nucl. Phys. B **133**, 275 (1978)
32. W. Siegel, Superfields in higher dimensional space-time. Phys. Lett. B **80**, 220 (1979)
33. M. Sohnius, K.S. Stelle, P.C. West, Off mass shell formulation of extended supersymmetric gauge theories. Phys. Lett. B **92**, 123 (1980)
34. E. Cremmer, J. Scherk, S. Ferrara, $SU(4)$ invariant supergravity theory. Phys. Lett. B **74**, 61 (1978)
35. E. Cremmer, B. Julia, J. Scherk, Supergravity theory in eleven-dimensions. Phys. Lett. B **76**, 409 (1978)
36. E. Cremmer, B. Julia, The $SO(8)$ supergravity. Nucl. Phys. B **159**, 141 (1979)
37. M.B. Green, J.H. Schwarz, Extended supergravity in ten-dimensions. Phys. Lett. B **122**, 143 (1983)
38. P.S. Howe, P.C. West, The complete $N = 2$, $D = 10$ supergravity. Nucl. Phys. B **238**, 181 (1984)
39. L. Brink, M. Gell-Mann, P. Ramond, J.H. Schwarz, Supergravity as geometry of superspace. Phys. Lett. B **74**, 336 (1978)
40. N. Dragon, Torsion and curvature in extended supergravity. Z. Phys. C **2**, 29 (1979)
41. L. Brink, P.S. Howe, The N = 8 supergravity in superspace. Phys. Lett. B **88**, 268 (1979)
42. L. Brink, P.S. Howe, Eleven-dimensional supergravity on the mass-shell in superspace. Phys. Lett. B **91**, 384 (1980)
43. E. Cremmer, S. Ferrara, Formulation of eleven-dimensional supergravity in superspace. Phys. Lett. B **91**, 61 (1980)

44. B.E.W. Nilsson, Simple ten-dimensional supergravity in superspace. Nucl. Phys. B **188**, 176 (1981)
45. P.S. Howe, Supergravity in superspace. Nucl. Phys. B **199**, 309 (1982)
46. E. Cartan, *Leçons sur la théorie des spineurs* (Hermann, Paris, 1937)
47. C. Chevalley, The algebraic theory of spinors and Clifford algebras (Springer, Berlin, 1996). Collected works.
48. J.-I. Igusa, A classification of spinors up to dimension twelve. Am. J. Math. **92**, 997 (1970)
49. X.W. Zhu, The classification of spinors under GSpin_{14} over finite fields. Trans. Am. Math. Soc. **333**, 93 (1992)
50. V.L. Popov, Classification of spinors of dimension fourteen. Trans. Moscow Math. Soc. **1**, 181 (1980)
51. B.E.W. Nilsson, Pure spinors as auxiliary fields in the ten-dimensional supersymmetric yang-mills theory. Class. Quant. Grav. **3**, L41 (1986)
52. P.S. Howe, Pure spinors lines in superspace and ten-dimensional supersymmetric theories. Phys. Lett. B 258, 141 (1991) [Addendum-ibid. B 259 (1991) 511].
53. P.S. Howe, Pure spinors, function superspaces and supergravity theories in ten-dimensions and eleven-dimensions. Phys. Lett. B **273**, 90 (1991)
54. N. Berkovits, Super Poincaré covariant quantization of the superstring. JHEP **0004**, 018 (2000). arXiv:hep-th/0001035
55. N. Berkovits, Covariant quantization of the superparticle using pure spinors. JHEP **0109**, 016 (2001). arXiv:hep-th/0105050
56. M. Cederwall, B.E.W. Nilsson, D. Tsimpis, The structure of maximally supersymmetric Yang-Mills theory: constraining higher order corrections. JHEP **0106**, 034 (2001). arXiv:hep-th/0102009
57. M. Cederwall, Superspace methods in string theory, supergravity and gauge theory. arXiv: hep-th/0105176
58. M. Cederwall, B.E.W. Nilsson, D. Tsimpis, Spinorial cohomology and maximally supersymmetric theories. JHEP **0202**, 009 (2002). arXiv:hep-th/0110069
59. M. Cederwall, B.E.W. Nilsson, D. Tsimpis, $D = 10$ super-Yang-Mills at $O(\alpha'^2)$. JHEP **0107**, 042 (2001). arXiv:hep-th/0104236
60. M. Cederwall, B.E.W. Nilsson, D. Tsimpis, Spinorial cohomology of abelian $D = 10$ super-Yang-Mills at $O(\alpha'^3)$. JHEP **0211**, 023 (2002). arXiv:hep-th/0205165
61. M. Cederwall, U. Gran, M. Nielsen, B.E.W. Nilsson, Manifestly supersymmetric M theory. JHEP **0010**, 041 (2000). arXiv:hep-th/0007035
62. M. Cederwall, U. Gran, M. Nielsen, B.E.W. Nilsson, Generalized 11-dimensional supergravity. arXiv:hep-th/0010042
63. M. Cederwall, U. Gran, B.E.W. Nilsson, D. Tsimpis, Supersymmetric corrections to eleven-dimensional supergravity. JHEP **0505**, 052 (2005). arXiv:hep-th/0409107
64. P.S. Howe, D. Tsimpis, On higher order corrections in M theory. JHEP **0309**, 038 (2003). arXiv:hep-th/0305129
65. D. Tsimpis, R^4 in type II superstrings. Fortsch. Phys. **56**, 537 (2008)
66. G. Policastro, D. Tsimpis, R^4, purified. Class. Quant. Grav. **23**, 4753 (2006). arXiv:hep-th/0603165
67. V. Alexandrov, D. Krotov, A. Losev, V. Lysov, On pure spinor superfield formalism. JHEP **0710**, 074 (2007). arXiv:0705.2191 [hep-th]
68. M. Movshev, A.S. Schwarz, On maximally supersymmetric Yang-Mills theories. Nucl. Phys. B **681**, 324 (2004). arXiv:hep-th/0311132
69. M.Movshev,A. Schwarz, Supersymmetric deformations ofmaximally supersymmetric gauge theories. JHEP **1209**, 136 (2012). arXiv:0910.0620 [hep-th]
70. M. Cederwall, Towards a manifestly supersymmetric action for 11-dimensional supergravity. JHEP **1001**, 117 (2010). arXiv:0912.1814 [hep-th]
71. M. Cederwall, $D = 11$ supergravity with manifest supersymmetry. Mod. Phys. Lett. A **25**, 3201 (2010). arXiv:1001.0112 [hep-th]

72. M. Cederwall, $N = 8$ superfield formulation of the Bagger-Lambert-Gustavsson model. JHEP **0809**, 116 (2008). arXiv:0808.3242 [hep-th]
73. M. Cederwall, Superfield actions for $N = 8$ and $N = 6$ conformal theories in three dimensions. JHEP **0810**, 070 (2008). arXiv:0809.0318 [hep-th]
74. M. Cederwall, Pure spinor superfields, with application to $D = 3$ conformal models. arXiv: 0906.5490 [hep-th]
75. N. Berkovits, Pure spinor formalism as an $N = 2$ topological string. JHEP **0510**, 089 (2005). arXiv:hep-th/0509120
76. O.A. Bedoya, N. Berkovits, GGI lectures on the pure spinor formalism of the superstring. arXiv:0910.2254 [hep-th]
77. N. Berkovits, Multiloop amplitudes and vanishing theorems using the pure spinor formalism for the superstring. JHEP **0409**, 047 (2004). arXiv:hep-th/0406055
78. P.A. Grassi, P. Vanhove, Topological M theory from pure spinor formalism. Adv. Theor. Math. Phys. **9**, 285 (2005). arXiv:hep-th/0411167
79. I. Oda, M. Tonin, On the b-antighost in the pure spinor quantization of superstrings. Phys. Lett. B **606**, 218 (2005). arXiv:hep-th/0409052
80. N. Berkovits, C.R. Mafra, Some superstring amplitude computations with the non-minimal pure spinor formalism. JHEP **0611**, 079 (2006). arXiv:hep-th/0607187
81. N. Berkovits, N. Nekrasov, Multiloop superstring amplitudes from non-minimal pure spinor formalism. JHEP **0612**, 029 (2006). arXiv:hep-th/0609012
82. I. Oda, M. Tonin, Y-formalism and b ghost in the non-minimal pure spinor formalism of superstrings. Nucl. Phys. B **779**, 63 (2007). arXiv:0704.1219 [hep-th]
83. J. Hoogeveen, K. Skenderis, BRST quantization of the pure spinor superstring. JHEP **0711**, 081 (2007). arXiv:0710.2598 [hep-th]
84. C. Stahn, Fermionic superstring loop amplitudes in the pure spinor formalism. JHEP **0705**, 034 (2007). arXiv:0704.0015 [hep-th]
85. Y. Aisaka, E.A. Arroyo, N. Berkovits, N. Nekrasov, Pure spinor partition function and the massive superstring spectrum. JHEP **0808**, 050 (2008). arXiv:0806.0584 [hep-th]
86. O.A. Bedoya, Superstring sigma model computations using the pure spinor formalism. arXiv: 0808.1755 [hep-th]
87. C.R. Mafra, Superstring scattering amplitudes with the pure spinor formalism. arXiv:0902.1552 [hep-th]
88. Y. Aisaka, N. Berkovits, Pure spinor vertex operators in Siegel gauge and loop amplitude regularization. JHEP **0907**, 062 (2009). arXiv:0903.3443 [hep-th]
89. H. Gomez, One-loop superstring amplitude from integrals on pure spinors space. JHEP **0912**, 034 (2009). arXiv:0910.3405 [hep-th]
90. P.A. Grassi, P. Vanhove, Higher-loop amplitudes in the non-minimal pure spinor formalism. JHEP **0905**, 089 (2009). arXiv:0903.3903 [hep-th]
91. C.R. Mafra, C. Stahn, The one-loop open superstring massless five-point amplitude with the non-minimal pure spinor formalism. JHEP **0903**, 126 (2009). arXiv:0902.1539 [hep-th]
92. N. Berkovits, J. Hoogeveen, K. Skenderis, Decoupling of unphysical states in the minimal pure spinor formalism II. JHEP **0909**, 035 (2009). arXiv:0906.3371 [hep-th]
93. C.R. Mafra, O. Schlotterer, S. Stieberger, D. Tsimpis, Six open string disk amplitude in pure spinor superspace. Nucl. Phys. B **846**, 359 (2011). arXiv:1011.0994 [hep-th]
94. I.Y. Park, Pure spinor computation towards open string three-loop. JHEP **1009**, 008 (2010). arXiv:1003.5711 [hep-th]
95. H. Gomez, C.R. Mafra, The overall coefficient of the two-loop superstring amplitude using pure spinors. JHEP **1005**, 017 (2010). arXiv:1003.0678 [hep-th]
96. M. Tonin, Pure spinor approach to type IIA superstring sigma models and free differential algebras. JHEP **1006**, 083 (2010). arXiv:1002.3500 [hep-th]
97. C.R. Mafra, O. Schlotterer, S. Stieberger, Complete N-point superstring disk amplitude I. Pure spinor computation. Nucl. Phys. B **873**, 419 (2013). arXiv:1106.2645 [hep-th]
98. G. Alencar, M.O. Tahim, R.R. Landim, R.N. Costa Filho, RNS and pure spinors equivalence for type I tree level amplitudes involving up to four fermions. arXiv:1104.1939 [hep-th]

99. I. Oda, M. Tonin, Free differential algebras and pure spinor action in IIB superstring sigma models. JHEP **1106**, 123 (2011). arXiv:1103.5645 [hep-th]
100. H. Gomez, Notes on the overall coefficient of the two-loop superstring amplitude using pure spinor. Fortsch. Phys. **60**, 1030 (2012)
101. N. Berkovits, Towards covariant quantization of the supermembrane. JHEP **0209**, 051 (2002). arXiv:hep-th/0201151
102. M. Babalic, N. Wyllard, Towards relating the kappa-symmetric and pure-spinor versions of the supermembrane. JHEP **0810**, 059 (2008). arXiv:0808.3691 [hep-th]
103. P. Fré, P.A. Grassi, Pure spinors, free differential algebras, and the supermembrane. Nucl. Phys. B **763**, 1 (2007). arXiv:hep-th/0606171
104. N. Berkovits, D.Z. Marchioro, Relating the Green-Schwarz and pure spinor formalisms for the superstring. JHEP **0501**, 018 (2005). arXiv:hep-th/0412198
105. M. Matone, L. Mazzucato, I. Oda, D. Sorokin, M. Tonin, The superembedding origin of the Berkovits pure spinor covariant quantization of superstrings. Nucl. Phys. B **639**, 182 (2002). arXiv:hep-th/0206104
106. Y. Aisaka, Y. Kazama, Origin of pure spinor superstring. JHEP **0505**, 046 (2005). arXiv:hep-th/0502208
107. P.S. Howe, Weyl superspace. Phys. Lett. B **415**, 149 (1997). arXiv:hep-th/9707184
108. S.J. Gates Jr, K.S. Stelle, P.C. West, Algebraic origins of superspace constraints in supergravity. Nucl. Phys. B **169**, 347 (1980)
109. S.J. Gates Jr, W. Siegel, Understanding constraints in superspace formulations of supergravity. Nucl. Phys. B **163**, 519 (1980)
110. M. Cederwall, A. Karlsson, Pure spinor superfields and Born-Infeld theory. JHEP **1111**, 134 (2011). arXiv:1109.0809 [hep-th]
111. N. Berkovits, Cohomology in the pure spinor formalism for the superstring. JHEP **0009**, 046 (2000). arXiv:hep-th/0006003
112. N. Berkovits, N. Nekrasov, The character of pure spinors. Lett. Math. Phys. **74**, 75 (2005). arXiv:hep-th/0503075
113. M. Chesterman, Ghost constraints and the covariant quantization of the superparticle in ten-dimensions. JHEP **0402**, 011 (2004). arXiv:hep-th/0212261
114. M. Chesterman, On the pure spinor superparticle cohomology. Nucl. Phys. Proc. Suppl. **171**, 269 (2007)
115. M. Cederwall, J. Palmkvist, "Serre relations, constraints and partition functions" to appear
116. J. Bagger, N. Lambert, Modeling multiple M2's. Phys. Rev. D **75**, 045020 (2007). arXiv:hep-th/0611108
117. A. Gustavsson, Algebraic structures on parallel M2-branes. Nucl. Phys. B **811**, 66 (2009). arXiv:0709.1260 [hep-th]
118. J. Bagger, N. Lambert, Gauge symmetry and supersymmetry of multiple M2-branes. Phys. Rev. D **77**, 065008 (2008). arXiv:0711.0955 [hep-th]
119. O. Aharony, O. Bergman, D.L. Jafferis, J. Maldacena, N = 6 superconformal Chern-Simons-matter theories, M2-branes and their gravity duals. JHEP **0810**, 091 (2008). arXiv:0806.1218 [hep-th]
120. M. Cederwall, B.E.W. Nilsson, Pure spinors and $D = 6$ super-Yang-Mills. arXiv:0801.1428 [hep-th]
121. M. Cederwall, U. Gran, B.E.W. Nilsson, $D = 3$, $N = 8$ conformal supergravity and the Dragon window. JHEP **1109**, 101 (2011). arXiv:1103.4530 [hep-th]
122. B.E.W. Nilsson, A.K. Tollsten, The geometrical off-shell structure of pure $N = 1$ $D = 10$ supergravity in superspace. Phys. Lett. B **169**, 369 (1986)
123. A. Candiello, K. Lechner, Duality in supergravity theories. Nucl. Phys. B **412**, 479 (1994). arXiv:hep-th/9309143
124. M. Cederwall, The geometry of pure spinor space. JHEP **1201**, 150 (2012). arXiv:1111.1932 [hep-th]
125. N. Berkovits, S.A. Cherkis, Higher-dimensional twistor transforms using pure spinors. JHEP **0412**, 049 (2004). arXiv:hep-th/0409243

126. M. Cederwall, A. Karlsson, Loop amplitudes in maximal supergravity with manifest supersymmetry. JHEP **1303**, 114 (2013). arXiv:1212.5175 [hep-th]
127. L. Anguelova, P.A. Grassi, P. Vanhove, Covariant one-loop amplitudes in $D = 11$. Nucl. Phys. B **702**, 269 (2004). arXiv:hep-th/0408171
128. N.A. Nekrasov, Lectures on curved beta-gamma systems, pure spinors, and anomalies. arXiv: hep-th/0511008
129. I.A. Batalin, G.A. Vilkovisky, Gauge algebra and quantization. Phys. Lett. B **102**, 27 (1981)
130. M. Henneaux, C. Teitelboim, *Quantization of Gauge Systems* (Princeton University Press, Princeton, 1992)
131. A. Fuster, M. Henneaux, A. Maas, BRST quantization: a short review. Int. J. Geom. Meth. Mod. Phys. **2**, 939 (2005). arXiv:hep-th/0506098
132. S. Weinber, The Quantum Theory of Fields, vol. 2 (Princeton University Press, Princeton, 1996) (Modern applications)
133. N. Boulanger, T. Damour, L. Gualtieri, M. Henneaux, Inconsistency of interacting, multigraviton theories. Nucl. Phys. B **597**, 127 (2001). arXiv:hep-th/0007220
134. A. Karlsson, Loop amplitude diagrams in manifest, maximal supersymmetry. Talk presented at BUDS, March 2013, Frascati.
135. W. Siegel, Introduction to String Field Theory. arXiv:hep-th/0107094
136. O. Chandia, The b ghost of the pure spinor formalism is nilpotent. Phys. Lett. B **695**, 312 (2011). arXiv:1008.1778 [hep-th]
137. R. Lipinski Jusinskas, Nilpotency of the b ghost in the non-minimal pure spinor formalism. JHEP **1305**, 048 (2013). arXiv:1303.3966 [hep-th]
138. N. Berkovits, Dynamical twisting and the b ghost in the pure spinor formalism. arXiv:1305. 0693 [hep-th]
139. J. Björnsson, M.B. Green, 5 loops in 24/5 dimensions. JHEP **1008**, 132 (2010). arXiv:1004. 2692 [hep-th]
140. J. Björnsson, Multi-loop amplitudes in maximally supersymmetric pure spinor field theory. JHEP **1101**, 002 (2011). arXiv:1009.5906 [hep-th]
141. A. Coimbra, C. Strickland-Constable and D. Waldram, $E_{d(d)} \times R^+$ generalised geometry, connections and M theory. arXiv:1112.3989 [hep-th]
142. A. Coimbra, C. Strickland-Constable, D. Waldram, Supergravity as generalised geometry II: $E_{d(d)} \times R^+$ and M theory. arXiv:1212.1586 [hep-th]
143. M. Cederwall, J. Edlund, A. Karlsson, Exceptional geometry and tensor fields. JHEP. arXiv: 1302.6736 [hep-th]
144. M. Cederwall, Non-gravitational exceptional supermultiplets. JHEP. arXiv:1302.6737 [hep-th]

Chapter 5
Loop Amplitude Diagrams in Manifest, Maximal Supergravity

Anna Karlsson

The issue of finiteness of maximal supergravity has been subject to research for quite some time. Here, we approach that question through an examination of how to describe amplitude diagrams in $D = 11$ maximal supergravity from a field theory point of view. The strength of the formulation is the presence of manifest supersymmetry through the use of pure spinors. An initial analysis of what the subsequent characteristics turn out to be, partly in lower dimensions through dimensional reduction, gives at hand results that agree with previous work, pointing towards a first divergence for the 7-loop contribution to the 4-point amplitude in four dimensions. The text is mainly based on [1] and may be regarded as an introduction to the main points presented there.

5.1 Introduction

Maximally supersymmetric Yang–Mills theory was proven to be perturbatively finite in up to four dimensions, and no further, in the 1980s. The question of whether or not maximal supergravity provides a well-defined quantum theory on its own in four dimensions, possible to treat perturbatively, has proven far more difficult to settle, though not for a lack of attempts. Investigations through U-duality arguments have pointed at a first possible divergence at 7 loops in four dimensions [2, 3], but the presence and elimination of the counterterms that govern the ultraviolet divergences are uncertain when more than 6 loops are present in the amplitude diagrams.

For the investigations in question, supersymmetry and U-duality properties have proven crucial for simplifications to occur, in order for results to be obtainable. However, the presently most promising examinations are performed through explicit calculations for the 4-graviton amplitude in four dimensions [4–6], that so far have

A. Karlsson (✉)
Fundamental Physics, Chalmers University of Technology, SE-412 96 Gothenburg, Sweden
e-mail: karann@chalmers.se

S. Bellucci (ed.), *Breaking of Supersymmetry and Ultraviolet Divergences in Extended Supergravity*, Springer Proceedings in Physics 153,
DOI: 10.1007/978-3-319-03774-5_5,

reached four loops. They are important as they may provide the way of actually proving the existence of a divergence in the description of the amplitudes. No deviation from the case of maximally supersymmetric Yang–Mills theory has been shown yet, but such are sought for at five loops and, in the general opinion, expected to occur, as well as eventual divergences in the theory.

However, along the lines of utilising as much of the supersymmetric properties as possible, formulations with manifest supersymmetry, obtained through the use of pure spinors, have been investigated as well. In specific, this has been done from a string theory point of view in [7, 8], to begin with, and later in [9, 10], with results corresponding to the ones obtained for the U-duality examinations, as mentioned above.

The last raised the question of what a description of amplitude diagrams in maximal supergravity would look like from a field theory approach, constructed from the superfield action in $D = 11$, $\mathcal{N} = 1$ supergravity [11] which provides a way to examine manifest, maximal supergravity through the use of pure spinors. This is what was presented in [1], including remarks on significant amplitude characteristics, such as the ultraviolet behaviour in a dimensional reduction to $D = 4$, in agreement with the results in [10]. The following text presents the very same thing, in a slightly shorter format and with a focus on the most important concepts of the construction and the analysis that so far have been deduced. Section 5.2 concerns the formulation of manifest, maximal supergravity, Sect. 5.3 contains the field theory description of the amplitude diagrams and Sect. 5.4 describes a way to make a rough estimate of the ultraviolet behaviour. Finally, some desired improvements are mentioned in the outlook.

5.2 Maximal Supergravity with Manifest Supersymmetry

Conventional formulations of maximally supersymmetric theories, such as maximal supergravity, typically have supersymmetry variations that only close on-shell, when the equations of motion are fulfilled. Because of this, it is impossible to formulate an action in terms of superfields in such a formulation. However, a formulation with an action is highly desirable, as the presence of an action simplifies examinations of various characteristics of a theory. This is why a formulation where supersymmetry holds both on- and off-shell, a formulation with manifest supersymmetry, is very useful in many contexts.

5.2.1 Manifesting Maximal Supersymmetry with Pure Spinors

The only manner known so far in which a formulation with manifest, maximal supersymmetry can be obtained is through the introduction of pure spinors [12, 13]. The motivation is quite natural, as the description is built straight from a superspace formulation, modified in order to fit the conventional free, on-shell theory while allowing for an off-shell formulation as well. The construction can be performed for both supersymmetric Yang–Mills theory and supergravity, the latter which is what is discussed here, in a rather brief way. For a more extensive, general description, see for example [14], where more extensive references with respect to the Yang–Mills case, where the procedure was first discovered and developed, also are to be found.

5.2.1.1 The Superspace Formulation of a Free Theory

Typically, the superspace formulation of a maximally supersymmetric theory:

1. Lacks a spinor derivative that maps superfields into superfields, so that such a derivative (D_α) must be introduced into the formulation.
2. Contains too many field components of the same sort (double copies), a redundancy that must be removed through some condition on at least one of the field strengths.

A characteristic of the superspace formulations is that a suitable condition on one of the field strengths removes the unasked for redundancy in a way which respects the Bianchi identities. Moreover, a pattern for how to put fields on-shell and how to allow them to go off-shell in a physical manner emerges.

For maximal supergravity, the spinor derivative acting on the 3-form $C_{\alpha\beta\gamma}$ turns out to represent the equations of motion, with certain irreducible representations removed. This absence of components is exactly what is reproduced by a construction with an operator $Q = \lambda D$ acting on $\lambda^\alpha\lambda^\beta\lambda^\gamma C_{\alpha\beta\gamma}$ with λ^α being a pure spinor; a bosonic spinor of ghost number 1 obeying a constraint [15]:

$$\lambda^\alpha : \quad \lambda\gamma_a\lambda = 0 \tag{5.1}$$

As certain irreducible representations are left untouched as long as the equations of motion are not required to hold, off-shell degrees have been allowed as well. Yet more can be introduced by taking a step away from the original theory; by incorporating the 3-form into a more extensive pure spinor superfield:

$$\psi = \lambda^\alpha\lambda^\beta\lambda^\gamma C_{\alpha\beta\gamma}(x,\theta) + \cdots \tag{5.2}$$

In this way, by allowing the equation of motion for the BRST formulation to be $Q\psi = 0$, a lot of fields, antifields etc. have been allowed into the theory, at other powers of the pure spinor in the series expansion above, at the same time as the original, free theory is obtainable at ghost number zero. Nothing is lost, but a lot is gained, as the necessary components for a formulation with an action have been incorporated into the theory.

This overall construction is useful for a formulation of an action as Q is nilpotent due to the pure spinor condition, and as the equations of motion in this way are given by a BRST operator. The BRST formulation in turn has a natural extension to a theory of interactions in the Batalin–Vilkovisky formalism, so the free theory just described in the pure spinor formalism can be extended to a theory of interactions described by an action.

5.2.1.2 A Theory of Interactions: The Batalin–Vilkovisky Formulation

For a Batalin–Vilkovisky formulation (interactions included) [16, 17] of the BRST pure spinor formulation (a free theory), the symmetry operator Q which in the BRST formulation acts linearly on fields simply is replaced by a generalised action which acts on fields through an antibracket (nonlinearly):

$$(A, B) \sim \int \frac{\delta A}{\delta \psi} \frac{\delta B}{\delta \psi} [dZ] \tag{5.3}$$

The above is the only available option that is similar to a conventional Batalin–Vilkovisky antibracket and possible to form in the pure spinor formalism, as the formalism only contains one superfield, where both fields and antifields reside in a way which is not desirable to alter or split up into components in any way. The formulation has the equation of motion $(S, \psi) = 0$ and the action is obtained through the master equation $(S, S) = 0$, corresponding to the BRST nilpotency demand. For consistency, it is constructed from a starting point in the BRST action:

$$S \sim \int \psi Q \psi [dZ]. \tag{5.4}$$

5.2.2 The Action in the Pure Spinor Formalism

For a proper action, a few more things than the solution to the master equation must be known. For example, the presence of a well-defined integration has been presumed in the reasoning above, and a well-defined gauge fixing ought to be present as well. There is, however, another very important characteristic of this formalism, originating in the BRST equivalence, which is best discussed first, before we get to the parts of and the final expression for the action.

5.2.2.1 BRST Equivalence

Due to the fact that all calculations in maximal supergravity are performed between on-shell, free external states, there is a "hidden" freedom of the theory, such that any term is defined up to BRST equivalent terms, as the equation of motion is $Q\psi = 0$ and Q is nilpotent. In practice, this gives at hand that a factor of 1 or that of $1+\{Q, \chi\}$, with χ a fermion of appropriate dimension and ghost number, are treated on an equal basis. BRST equivalent terms can be added or removed at will, whenever that helps with the interpretation of the characteristics of the pure spinor formalism. The last is especially handy with respect to so called regulators, $e^{\{Q,\chi\}}$, that can be inserted into or removed from an expression at any time.

5.2.2.2 Gauge Fixing Through a Siegel Gauge

The process of gauge fixing normally is performed in order to eliminate the antifields of a formalism, so that a gauge fixing fermion expressed in the fields is given at hand. In the pure spinor formalism though, the single superfield ψ contains both fields and antifields, in a way which we do not want to alter by dividing it into components etc. Conventional gauge fixing is therefore abandoned for an imitation of gauge fixing in string theory through a Siegel gauge [18], which originally is performed for the scalar particle. This process gives at hand a propagator b/p^2 for the theory, which contains a nilpotent b-ghost, the expression for which can be found in [1]:

$$\{Q, b\} = \partial^2 \qquad b\psi_{\text{on-shell}} = 0. \tag{5.5}$$

5.2.2.3 Integration in the Non-minimal Formalism

As to the issue of defining a working integral measure for the formalism, the superspace formulation extended with the presence of the pure spinor falls somewhat short. The most intuitive way of constructing an integral measure would chop off the integrand at some order of (λ, θ) with respect to a series expansion in those variables, which is not wished for. The easiest way to deal with this is through an extension of the superspace to the so called non-minimal formalism, in which two more variables exist: counterparts to $(\lambda^\alpha, \theta^\alpha)$ which are $(\bar{\lambda}_\alpha, r_\alpha)$. In the presence of these "extra" variables, it is possible to construct a regulator so that a well-defined integral measure, in the non-minimal formalism, exists. The first of the "new" variables obey the same condition as the pure spinor, but has the opposite ghost number. The other is a fermionic spinor of 23 degrees of freedom through $r\gamma^a\bar{\lambda} = 0$ [19].

5.2.2.4 General Regularisations

With the extension of the superspace into the non-minimal formalism, the kinetic operator Q is enlarged so that it incorporates the non-minimal variables as well, but the most notable difference is that it is possible to form scalars out of the variables:

$$(\lambda\bar{\lambda}), \quad \left(\lambda\gamma^{ab}\lambda\right)(\bar{\lambda}\gamma_{ab}\bar{\lambda}) \tag{5.6}$$

Negative powers of these can show up in the integrands in a way as to create superficial divergences, and a closer look upon this matter shows that they in fact tend to do just that. However, any operator that is singular with respect to the bosonic spinors can be shown to be BRST equivalent to an operator which shows no such singularity. The process is quite complicated and involves the introduction of a new set of non-minimal superspace variables, or several, and the use of regulators with respect to these; a so called general regularisation. However, the important thing is that such an alteration can be performed. Moreover, it can be performed at any time of the analysis of an integrand, and as the process typically renders any examinations of an integrand impossible to perform or at least highly difficult to go through with, it typically is left implicit up to the point of integration. As the latter process is not performed in practice, though vital to the existence of an action and the examination of different characteristics of an integrand, the process of general regularisation falls within the very same category of utilisation: it is not put into use explicitly [7, 10].

5.2.2.5 The Action in Maximal Supergravity

The final expression for the supergravity action in $D = 11$, $\mathcal{N} = 1$ which is given by the master equation is [11]:

$$S = \frac{1}{\kappa^2} \int [\mathrm{d}Z] \Big(\frac{1}{2} \psi Q \psi + \frac{1}{6} (\lambda \gamma_{ab} \lambda) \left(1 - \frac{3}{2} T \psi \right) \psi R^a \psi R^b \psi \Big) \tag{5.7}$$

Here, the first term originates in the free theory, whereas the second and third term correspond to 3- and 4-point couplings, respectively. A few things can be noted about these terms. To begin with, the couplings show up as there in $D = 11$ maximal supergravity exists one more pure spinor field than ψ, which is expressible in terms of ψ through the operator R^a. This lays the foundation for the description of a 3-point coupling as shown above, and allows for the construction of a 4-point coupling through the master equation, where the operator T turns out to be nilpotent, thus putting an end to what might go into the action, again with respect to the master equation. These terms and operators all have implications for the description of the theory and the amplitude diagrams in it, some of which will be discussed later. However, for a detailed description of the action and its constituent parts, see [1, 11, 15].

5.3 Field Theory Construction of Amplitude Diagrams

In order to capture the field theory description of amplitude diagrams in maximal supersymmetry, as well as to incorporate the desirable characteristic of manifest supersymmetry into the investigations, we start off from the pure spinor action in $\mathcal{N} = 1$, $D = 11$ supergravity, displayed in (5.7). In principle, it tells us what kind of vertices that are part of the theory straight off, and allows us to define the constituent parts of the tree diagrams. In order to capture loop characteristics, that description then needs to be extended slightly, before any amplitude characteristics can be analysed.

5.3.1 Building Blocks from the Action

The first term in the action merely tells us the free behaviour of the pure spinor superfield. When it is off-shell, it propagates with a propagator proportional to b/p^2, as described in Sect. 5.2.2. Moreover, as each end of the propagator is supposed to connect to vertices etc. as a superfield of ghost number 3, the propagator must have ghost number 6. It also needs to connect the sets of variables in the fields, that is, it needs to contain a delta function with respect to the superspace variables. The last has ghost number 7 and the b-ghost has ghost number -1, so the addition gives at hand the correct propagator in a very natural way. The relation in (5.5) also gives at hand a propagator identity:

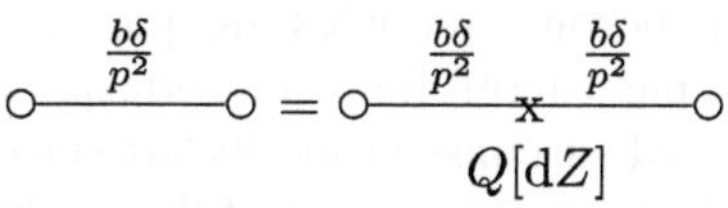

(5.8)

Here, it is implicit that the two propagators on the right hand side depend on two different sets of variables.

As to the second term in the action, it describes a vertex, which in contrast to the propagator can connect to on-shell fields as well as off-shell fields. In specific, it is a 3-point coupling with two operators acting on two of the connected fields. Which ones can be chosen freely, but both operators cannot act on the same field:

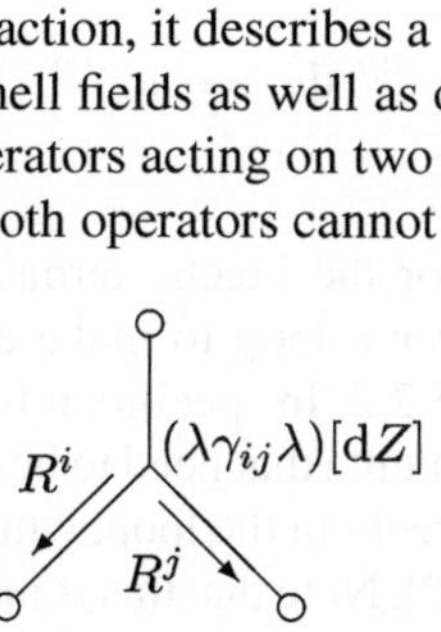

(5.9)

The third term describes a 4-point coupling, which resembles the 3-point coupling except for that it connects to a fourth field and has one more operator acting on one of the fields, different from the other operators:

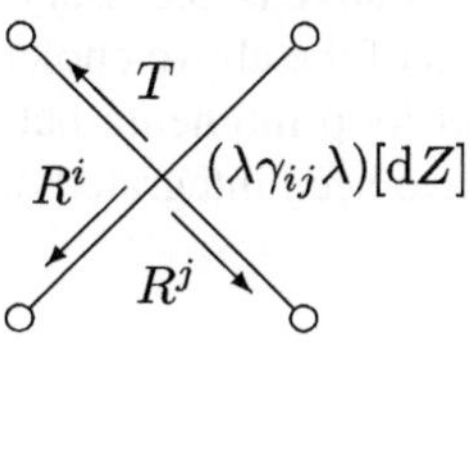

(5.10)

These tree components, the propagator and the two different vertices, are easily combined into any tree diagram. Vertices are connected through propagators, and external fields are connected to vertices. By construction, the total ghost number of any amplitude diagram, may it be a tree or a loop diagram, is zero, as ought to be for a physical quantity.

5.3.2 Loop Regularisation

At an extension of the field theory tree amplitude description to one including loops, the field theory description fails to describe the new degrees of freedom of momenta which are allowed when loops are formed. The propagator expressed in terms of the b-ghost is too local in terms of some of the variables, for example in the pure spinor, where the dependence corresponds to a function $\delta(\lambda_\alpha - \lambda'_\alpha)$ in terms of the two sets of variables that the propagator connects. In a phase space description of loops, this results in a nonsensical integration over $\delta(0)$, and the traditional way of circumventing this problem, by moving to a description in momentum space, is not feasible for two of the bosonic variables: the pure spinor and its non-minimal counterpart. There are no kinetic terms for them, corresponding to what is given for x by the gauge fixing, the last which consequentially has a propagator in the momentum space which is expressible as an exponential of the Laplacian, corresponding to a Gaussian curve in phase space:

$$\frac{1}{p^2} \sim \int_0^\infty e^{-ap^2} \mathrm{d}a \leftrightarrow e^{-a(x-y)^2}, \quad a > 0 \tag{5.11}$$

With no specific description for the kinetic terms, regulators must be introduced in order for the integration over a loop to make any sense. This is performed in the manner described in Sect. 5.2.2. In specific, it is performed in combination with the introduction of the new momenta that need to be present in a loop description: the momenta that may propagate freely in the loop, which also need to be integrated out for a full, final description [7, 10]. Note that this is just one example of making the too local propagator less so. Effectively, the regulators play the role of kinetic terms, but it may also be possible to change the expression for the b-ghost into something BRST equivalent, yet less local. Also, the general regularisation described in Sect. 5.2.2 might do the very same thing, yet as the result of such a regularisation is very difficult to interpret, another alternative is necessary.

There are a few consequences of the above choice of a regularisation of the loop amplitudes, noticeable when all loop momenta but the ones corresponding to the original superspace variables have been integrated out:

- In each loop, there are additional degrees of freedom of loop momenta p_a and loop spinor momenta D_α. Each original momentum becomes split between the different loops the propagator it acts along is part of.
- For each loop integration, 9 spinor derivatives belonging to the loop in question need to be claimed for a non-zero result, the rest may combine into momenta p_a.
- The regulators allow for a transformation of one of the non-minimal variables, the fermionic r_α with 23 degrees of freedom:

$$r_\alpha \rightarrow \begin{cases} (\lambda D)\bar{\lambda}_\alpha \\ (\lambda\gamma_{ab}D)(\bar{\lambda}\gamma^{ab})_\alpha \end{cases} \tag{5.12}$$

- At the final integration over the loop momenta p, the combined terms (with the modifications described above) threaten to have too high a power of p^2 so that the result diverges in the ultraviolet limit. This will be discussed in more detail in Sect. 5.4.

It is also important to note that this regularisation differs greatly from the general regularisation in Sect. 5.2.2 since it alters the behaviour of the constituent parts, as illustrated in (5.12). For example, any term with $r^x\colon x > 23$ is not identically equal to zero, but needs to be transformed for a correct interpretation. Moreover, this cannot be done after antisymmetric properties between these variables have been used, at least not for more than 23 of them, for which the BRST equivalence holds. This matter complicates analyses of the amplitudes slightly, as the constituent parts typically combine to a high power of r_α, so that "new" spinor derivatives that can combine in many different ways show up in the descriptions.

5.3.3 *Amplitude Characteristics*

Despite the complications brought on by the loop regularisations, simple characteristics of the amplitudes can be examined quite easily. Through a look at how the components in the operators in the vertices, the propagator and the loop regularisations may interact to create a final expression for a part of an amplitude, certain characteristics can be discerned.

5.3.3.1 No Bubbles or Triangles

Notably, the no bubble or triangle observation which has been made for amplitude diagrams in maximal supergravity follows directly from the loop regularisation requirement of that 9 spinor derivatives, part of the loop in question, must be claimed for each loop integration. Loops with less than four vertices connected to them simply cannot fulfil this requirement, so no bubbles or triangles can exist.

5.3.3.2 No More than One 4-Point Vertex

A second characteristic that is obvious already from the building blocks is that an amplitude diagram at most can contain one 4-point vertex. This occurs due to the fact that the operator T that is present in the 4-point vertex is nilpotent, so that the presence of two such operators in the same diagram renders the entire expression to zero. Of course, loops with four vertices are still allowed, and after an integration of such a loop, something similar to a 4-point vertex shows up, but as to the number of 4-point vertices that can be present without any impact on the number of loops in the diagram in question, that is, pure vertices, its upper limit is one.

5.3.3.3 Splitting and Combination of Momenta in Different Diagrams

As mentioned above, the introduction of loop momenta transforms for example the momentum p_a according to

$$p_a \to p_a + \sum_I p_a^I \tag{5.13}$$

where the sum is over the different loops the propagator the momentum acts along is part of. The corresponding happens for the other momenta as well, which all of them, apart from the spinor derivative, can be integrated out easily.

Whereas structures with no more than 23 r's present (in a superfluous way, not required to transform), such as bubbles and triangles, easily can be examined, the transformations in (5.12) need to be taken into account otherwise, with BRST equivalence in mind. This represents a quite complicated process for diagrams with a high number of loops, and needs further investigation, especially as it has a decisive impact on the ultraviolet behaviour of the diagrams, the currently discernible one which will be presented in the next section. For example, it would be interesting to know:

- How the operators in the vertices behave in combination with each other.
- If and how the structure of a loop diagram of a certain genus (how the loops are placed with respect to each other) changes its properties with respect to integration over the loop variables.

The last issue, the possibility of transformations between diagrams with the same number of loops, is very interesting. The structures have the same number of constituent components, which seemingly would not allow for the same type of combinations, in a perhaps deceptive way. Note though, that this refers to structures of loops that cannot be divided into blocks with a single propagator in between. Such constructions can be analysed separately, as can tree structures connected to a loop structure.

5.4 Behaviour in the Ultraviolet Limit

The ultraviolet divergences in maximal supergravity expressed in the pure spinor formalism show up due to the integrations over the loop momenta. These are performed in D dimensions for each loop, where contributions show up from the denominators in the propagators and through spinor derivatives D_α that have combined into powers of p^2. What this looks like in the infrared limit is not yet determined, but examinations can be performed in the ultraviolet region. The relevant requirement for finiteness in the ultraviolet limit originates in that the power of momenta in the total expression must be negative, and corresponds to:

$$LD - 2m + 2n < 0, \quad \begin{cases} L\text{: number of loops} \\ m\text{: number of propagators} \\ n\text{: number of } p^2\text{'s made out of } D\text{'s, } n \geq 0 \end{cases} \tag{5.14}$$

Of course, the various combinations are restricted by the characteristics of the loop structures, which are not completely known, but a worst case scenario with the known restrictions can easily be examined.

The procedure for discerning a "worst case scenario" simply consists of a power counting of the r's and D's present in a loop structure, as they can be treated on an equal footing according to (5.12) and $\{D, D\} \propto p$ so that:

$$r \sim D \sim p^{1/2} \tag{5.15}$$

For example, an R^a contains the equivalent of r^2 and the b-ghost r^3. Moreover, a loop structure (no tree structures attached) only constructed out of 3-point vertices contains $3(L-1)$ propagators and $2(L-1)$ inner vertices as well as some required vertices connecting to external states, for example via tree structures.

As many as possible of the r's in an amplitude diagram are then assumed to combine into powers of loop momenta p^2, apart from the ones required for loop integrations and the 23 already allowed. The known characteristics are also taken into account, for example the absence of an impact from:

- The operators in the vertices which without a loss of generality can be assumed to act out of the loop structure, without any impact on the loop integrations.
- Structures with a 4-point vertex, which can be shown not to have any worse an impact than structures made solely out of 3-point vertices.
- Connections to external states or other loop structures through tree diagrams.

In total, this gives at hand the same result for the ultraviolet divergences as given in [9, 10], though the relevant counterterms are not specified. For a high number of loops, the procedures of the two different approaches are very similar, though a bit more straightforward in the field theory approach, as to what goes into the calculation. The resulting demand for ultraviolet finiteness is dependent on the number of loops L in a loop structure and identical to the case of Yang–Mills theory up until the n in (5.14) exceeds zero. For $n \geq 0$ the condition is

$$D < 2 + \frac{14}{L}, \quad L \geq 4 \tag{5.16}$$

which implies a first divergence at 7 loops in 4 dimensions. However, the properties of the transformations in (5.12) do not set in until at 5 loops. What precisely this means for the combination of the loop momenta into p^2 thus remains an open question, so that the real result as of yet is uncertain.

Worthwhile to notice its that the exact same type of investigation for the case of maximally supersymmetric Yang–Mills theory, which do not have operators in the vertices, apart from quite naturally showing the same result up to 4 loops, also shows the well-established requirement with respect to finiteness and dimension versus number of loops present. That is, without the necessity of any further investigations of characteristics of the amplitude diagrams etc. The complications in maximal supergravity seem to be brought on by nothing but the operators in the vertices, and the way they behave.

5.5 Outlook

The results of the examinations of the loop amplitudes in maximal supergravity through a formulation with manifest supersymmetry has as of yet provided no results other than those already known, perhaps with the exception of its advantages of constituting a field theory description. However, further investigations with respect to how the components of the amplitude diagrams may combine are of interest, as well as other examinations.

To begin with, other characteristics may come into play for the behaviour of the amplitude diagrams, for example U-duality. Such arguments are commonplace when discussing these questions in general, but it is not known how to incorporate U-duality into a formulation with pure spinors. A most desirable formulation would have both U-duality and supersymmetry as manifest properties.

Lastly, the more detailed calculations of Bern et al. may give hints at integral, basic properties once the amplitude with 5 loops has been examined in full. The possible outcome ranges between a scenario similar to that which takes place in Yang–Mills theory and the above proposed one, and the result might make it possible to predict the ultraviolet behaviour of the amplitudes in maximal supergravity with more reliability.

References

1. M. Cederwall, A. Karlsson, JHEP **03**, 114 (2013)
2. P. Vanhove, (2010). arXiv:1004.1392 [hep-th]
3. N. Beisert, H. Elvang, D.Z. Freedman et al., Phys. Lett. B **694**, 265 (2010)
4. Z. Bern, L.J. Dixon, D.C. Dunbar et al., Nucl. Phys. B **530**, 401 (1998)
5. Z. Bern, J.J. Carrasco, L.J. Dixon et al., Phys. Rev. Lett. **98**, 161303 (2007)

6. Z. Bern, J.J. Carrasco, L.J. Dixon et al., Phys. Rev. Lett. **103**, 081301 (2009)
7. N. Berkovits, N. Nekrasov, JHEP **12**, 029 (2006)
8. Y. Aisaka, N. Berkovits, JHEP **07**, 062 (2009)
9. J. Björnsson, M.B. Green, JHEP **08**, 132 (2010)
10. J. Björnsson, JHEP **01**, 002 (2011)
11. M. Cederwall, Mod. Phys. Lett. A **25**, 3201 (2010)
12. N. Berkovits, JHEP **04**, 018 (2000)
13. N. Berkovits, JHEP **09**, 016 (2001)
14. M. Cederwall (2013). arXiv:1307.1762 [hep-th]
15. M. Cederwall, JHEP **01**, 117 (2010)
16. I.A. Batalin, G.A. Vilkovisky, Phys. Lett. B **102**, 27 (1981)
17. A. Fuster, M. Henneaux, A. Maas, Int. J. Geom. Meth. Mod. Phys. **2**, 939 (2005)
18. W. Siegel, *Introduction to String Field Theory* (World Scientific, Singapore, 1988)
19. N. Berkovits, JHEP **10**, 089 (2005)

Chapter 6
Perturbative Ultraviolet Calculations in Supergravity

Tristan Dennen

There has been an abundance of recent progress in calculating and understanding the ultraviolet properties of supergravity theories. On the calculation side, the duality between color and kinematics proposed in [1] has opened new avenues for constructing supergravity amplitude integrands at relatively high loop order. In this talk, I explain how the color-kinematics duality is used to construct gravity integrands, and I detail how to extract ultraviolet divergences once the integrands are known. In the specific case of the three-loop four-point amplitude of $\mathcal{N} = 4$ supergravity, this leads to a surprising cancellation of the coefficient of the R^4 counterterm, whose presence was expected based on symmetry arguments [2].

6.1 Ultraviolet Divergences via the Double Copy

An n-point tree-level Yang-Mills amplitude can always be written in the form

$$\mathcal{A}_n = g^{n-2} \sum_i \frac{n_i c_i}{D_i}, \tag{6.1}$$

where the sum is over all trivalent diagrams i, n_i are kinematic numerator factors, c_i are f^{abc} color factors, and D_i are the propagators of diagram i. This is achieved simply by regrouping kinematic terms according to their associated color factors and multiplying numerator and denominator by inverse propagators where necessary.

By a further rearrangement, it is also possible for the kinematic numerator factors to satisfy linear identities in one-to-one correspondence with the color Jacobi identities [1],

T. Dennen (✉)
Niels Bohr International Academy and Discovery Center, Niels Bohr Institute,
University of Copenhagen, Blegdamsvej 17, 2100 Copenhagen Ø, Denmark
e-mail: tdennen@nbi.ku.dk

S. Bellucci (ed.), *Breaking of Supersymmetry and Ultraviolet Divergences in Extended Supergravity*, Springer Proceedings in Physics 153,
DOI: 10.1007/978-3-319-03774-5_6,

$$c_i \pm c_j \pm c_k = 0 \quad \leftrightarrow \quad n_i \pm n_j \pm n_k = 0. \tag{6.2}$$

For the four-point amplitude, simply writing down the Feynman diagrams in Feynman gauge suffices. For five points and above, the required rearrangement is nontrivial.

As conjectured in [1] and proven in [3], once the numerators are put into this color-dual form, one can replace the color factors in the amplitude with another copy of the kinematic numerators to obtain a gravity amplitude,

$$\mathcal{M}_n = i\left(\frac{\kappa}{2}\right)^{n-2}\sum_i \frac{n_i \tilde{n}_i}{D_i}. \tag{6.3}$$

This is called the double-copy method of constructing the gravity amplitude. The two sets of numerators n_i and $\tilde{n}_i$ can come from different Yang-Mills theories, and different choices will lead to different gravity theories. In the later part of this talk, we are concerned with $\mathcal{N} = 4$ supergravity, which is formed as a double copy of pure YM and $\mathcal{N} = 4$ SYM.

We are also primarily interested in multi-loop amplitudes. Luckily, the double-copy method still works to construct multi-loop gravity amplitudes. Again we begin by regrouping Yang-Mills amplitudes into sums over trivalent diagrams,

$$\mathcal{A}_n^{(L)} = i^L g^{n-2+2L}\sum_i \int \prod_{l=1}^{L} \frac{d^D \ell_l}{(2\pi)^D}\frac{1}{S_i}\frac{n_i c_i}{D_i}, \tag{6.4}$$

where the only new ingredients are the loop momentum dependence and the symmetry factors S_i. If one is able to rearrange the kinematic numerators in this expression into a form where they satisfy the same identities as the color factors, then it is possible to replace the color factors c_i with another set of Yang-Mills kinematic numerators $\tilde{n}_i$ to obtain a corresponding multi-loop gravity amplitude [4],

$$\mathcal{M}_n^{(L)} = i^{L+1}\left(\frac{\kappa}{2}\right)^{n-2+2L}\sum_i \int \prod_{l=1}^{L} \frac{d^D \ell_l}{(2\pi)^D}\frac{1}{S_i}\frac{n_i \tilde{n}_i}{D_i}. \tag{6.5}$$

The path to calculating ultraviolet divergences in supergravity is now clear:

1. Find a representation of the SYM amplitude that satisfies the color-kinematics duality.
2. Construct the integrand for the gravity amplitude using the double copy method.
3. Extract the ultraviolet divergences from the integrals.

In the next section, we will discuss some recent advancements regarding point 1. In the subsequent section, we will discuss point 3 in more detail.

6.2 Obtaining Color-Dual Numerators

Finding a representation of a multi-loop SYM amplitude that satisfies the color-kinematics duality is generally a difficult task. Such a representation has been found for the four-point amplitude through four loops [4, 5], and for the five-point amplitude through three loops [6]. Progress on higher-point amplitudes requires a better finesse of the Jacobi identities between the kinematic numerators, and until recently little was known above five points, except in restricted kinematic regimes [7–9].

At one loop, however, it is possible to see explicitly how to use the Jacobi identities to reconstruct the loop-momentum dependence of the numerators from a relatively small ansatz for box numerators. This is well illustrated in the example of the five-point amplitude. Consider the pentagon numerator $n_{\text{pent}}(1, 2, 3, 4, 5; \ell)$, where the loop momentum ℓ flows between legs 5 and 1. By using Jacobi identities to commute leg 1 all the way around the loop, it is possible to obtain an equation for the pentagon numerator in terms of box numerators:

$$n_{\text{pent}}(1, 2, 3, 4, 5; \ell) - n_{\text{pent}}(1, 2, 3, 4, 5; \ell - k_1) = \sum n_{\text{box}} \tag{6.6}$$

This has the structure of a finite difference equation for n_{pent}, and assuming the box numerators are independent of loop momentum (as is the case for $\mathcal{N} = 4$ SYM), it can be solved by a function linear in loop momentum,

$$n_{\text{pent}}(\ldots; \ell) = n_{0,\text{pent}}(\ldots) + n_{1,\text{pent}}(\ldots)_\mu \ell^\mu. \tag{6.7}$$

With this form, the loop momentum drops out of the difference equation, and we are left with

$$k_1 \cdot n_{1,\text{pent}}(1, 2, 3, 4, 5) = \sum n_{\text{box}}. \tag{6.8}$$

It is straightforward to reconstruct $n_{1,\text{pent}}$ from any four independent projections of itself, which gives $n_{1,\text{pent}}$ in terms of box numerators. More generally, if the $(m-1)$-gon numerators are known for a given amplitude, it is possible to reconstruct all of the loop-momentum dependence of the m-gon numerators in the same way (see [10] for further details).

Similarly, it is possible to use the reflection symmetry of the pentagon to obtain $n_{0,\text{pent}}$ in terms of box numerators. Using this method, we have explicitly constructed color-dual numerators for six- and seven-point $\mathcal{N} = 4$ SYM numerators in the MHV and NMHV sectors. We have conjectured that this construction will work at any multiplicity. This also points towards a strategy for obtaining color-dual numerators at higher loops; one can hope to use global properties of the system of Jacobi identities to constrain the loop momentum dependence of numerators in terms of simpler quantities.

6.3 Extracting Ultraviolet Divergences

Once the appropriate kinematic numerators have been rearranged into a color-dual form, multi-loop $\mathcal{N}=4$ gravity amplitudes can be constructed via the double copy method, which yields the amplitude in the form

$$\mathcal{M}_n^{(L)} = i^{L+1}\left(\frac{\kappa}{2}\right)^{n-2+2L}\sum_i \int \prod_{l=1}^{L} \frac{d^D \ell_l}{(2\pi)^D}\frac{1}{S_i}\frac{n_i \tilde{n}_i}{D_i} \tag{6.9}$$

where n_i are numerators from the pure Yang-Mills theory, and $\tilde{n}_i$ are numerators from $\mathcal{N}=4$ SYM. The $\mathcal{N}=4$ SYM copy satisfies the BCJ duality, and was constructed at three loops in [4]. The pure YM copy does not need to satisfy the BCJ duality, so we construct its numerators using Feynman diagrams in Feynman gauge. This has the advantage of being straightforward and easy to pipeline, as well as being D-dimensional. On the other hand, there are a lot of diagrams, so time and memory constraints need to be considered.

Once the integrand of the gravity amplitude has been constructed, the extraction of the ultraviolet divergences proceeds in five steps:

1. Series expand the integrand and select the logarithmic terms.
2. Reduce all of the tensor integrals to scalar integrals using Lorentz invariance.
3. Regulate the infrared divergences with a uniform mass.
4. Subtract away all subdivergences.
5. Evaluate the vacuum integrals.

This procedure is explained in more detail in [14], but we summarise the salient points here.

Each integral in the problem contributes a polynomial to the overall ultraviolet divergence, so we can count up its degree using the differential operator

$$\Delta = \sum k^\mu \frac{\partial}{\partial k^\mu}. \tag{6.10}$$

For example,

$$\Delta \int \frac{d^{6-2\epsilon}\ell}{\ell^2(\ell+k)^2} = 2\int \frac{d^{6-2\epsilon}\ell}{\ell^2(\ell+k)^2}. \tag{6.11}$$

But because Δ is a differential operator, it serves to explicitly extract the polynomial dependence of the integral on the external momenta. After all of the dependence on external momenta has been extracted in this way, we can drop k from the propagators, and we get

$$\int \frac{d^{6-2\epsilon}\ell}{\ell^2(\ell+k)^2} = \int d^{6-2\epsilon}\ell\left\{\frac{4(k\cdot\ell)^2}{[\ell^2]^4} - \frac{k^2}{[\ell^2]^3}\right\} + \mathcal{O}(\epsilon^0). \tag{6.12}$$

This is equivalent to performing a simple series expansion on the integrand and keeping only those terms that are logarithmically divergent by power counting.

The next step in the extraction of the ultraviolet divergence is to reduce the tensor integrals to scalar integrals using the Lorentz invariance of the integral. For example, a rank-two tensor integral must be proportional to the metric tensor,

$$\int d^{6-2\epsilon}\ell \frac{\ell^\mu \ell^\nu}{[\ell^2]^4} = A\eta^{\mu\nu}. \tag{6.13}$$

The scalar integral A is found by contracting both sides with the metric,

$$\int d^{6-2\epsilon}\ell \frac{\ell^2}{[\ell^2]^4} = A(6-2\epsilon). \tag{6.14}$$

In this way, the rank-two tensor integral is given in terms of a logarithmically divergent scalar integral, which is easier to evaluate directly. This generalises straightforwardly to higher-rank tensors.

As described long ago in [11], the infrared divergences can be sidestepped by introducing a uniform mass regulator in all of the integrals. With a mass regulator, the integrals will begin at ϵ_{UV}^{-L} instead of the $\epsilon_{UV}^{-L}\epsilon_{IR}^{-L}$ that comes with a dimensional regulator, and they are thus significantly easier to evaluate. The down side is that the integral-by-integral subdivergences are not guaranteed to cancel from the final answer—in other words, the subleading UV divergences of the integrals are individually regulator-dependent, and so the subdivergences must be subtracted at the level of individual integrals.

The subtraction of subdivergences proceeds recursively,

$$\begin{aligned} S\left[\int \prod_{i=1}^{L} dp_i I\right] &= \text{Div}\left[\int \prod_{i=1}^{L} dp_i I\right] \\ &\quad - \sum_{l=1}^{L-1} \times \sum_{\substack{l\text{-loop} \\ \text{subloops}}} \text{Div}\left[\int \prod_{j=l+1}^{L} dp'_j S\left[\int \prod_{i=1}^{l} dp'_i I\right]\right], \end{aligned} \tag{6.15}$$

where $S[\ldots]$ is a subtracted divergence. This is in direct analogy with how counterterms work, but applied to individual integrals. After subtracting all of the subdivergences, the quantity $S[\ldots]$ is free of infrared regulator dependence.

At the final stage, all of the remaining integrals are scalar, single-scale vacuum integrals. These can be evaluated by reducing them to a basis using IBP relations [12] and Mellin-Barnes techniques [13]. At three loops, there are about 600 such vacuum integrals.

By performing the three-loop four-point ultraviolet calculation in $\mathcal{N} = 4$ supergravity in the manner described here, in [14] we have demonstrated that the R^4 type counterterm is not present. This result was largely unexpected by the supergravity community based on symmetry arguments. Since its appearance, a number of possible explanations have appeared [15–19]. To disentangle these arguments, the calculation of the four-loop divergence will be crucial; this calculation is currently underway.[1]

Acknowledgments The work described in this talk was done in collaboration with Zvi Bern, Scott Davies, and Yu-tin Huang at UCLA, and with Emil Bjerrum-Bohr, Ricardo Monteiro, and Donal O'Connell at the Niels Bohr International Academy.

References

1. Z. Bern, J.J.M. Carrasco, H. Johansson, Phys. Rev. D 78, 085011 (2008). arXiv:0805.3993 [hep-ph]
2. G. Bossard, P.S. Howe, K.S. Stelle, P. Vanhove, Class. Quant. Grav. **28**, 215005 (2011). arXiv: 1105.6087 [hep-th]
3. Z. Bern, T. Dennen, T. Huang, M. Kiermaier, Phys. Rev. D **82**, 065003 (2010). arXiv:1004. 0693 [hep-th]
4. Z. Bern, J.J.M. Carrasco, H. Johansson, Phys. Rev. Lett. **105**, 061602 (2010). arXiv:1004.0476 [hep-th]
5. Z. Bern, J.J.M. Carrasco, L.J. Dixon, H. Johansson, R. Roiban, Phys. Rev. D **85**, 105014 (2012). arXiv:1201.5366 [hep-th]
6. J.J. Carrasco, H. Johansson, Phys. Rev. D **85**, 025006 (2012). arXiv:1106.4711 [hep-th]
7. R. Monteiro, D. O'Connell, JHEP **1107**, 007 (2011). arXiv:1105.2565 [hep-th]
8. N.E.J. Bjerrum-Bohr, P.H. Damgaard, R. Monteiro, D. O'Connell, JHEP **1206**, 061 (2012). arXiv:1203.0944 [hep-th]
9. R.H. Boels, R.S. Isermann, R. Monteiro, D. O'Connell, JHEP **1304**, 107 (2013). arXiv:1301. 4165 [hep-th]
10. N.E.J. Bjerrum-Bohr, T. Dennen, R. Monteiro, D. O'Connell, Integrand Oxidation and One-Loop Colour-Dual Numerators in N=4 Gauge Theory. JHEP **1307**, 092 (2013). arXiv:1303. 2913 [hep-th]
11. N. Marcus, A. Sagnotti, Nuovo Cim. A **87**, 1 (1985)
12. A.V. Smirnov, JHEP **0810**, 107 (2008). arXiv:0807.3243 [hep-ph]
13. M. Czakon, Comput. Phys. Commun. **175**, 559 (2006). arXiv:hep-th/0511200
14. Z. Bern, S. Davies, T. Dennen, Y.-T. Huang, Phys. Rev. Lett. **108**, 201301 (2012). arXiv:1202. 3423 [hep-th]
15. P. Tourkine, P. Vanhove, Class. Quant. Grav. **29**, 115006 (2012). arXiv:1202.3692 [hep-th]
16. R. Kallosh, Phys. Rev. D **85**, 081702 (2012). arXiv:1202.4690 [hep-th]
17. S. Ferrara, R. Kallosh, A. Van Proeyen, Phys. Rev. D **87**, 025004 (2013). arXiv:1209.0418 [hep-th]
18. G. Bossard, P.S. Howe, K.S. Stelle, Phys. Lett. B **719**, 424 (2013). arXiv:1212.0841 [hep-th]
19. G. Bossard, P.S. Howe, K.S. Stelle, Invariants and divergences in half-maximal supergravity theories. JHEP **1307**, 117 (2013). arXiv:1304.7753 [hep-th]
20. Z. Bern, S. Davies, T. Dennen, A.V. Smirnov, V.A. Smirnov, The Ultraviolet Properties of N=4 Supergravity at Four Loops, Phys. Rev. Lett. **111**, 231302 (2013). arXiv:1309.2498 [hep-th]

[1] Since this contribution was written, the four-loop calculation has been completed. See ref. [20].

Chapter 7
Scalars with Higher Derivatives in Supergravity and Cosmology

Michael Koehn, Jean-Luc Lehners and Burt A. Ovrut

We construct $\mathcal{N} = 1$ supergravity extensions of scalar field theories with higher-derivative kinetic terms. Special attention is paid to the auxiliary fields, whose elimination leads not only to corrections to the kinetic terms, but to new expressions for the potential energy as well. Our formalism allows one to write a supergravity extension of any higher-derivative scalar field theory and therefore has applications to both particle physics and cosmological model building. For instance, the ghost condensate vacuum spontaneously breaks local supersymmetry without the super-Higgs effect taking place. Supersymmetric cubic Galileons are shown to imply equations of motion of higher than second order, thus leading to the appearance of ghosts.

7.1 Introduction

Supersymmetry [1–3] is believed to be a symmetry of particle physics at high energies. This is based on the result that the supersymmetry algebra is the only graded Lie algebra of S-matrix symmetries that is consistent with relativistic quantum field

---

Proceedings of the workshop "Breaking of supersymmetry and ultraviolet divergences in extended supergravity", held in March 2013 at the INFN in Frascati. Michael Koehn: Speaker

---

M. Koehn (✉) · J.-L. Lehners
Max Planck Institute for Gravitational Physics (Albert Einstein Institute),
Am Mühlenberg 1,14476 Potsdam, Germany
e-mail: michael.koehn@aei.mpg.de

J.-L. Lehners
e-mail: jean-luc.lehners@aei.mpg.de

B. A. Ovrut
Department of Physics, University of Pennsylvania, 209 South 33rd Street,
Philadelphia, PA19104–6395, USA
e-mail: ovrut@elcapitan.hep.upenn.edu

S. Bellucci (ed.), *Breaking of Supersymmetry and Ultraviolet Divergences in Extended Supergravity*, Springer Proceedings in Physics 153,
DOI: 10.1007/978-3-319-03774-5_7,

theory [4]. Representations of the supersymmetry algebra contain bosonic and fermionic degrees of freedom in equal numbers. Moreover, particles belonging to the same representation have equal mass. Since superpartners with the same mass as conventional particles have not been observed, four-dimensional supersymmetry cannot be an unbroken low-energy symmetry. But supersymmetry—particularly four-dimensional $\mathcal{N} = 1$ supersymmetry—might be relevant at higher energies. For example, when $\mathcal{N} = 1$ supersymmetry is taken into account, the gauge couplings of the electroweak and strong forces unite to good precision at high energies [5], suggesting the existence of supersymmetric grand unification. Moreover, supersymmetric theories enjoy special finiteness properties that help to explain the hierarchy between the electroweak and the unification/gravitational scales [6, 7]. Last, but not least, $\mathcal{N} = 1$ supersymmetry is a central feature of phenomenologically realistic string theories—see, for example [8, 9].

Hence it is of natural interest to study the cosmology of the early universe within the context of $\mathcal{N} = 1$ supersymmetry. Since cosmology quintessentially involves gravitation, such theories must be constructed using "local" supersymmetry, i.e. $\mathcal{N} = 1$ supergravity in this case. This has been done within the context of two-derivative kinetic theories, both in local quantum field theory and superstrings. More recently, however, it has become clear that higher-derivative theories of cosmology are potentially important, with applications including DBI inflation [10], ekpyrotic theories with brane collisions [11, 12] and ghost condensation [13–15]. These proceedings are intended to report recent results of the authors concerning the development of a framework for constructing higher-derivative kinetic theories of chiral superfields coupled to $\mathcal{N} = 1$ supergravity [16] (building on previous work [17, 18] concerning it global supersymmetry). As an application of these results, we study supergravitational DBI inflation [19], and show how supersymmetric cubic Galileons [20–22] necessarily imply the existence of ghosts [23]. For the application of our framework to ghost condensation [24], we refer the reader to the article by Professor Ovrut, also in the proceedings of this workshop.

We neglect fermions because (a) they are typically unimportant in models of early universe cosmology and (b) since their inclusion greatly complicates all equations. Instead, we focus on the physics of the scalar bosons and the associated auxiliary fields, and refer to [24]. When the fermions are set to zero, our supergravity extension of $(\partial\phi)^4$ has a special—perhaps unique—property; namely, it can be multiplied by an arbitrary function of the scalar fields and their spacetime derivatives, while not altering the pure supergravity sector of the Lagrangian. Because this multiplicative factor is arbitrary, our formalism allows one to write a supergravity extension of *any* higher-derivative Lagrangian built out of scalar fields and their spacetime derivatives.

There are many further potential applications of our results, particularly in early-universe cosmology. For example, cosmological models that are constructed in—or inspired by—string theory should admit an effective $\mathcal{N} = 1$ supergravity description in four-dimensions. These theories typically have scalar fields arising as the moduli associated with branes [25], flux [26, 27] or the compactification manifold. For most—if not all—of these models, whether they are of DBI inflation [10], k-inflation [28], k-essence [28], ekpyrotic/cyclic cosmology [11, 13, 29, 30], effective theories of Galileons [31] or higher-derivative induced cosmic bounces [32–34], the proper setting is supergravity—and all contain phases where the dynamic description includes scalar higher-derivative terms.

The plan of this talk is the following. We begin by reviewing the construction of higher-derivative kinetic terms for chiral multiplets in global supersymmetry; that is, when gravity is neglected. Then, we show how this construction can be generalized to supergravity. We proceed by eliminating the auxiliary fields, where the auxiliary fields F^i of the chiral multiplets require special attention. We outline the implications of our formalism for DBI inflation. It turns out that in order for DBI inflation to work, additional modifications of the theory are necessary. The rest of the talk is devoted to a study of cubic Galileons. We show that their supersymmetric generalizations contain ghosts, and discuss the realm of this ghost. The notation and conventions of the book by Wess and Bagger [35] are used throughout the paper.

7.2 Higher-Derivative Chiral Superfield Actions in $\mathcal{N} = 1$ Supergravity

7.2.1 Higher-Derivative Chiral Superfields in Flat Superspace

We begin by considering global $\mathcal{N} = 1$ supersymmetry in flat four-dimensional spacetime. The associated supersymmetry algebra is given by

$$\{Q_\alpha, \bar{Q}_{\dot{\alpha}}\} = -2\sigma^m_{\alpha\dot{\alpha}} P_m, \tag{7.1}$$

where Q_α, $\bar{Q}_{\dot{\alpha}}$ and $P_m = -\mathrm{i}\partial_m$ generate supersymmetry and translations respectively. Here $\alpha, \beta, \ldots$ and $\dot{\alpha}, \dot{\beta}, \ldots$ are the conjugate indices of two-component Weyl spinors and $m, n \ldots$ are spacetime indices. To construct supersymmetric Lagrangians in this context, it is useful to work in flat superspace where, in addition to the four ordinary spacetime dimensions (with coordinates x^m), one adds four fermionic, Grassmann-valued dimensions (with coordinates $\theta_\alpha, \bar{\theta}_{\dot{\alpha}}$). In terms of these coordinates, the supersymmetric generators are represented by the superspace derivatives

$$D_\alpha = \frac{\partial}{\partial\theta^\alpha} + \mathrm{i}\sigma^m_{\alpha\dot{\alpha}}\bar{\theta}^{\dot{\alpha}}\partial_m, \quad \bar{D}_{\dot{\alpha}} = -\frac{\partial}{\partial\bar{\theta}^{\dot{\alpha}}} - \mathrm{i}\theta^\alpha\sigma^m_{\alpha\dot{\alpha}}\partial_m \tag{7.2}$$

which satisfy the algebra

$$\{D_\alpha, \bar{D}_{\dot{\alpha}}\} = -2\mathrm{i}\sigma^m_{\alpha\dot{\alpha}}\partial_m. \tag{7.3}$$

Any supermultiplet can be obtained as an expansion of a superfield, appropriately constrained, in the anti-commuting coordinates $\theta, \bar{\theta}$. The expansion terminates at order $\theta\theta\bar{\theta}\bar{\theta}$ because of the Grassmann nature of these coordinates. For example, a chiral superfield Φ, defined by the constraint

$$\bar{D}\Phi = 0, \tag{7.4}$$

has the expansion

$$\begin{aligned}\Phi &= A(x) + \sqrt{2}\theta\chi(x) + \theta\theta F(x) \\ &+ \mathrm{i}\theta\sigma^m\bar{\theta}\partial_m A(x) - \frac{\mathrm{i}}{\sqrt{2}}\theta\theta\partial_m\chi(x)\sigma^m\bar{\theta} + \frac{1}{4}\theta\theta\bar{\theta}\bar{\theta}\Box A(x),\end{aligned} \tag{7.5}$$

where A is a complex scalar, χ_α is a spin-$\frac{1}{2}$ fermion and F is a complex auxiliary field—which, for Lagrangians with canonical kinetic energy, is not a dynamical degree of freedom. (A, χ, F) are the component fields of the chiral supermultiplet. The component expansion (7.5) can be simplified by using the coordinates $y^m = x^m + \mathrm{i}\theta\sigma^m\bar{\theta}$, in terms of which

$$\Phi = A(y) + \sqrt{2}\theta\chi(y) + \theta\theta F(y). \tag{7.6}$$

This form of the component expansion has a straightforward generalization to curved superspace, as we will see shortly. It also suggests an alternative way of defining component fields, which turns out to be more useful in supergravity. Consider, for example, the chiral supermultiplet Φ. We note that one can also define the components of Φ as

$$A \equiv \Phi \mid \tag{7.7}$$

$$\chi_\alpha \equiv \frac{1}{\sqrt{2}} D_\alpha \Phi \mid \tag{7.8}$$

$$F \equiv -\frac{1}{4} D^2 \Phi \mid \tag{7.9}$$

where | denotes taking the lowest component. It is straightforward to check that these fields are identical to those in the θ, $\bar{\theta}$ expansion (7.5).

A general feature of superspace is that the highest component (that is, the $\theta\theta\bar{\theta}\bar{\theta}$ component) transforms under supersymmetry into a total spacetime derivative. Thus, the highest component of a superfield can be used to construct a supersymmetric Lagrangian. Because of the Grassmann nature of the fermionic coordinates, one can isolate the top component by integrating over superspace with $\mathrm{d}^2\theta\mathrm{d}^2\bar{\theta}$. Moreover, one can replace the $\mathrm{d}^2\theta\mathrm{d}^2\bar{\theta}$ integral over all superspace by a chiral integral $-\frac{1}{4}\mathrm{d}^2\theta\bar{D}^2$ using the chiral projector $\bar{D}^2$. This follows from the flat superspace relation $\bar{D}^3 = 0$.

In [17], it was shown how to construct supersymmetric actions involving higher derivatives of chiral superfields. The construction is based on a particular supersymmetric extension of the scalar-field Lagrangian $(\partial\phi)^4$ given by $D^\alpha\Phi D_\alpha\Phi\bar{D}_{\dot{\alpha}}\Phi^\dagger \bar{D}^{\dot{\alpha}}\Phi^\dagger$. Ignoring the fermion χ, this superfield contains only the $\theta\theta\bar{\theta}\bar{\theta}$ component

$$D^{\alpha}\Phi D_{\alpha}\Phi\bar{D}_{\dot{\alpha}}\Phi^{\dagger}\bar{D}^{\dot{\alpha}}\Phi^{\dagger} = \theta\theta\bar{\theta}\bar{\theta}\Big(16(\partial A)^2(\partial A^*)^2 - 32|\partial A|^2|F|^2 + 16|F|^4\Big), \tag{7.10}$$

where the complex scalar A is composed of two real scalars ϕ, ξ as

$$A = \frac{1}{\sqrt{2}}(\phi + \mathrm{i}\xi) \tag{7.11}$$

and $|\partial A|^2 \equiv \partial A \cdot \partial A^*$. Thus, the superspace integral of the superfield (7.10) yields the term

$$16(\partial A)^2(\partial A^*)^2 = 4(\partial\phi)^4 + 4(\partial\xi)^4 - 8(\partial\phi)^2(\partial\xi)^2 + 16(\partial\phi\cdot\partial\xi)^2 \tag{7.12}$$

plus terms involving the auxiliary field F. Hence, (7.10) constitutes a possible supersymmetric extension of $(\partial\phi)^4$. The superfield expression (7.10) possesses several particularly useful properties[1]:

- It constitutes a supersymmetric extension of the precise expression $(\partial\phi)^4$, and does not contain other terms involving ϕ alone.
- Despite the higher-derivative nature of the superfield, the auxiliary field F does not obtain a kinetic energy. This is non-trivial, as on dimensional grounds a term such as $|A|^2|\partial F|^2$ could have arisen, and implies that F remains truly auxiliary.
- As pointed out in [17], the auxiliary field now appears at quartic order in the action and, thus, its equation of motion is cubic. Hence, in contrast to the usual two-derivative supersymmetric theories, there exist now up to three different solutions for F. We will explore this issue much further in Sect. 7.3.
- Finally, the most crucial property for our present purposes is the fact that the bosonic part of $D^{\alpha}\Phi D_{\alpha}\Phi\bar{D}_{\dot{\alpha}}\Phi^{\dagger}\bar{D}^{\dot{\alpha}}\Phi^{\dagger}$, given in (7.10), only contains a non-zero top $\theta\theta\bar{\theta}\bar{\theta}$ component—all lower components vanish. It follows that if one multiplies this superfield with any function T of Φ, $\Phi^{\dagger}$ and (an arbitrary number of) their spacetime derivatives, then the component expansion will be given by (7.10) times $T|$, where inside $T|$ the chiral superfield Φ is simply replaced by its lowest component A. This allows one to easily construct a supersymmetric extension of *any* higher-derivative scalar Lagrangian containing $(\partial\phi)^4$ as a factor, simply by performing the replacement $\phi \to \sqrt{2}A \to \sqrt{2}\Phi$ in the co-factor.

This last property was used in [17] to construct a supersymmetric extension of theories with Lagrangian $P(X, \phi)$, where $X \equiv -\frac{1}{2}(\partial\phi)^2$. Specifically, for

$$P(X, \phi) = \sum_{n\geq 1} a_n(\phi)X^n \tag{7.13}$$

[1] There exists a different supersymmetric extension of $(\partial\phi)^4$ [36, 37], which however doesn't live up to every point of this list of useful properties that (7.10) has, and which thus feels much less natural, particularly so when it comes to the extension to supergravity.

it was shown that the higher-derivative terms in the supersymmetric generalization are the $d^2\theta d^2\bar{\theta}$ integral of

$$\frac{1}{16} D\Phi D\Phi \bar{D}\Phi^\dagger \bar{D}\Phi^\dagger\, T(\Phi, \Phi^\dagger, \partial_m\Phi, \partial_n\Phi^\dagger), \tag{7.14}$$

where

$$\begin{aligned} T(\Phi, \Phi^\dagger, \partial_m\Phi, \partial_n\Phi^\dagger) &= \sum_{n\geq 2} a_n \left(\frac{1}{32}\{D, \bar{D}\}(\Phi + \Phi^\dagger)\{D, \bar{D}\}(\Phi + \Phi^\dagger)\right)^{n-2} \\ &= \sum_{n\geq 2} a_n \left(\frac{1}{4}\partial^m(\Phi + \Phi^\dagger)\partial_m(\Phi + \Phi^\dagger)\right)^{n-2}, \end{aligned} \tag{7.15}$$

$a_n = a_n\left(\frac{\Phi+\Phi^\dagger}{\sqrt{2}}\right)$ and we have made use of (7.3) to write $\{D, \bar{D}\} \propto \partial_m$.

Particular applications were a supersymmetric form of the DBI action, as well as a supersymmetric ghost condensate theory—both in flat spacetime. However, the most interesting phenomenological consequences occur when these models are coupled to gravity—for example, inflation driven by the DBI part of the action or cosmic bounces induced by a ghost condensate. It is, therefore, of interest to include gravity in the analysis. In a supersymmetric context, this means extending the above construction to *curved* superspace. This will be the topic of the next section.

7.2.2 Higher-Derivative Kinetic Terms in Supergravity

In [16], we showed how to couple chiral superfields with higher-derivative kinetic terms to four-dimensional $\mathcal{N} = 1$ supergravity.[2] To keep the equation length reasonable in this contribution, and since we are interested in cosmological applications, fermionic component fields will be ignored throughout. The construction takes place in curved superspace [35], which is the most natural setting for writing actions invariant under local supersymmetry transformations. A chiral superfield Φ then admits the expansion

$$\Phi = A + \Theta^\alpha \Theta_\alpha F, \tag{7.16}$$

where A is a complex scalar field and F is a complex auxiliary field. The Θ coordinates are Grassmann-valued and carry local Lorentz indices (α denotes the index of a two-component Weyl spinor). They extend ordinary spacetime to curved superspace, and are defined precisely so that A and F arise as the components of Φ in the above expansion. In curved superspace, supersymmetric Lagrangians can be constructed from the chiral integrals

[2] Also see [38], where related results were obtained. Earlier work of interest includes [17, 18, 39–44].

$$\int d^2\Theta(\bar{\mathcal{D}}^2 - 8R)L, \tag{7.17}$$

where L is a scalar, Hermitean function. The chiral projector in curved superspace is $\bar{\mathcal{D}}^2 - 8R$, where $\bar{\mathcal{D}}_{\dot{\alpha}}$ is a spinorial component of the curved superspace covariant derivative $\mathcal{D}_A = \{\mathcal{D}_a, \mathcal{D}_\alpha, \bar{\mathcal{D}}_{\dot{\alpha}}\}$ and R is the curvature superfield. In its component expansion, R contains the Ricci scalar $\mathcal{R}$ as well as the auxiliary fields of supergravity—namely a complex scalar M and a real vector b_m. The purely bosonic components in the Θ expansion of R are

$$R = -\frac{1}{6}M + \Theta^2\left(\frac{1}{12}\mathcal{R} - \frac{1}{9}MM^* - \frac{1}{18}b_m b^m + \frac{1}{6}\mathrm{i}e_a{}^m\mathcal{D}_m b^a\right). \tag{7.18}$$

Another superfield that we will need is the chiral density $\mathcal{E}$ with expansion

$$2\mathcal{E} = e(1 - \Theta^2 M^*), \tag{7.19}$$

where e is the determinant of the vierbein. Note that the tangent space Lorentz indices $A = \{a, \alpha, \dot{\alpha}\}$ are related to the spacetime indices $M = \{m, \mu, \dot{\mu}\}$ via the supervielbein $E_M{}^A$ and its inverse, with $E_m{}^a = e_m{}^a$ being the ordinary vierbein. For a complete discussion of curved superspace we refer the reader to [35].

We start by introducing an Hermitean *Kähler potential* $K(\Phi^i, \Phi^{\dagger k*})$ of the chiral superfields Φ^i (where $i = 1, 2, \ldots$ enumerates the fields), along with a holomorphic *superpotential* $W(\Phi^i)$. The associated Lagrangian is given by

$$\frac{1}{e}\mathcal{L} = \frac{1}{e}\int d^2\Theta 2\mathcal{E}\left[\frac{3}{8}(\bar{\mathcal{D}}^2 - 8R)e^{-K(\Phi^i,\Phi^{\dagger k*})/3} + W(\Phi^i)\right] + h.c. \tag{7.20}$$

$$\begin{aligned} &= e^{-K/3}\left(-\frac{1}{2}\mathcal{R} - \frac{1}{3}MM^* + \frac{1}{3}b^a b_a\right) \\ &\quad + 3\left(\frac{\partial^2 e^{-K/3}}{\partial A^i \partial A^{k*}}\right)\left(\partial A^i \cdot \partial A^{k*} - F^i F^{k*}\right) \\ &\quad + \mathrm{i}b^m\left(\partial_m A^i \frac{\partial e^{-K/3}}{\partial A^i} - \partial_m A^{k*}\frac{\partial e^{-K/3}}{\partial A^{k*}}\right) + MF^i\frac{\partial e^{-K/3}}{\partial A^i} \\ &\quad + M^* F^{k*}\frac{\partial e^{-K/3}}{\partial A^{k*}} - WM^* - W^*M + \partial W_i F^i + \partial W^*_{k*}F^{k*}, \end{aligned} \tag{7.21}$$

where $\partial W_i = \frac{\partial W}{\partial A^i}$. This Lagrangian is obtained after integration by parts; it is meant to be integrated over spacetime to yield an action.

We now add the higher-derivative kinetic terms for the chiral superfields in a manifestly diffeomorphism invariant manner. Specifically, we introduce

$$\begin{aligned} \mathcal{L}_{\text{h-d}} &= -\frac{1}{8}\int d^2\Theta 2\mathcal{E}(\bar{\mathcal{D}}^2 - 8R)\mathcal{D}\Phi^i\mathcal{D}\Phi^j\bar{\mathcal{D}}\Phi^{\dagger k*}\bar{\mathcal{D}}\Phi^{\dagger l*}\,T_{ijk*l*} + h.c. \\ &= 16\,e(\partial A^i \cdot \partial A^j)(\partial A^{k*} \cdot \partial A^{l*})\,T_{ijk*l*}| \end{aligned}$$

$$- 32\, eF^i F^{k*} (\partial A^j \cdot \partial A^{l*})\, T_{ijk*l*}|$$
$$+ 16 e F^i F^j F^{k*} F^{l*}\, T_{ijk*l*}|, \qquad (7.22)$$

where $T_{ijk*l*}|$ is the lowest component of the tensor superfield T_{ijk*l*}. Let us clarify the meaning of T_{ijk*l*}. First, this superfield transforms as a four-index tensor on the Kähler manifold in which the scalar fields take their values and, thus, ensures target space diffeomorphism invariance. Second, T_{ijk*l*} is required to be Hermitean and symmetric in the pair of indices i, j as well as in k^*, l^*. Third, any tensor satisfying these constraints can be multiplied by an arbitrary real function of the chiral superfields and an unlimited number of their $\mathcal{D}_m$ covariant derivatives, as long as all indices stemming from the covariant derivatives are contracted. Examples of $T_{ijk*l*}|$ include $\frac{1}{2}(g_{ik*}g_{jl*} + g_{il*}g_{jk*})$, where g_{ij^*} is the Kähler metric, and the Riemann tensor R_{ik*jl*}. However, more general—non-geometric—choices respecting the required symmetries are equally possible.[3] The fact that one can multiply this tensor with an arbitrary function of the chiral superfields and their spacetime derivatives means that we can obtain a supergravity extension of any term that involves $(\partial\phi)^4$ as a factor and, thus, by dividing out by $(\partial\phi)^4$ if necessary, of *any* higher-derivative scalar Lagrangian (e.g. DBI action).

The sum of the two actions (7.21) + (7.22) does not lead to ordinary Einstein frame gravity but, rather, to a scalar-gravity theory of the form $e^{-K/3}\mathcal{R}$. One can transform the action into Einstein frame by performing the Weyl rescaling

$$e_n{}^a \to e_n{}^a e^{K/6}. \qquad (7.23)$$

Note that the higher-derivative term does not contribute to the gravity-scalar coupling and, hence, we can perform the same Weyl rescaling as in ordinary chiral supergravity without higher-derivatives. This is a non-trivial feature of our framework, which greatly facilitates subsequent calculations. After Weyl-rescaling and elimination of the auxiliary fields b_m and F, the action reads

$$\begin{aligned}\frac{1}{e}\mathcal{L}_{\text{Weyl}} = &-\frac{1}{2}\mathcal{R} - g_{ik*}\partial A^i \cdot \partial A^{k*} + g_{ik*}e^{K/3} F^i F^{k*} \\ &+ e^{2K/3}\left[F^i (D_A W)_i + F^{k*}(D_A W)^*_{k*}\right] + 3e^K W W^* \\ &+ 16(\partial A^i \cdot \partial A^j)(\partial A^{k*} \cdot \partial A^{l*})\, T_{ijk*l*\text{Weyl}}| \\ &- 32\, e^{K/3} F^i F^{k*}(\partial A^j \cdot \partial A^{l*})\, T_{ijk*l*\text{Weyl}}| \\ &+ 16 e^{2K/3} F^i F^j F^{k*} F^{l*}\, T_{ijk*l*\text{Weyl}}|. \end{aligned} \qquad (7.24)$$

We next move on to the remaining auxiliary field F, which reveals some subtleties of the higher-derivative theory.

---

[3] In all examples in this paper, we will, for specificity, choose $T_{ijk*l*}|$ to be proportional to $\frac{1}{2}(g_{ik*}g_{jl*} + g_{il*}g_{jk*})$.

7.3 New Potentials from the Equation for the Auxiliary Field F

We now consider the most interesting of the auxiliary fields, namely F. The equation of motion for F is easily derived from the action (7.24) and reads

$$g_{ik*}F^i + e^{K/3}(D_A W)^*_{k*} + 32F^i(e^{K/3}F^j F^{l*} - \partial A^j \cdot \partial A^{l*})T_{ijk*l*\text{Weyl}}| = 0. \quad (7.25)$$

This equation is now cubic in F and, thus, it can have up to three inequivalent solutions. As we will see, these different solutions lead to *different* theories! From now on, we will restrict our analysis to a single chiral superfield $\Phi^1 = \Phi$, the extension to multiple superfields being straightforward to implement. In this case, the equation of motion for F becomes

$$K_{,AA^*}F + e^{K/3}(D_A W)^* + 32F(e^{K/3}|F|^2 - |\partial A|^2)\mathcal{T} = 0, \quad (7.26)$$

where

$$|\partial A|^2 = \partial A \cdot \partial A^* = g^{mn}\partial_m A \partial_n A^* \quad (7.27)$$

and where we use the simplified notation

$$\mathcal{T} \equiv T_{111*1*\text{Weyl}}|. \quad (7.28)$$

Note that $\mathcal{T}$ is effectively an arbitrary real scalar function of A, A^* and their spacetime covariant derivatives $\mathcal{D}_m \ldots \partial_n A$, $\mathcal{D}_m \ldots \partial_n A^*$. Multiplying (7.26) with F^* shows that $(D_A W)^* F^*$ must be real. Thus, one can relate F and F^* via

$$F^* = \frac{D_A W}{(D_A W)^*}F \quad (7.29)$$

as long as $D_A W \neq 0$, which we now assume. One can use this relation to obtain a cubic equation for F alone. This is given by

$$K_{,AA^*}F + e^{K/3}(D_A W)^* + 32\left(e^{K/3}\frac{D_A W}{(D_A W)^*}F^3 - |\partial A|^2 F\right)\mathcal{T} = 0. \quad (7.30)$$

In general, this equation admits *three* distinct solutions—which we denote by F_1, F_2, F_3—leading to three different theories. One can find these solutions using Cardano's formula. Define

$$p = e^{-K/3}\frac{(D_A W)^*}{D_A W}\left(\frac{K_{,AA^*}}{32\mathcal{T}} - |\partial A|^2\right), \quad (7.31)$$

$$q = \frac{1}{32\mathcal{T}} \frac{(D_A W)^{*2}}{D_A W}, \tag{7.32}$$

$$\begin{aligned} D &= \left(\frac{q}{2}\right)^2 + \left(\frac{p}{3}\right)^3 \\ &= \frac{1}{(64\mathcal{T})^2} \frac{(D_A W)^{*4}}{(D_A W)^2} + \frac{1}{27e^K} \frac{(D_A W)^{*3}}{(D_A W)^3} \left(\frac{K_{,AA^*}}{32\mathcal{T}} - |\partial A|^2\right)^3. \end{aligned} \tag{7.33}$$

Then the solutions are given by

$$F_{k+1} = \omega^k F_+ + \omega^{-k} F_-, \tag{7.34}$$

where $k = 0, 1, 2$, $\omega = e^{2\pi i/3} = -\frac{1}{2} + i\frac{\sqrt{3}}{2}$ is a cube root of unity and

$$F_+ = \left(-\frac{q}{2} + D^{1/2}\right)^{1/3}, \quad F_- = \left(-\frac{q}{2} - D^{1/2}\right)^{1/3}. \tag{7.35}$$

Substituting these back into the action generates three different branches of the theory. We call the theory that results from substituting F_1 the *ordinary branch*, as only this solution approaches the usual solution when $\mathcal{T} \to 0$. The new branches are discussed in detail in [16].

The higher-derivative terms are all proportional to the $\mathcal{T}$ tensor. Therefore, by assuming that $\mathcal{T}$ contains a factor that can be tuned to be small, one can treat such terms as sub-leading. The $\mathcal{T} \to 0$ limit then corresponds to $q \ll p^{3/2}$, and hence

$$\begin{aligned} F_1 = &-K^{,AA^*} e^{K/3} (D_A W)^* \\ &+ 32\mathcal{T} e^{4K/3} (K^{,AA^*})^4 (D_A W)^{*2} D_A W \\ &- 32\mathcal{T} e^{K/3} (K^{,AA^*})^2 (D_A W)^* |\partial A|^2 + \mathcal{O}(\mathcal{T}^2). \end{aligned} \tag{7.36}$$

Note that this corresponds to a small correction to the usual solution for the auxiliary field F in the presence of a superpotential. Correspondingly, we obtain small corrections in the Lagrangian by substituting this solution for F. To first order in the higher-derivative terms, the Lagrangian becomes

$$\begin{aligned} \frac{1}{e}\mathcal{L}_{\text{ordinary},\mathcal{T}\to 0} = &-\frac{1}{2}\mathcal{R} - K_{,AA^*}|\partial A|^2 - e^K (K^{,AA^*}|D_A W|^2 - 3|W|^2) \\ &- 32\, e^K K^{,AA^*} |D_A W|^2 K^{,AA^*} |\partial A|^2 \mathcal{T} \\ &+ 16\, (\partial A)^2 (\partial A^*)^2 \mathcal{T} \\ &+ 16 e^{2K} (K^{,AA^*} |D_A W|^2)^2 \, (K^{,AA^*})^2 \mathcal{T}. \end{aligned} \tag{7.37}$$

An interesting feature is that both the kinetic terms and the potential get corrected. The potential now becomes

$$V = e^K (K^{,AA^*}|D_A W|^2 - 3|W|^2)$$
$$-16(e^K K^{,AA^*}|D_A W|^2)^2 (K^{,AA^*})^2 \mathcal{T}_{\text{no der.}}, \tag{7.38}$$

where $\mathcal{T}_{\text{no der.}}$ stands for the part of $\mathcal{T}$ that does not contain spacetime derivatives. Note that all the correction terms in the Lagrangian above are invariant under Kähler transformations.

As an example, consider the case where $K = \Phi\Phi^\dagger$, $\mathcal{T} = \tau(K_{,AA^*})^2$ is of canonical form with τ a small parameter and $W = \Phi$. Near the minimum at $A = 0$, to first order in τ, the potential is given by $V = \bar{V} + \delta V$ and can be approximated by

$$\bar{V}_{n=1} \approx 1 + \frac{1}{2}|A|^4 + \cdots . \tag{7.39}$$

Note that the $|A|^2 = \phi^2 + \xi^2$ term cancels in the expansion. Therefore, this potential is very flat near the origin, rising only quartically as $(\phi^2 + \xi^2)^2$. The leading order correction to this potential is given by

$$\delta V_{n=1} \approx -16\tau(1 + 6|A|^2 + 16|A|^4 + \cdots). \tag{7.40}$$

For $\frac{1}{128} > \tau > 0$, the minimum at $A = 0$ becomes a local maximum. The potential is now minimized along a circle defined by $|A|^2 = 12\tau/(1 - 128\tau)$. In other words, the potential changes from a slowly rising quartic potential with a minimum at the origin to a "Mexican hat".

7.3.1 Supergravitational DBI Inflation: Large Higher-Derivative Terms

Although our work concers purely the supergravity context, the motivation stems from string theory. There, the dynamics of D-branes and M5-branes are described by the Dirac-Born-Infeld (DBI) action [45].[4] This action is unusual in that it contains higher-derivative terms which are essential to understanding its dynamics.[5] Furthermore, interactions between branes (and anti-branes) can generate an effective potential [26, 48–51]. In such a setting, inflationary models based on the DBI action, in which the inflaton field is identified with a position modulus of the brane, have been constructed and shown to lead to interesting observational predictions—such as equilateral non-Gaussianities [10, 52]. These models have mainly been analyzed in non-supersymmetric effective field theory. However, realistic string

---

[4] The effective description in terms of the DBI action is valid at arbitrary velocity, but only as long as the proper acceleration of the branes is small.

[5] Higher-derivative terms involving the extrinsic and intrinsic brane curvatures—such as those discussed in [46, 47]—can arise as well. We will not consider these couplings here, but note that they might be significant in certain applications.

compactifications typically preserve minimal supersymmetry in four dimensions—see, for example [8, 9]. It is of interest, therefore, to re-formulate these models within the context of four-dimensional, $\mathcal{N}=1$ supergravity.

By choosing the function T in (7.24) appropriately, we can write a supergravity version of the single real scalar field DBI action. It turns out that we need to consider a Kähler potential with the property

$$\frac{\partial^2 K}{\partial \Phi \partial \Phi^\dagger}| = K_{,AA^*} = 1 \tag{7.41}$$

and a tensor superfield [16, 53]

$$16T = \frac{f(\Phi, \Phi^\dagger)}{1 + f\partial\Phi \cdot \partial\Phi^\dagger e^{K/3} + \sqrt{(1 + f\partial\Phi \cdot \partial\Phi^\dagger e^{K/3})^2 - f^2(\partial\Phi)^2(\partial\Phi^\dagger)^2 e^{2K/3}}}. \tag{7.42}$$

Here $f(\Phi, \Phi^\dagger)$ is an arbitrary hermitian function and we have used the notation that $\partial\Phi \cdot \partial\Phi^\dagger = g^{mn}\mathcal{D}_m\Phi\mathcal{D}_n\Phi^\dagger$. In a brane setting, the lowest component of the f function can be identified with the warp factor of the direction in which the brane moves. Then the Lagrangian reduces to

$$\begin{aligned}\frac{1}{e}\mathcal{L} = &-\frac{1}{2}\mathcal{R} + 3e^K|W|^2 \\ &-\frac{1}{f}\left(\sqrt{1 + 2f\,\partial A \cdot \partial A^* + f^2\,(\partial A \cdot \partial A^*)^2 - f^2\,(\partial A)^2(\partial A^*)^2} - 1\right) \\ &+ e^{K/3}|F|^2 + e^{2K/3}\left(F(D_A W) + F^*(D_A W)^*\right) \\ &- 32\,e^{K/3}|F|^2 \partial A \cdot \partial A^*\,\mathcal{T} + 16e^{2K/3}|F|^4\,\mathcal{T}.\end{aligned} \tag{7.43}$$

Here $\mathcal{T}$, which is the Weyl rescaled lowest component of T, is given by

$$16\mathcal{T} = \frac{f}{1 + f\,\partial A \cdot \partial A^* + \sqrt{(1 + f\,\partial A \cdot \partial A^*)^2 - f^2\,(\partial A)^2(\partial A^*)^2}} \tag{7.44}$$

with $f = f(A, A^*)$. The second line of (7.43) can be recognized as the DBI action for the *two* real scalar fields ϕ, ξ that make up the complex scalar A [53]. That is, the simplest $\mathcal{N}=1$ supergravity generalization of the single real scalar DBI action naturally produces a DBI theory for both real scalar component fields. As can be seen from the action, when the fields depend only on time there exists an upper bound on the velocity of A given by

$$|\dot{A}|^2 \leq \frac{1}{2f}. \tag{7.45}$$

The so-called *relativistic regime* corresponds to the situation where this bound is (almost) saturated. Models of DBI inflation [10] exploit this inequality. As the brane moves towards a region of large f, the scalars are automatically constrained to move slowly—allowing for inflation to occur on potentials that would otherwise be too steep.

When f is small, then so is $\mathcal{T}$, and F approaches the usual solution

$$F \approx -e^{K/3}(D_A W)^*. \quad (f \text{ small}) \tag{7.46}$$

In this *non-relativistic* limit, after substituting for F one obtains the usual potential

$$V_{\text{non-rel.}} = e^K \big(|D_A W|^2 - 3|W|^2 \big). \tag{7.47}$$

Note that this expression is only valid as long as the higher-derivative terms in A are irrelevant.

More interesting for our purposes is the *relativistic limit*, where f is large and $|\dot{A}|^2$ correspondingly small, with $\mathcal{T} \approx f/8$. In that case, the solution for F approaches

$$F \approx -\left(\frac{(D_A W)^{*2}}{4f\, D_A W}\right)^{1/3}. \quad (f \text{ large}) \tag{7.48}$$

After substituting for F in the relativistic limit, the Lagrangian becomes

$$\begin{aligned} \frac{1}{e}\mathcal{L}_{\text{rel.}} = &-\frac{1}{2}\mathcal{R} + 3e^K|W|^2 - \frac{3}{2}\frac{e^K|D_A W|^2}{\left(4f\, e^K |D_A W|^2\right)^{1/3}} \\ &-\frac{1}{f}\left(\sqrt{1 + 2f\, \partial A \cdot \partial A^* + f^2\, (\partial A \cdot \partial A^*)^2 - f^2\, (\partial A)^2 (\partial A^*)^2} - 1\right) \\ &+ \mathcal{O}(f^{-2/3}). \end{aligned} \tag{7.49}$$

Thus, to leading order the potential is given by

$$V_{\text{rel.}} = -3e^K|W|^2, \tag{7.50}$$

which is *negative* for any choice of superpotential. The term arising from eliminating F is sub-leading. It is evident, therefore, that inflation cannot occur since a phase of de-Sitter-like expansion requires a positive energy density in the universe. Thus, supergravitational relativistic DBI inflation with a single chiral superfield does not work!

Let us now extend this theory by coupling it to a second chiral superfield S with component expansion

$$S = B + \Theta^\alpha \Theta_\alpha F_B. \tag{7.51}$$

Here B is a complex scalar and F_B the complex auxiliary field associated with S. We will assume that this second field has a two-derivative action.[6] Then, choosing a Kähler potential such that

$$K_{,AA^*} = 1\,, \tag{7.52}$$

$$K_{,AB^*} = 0 = K_{,A^*B}\,, \tag{7.53}$$

and after the same manipulations as in the previous section—for example, Weyl rescaling the action and eliminating the auxiliary fields b_m, M—we obtain the Lagrangian

$$\begin{aligned}\frac{1}{e}\mathcal{L} = &-\frac{1}{2}\mathcal{R} + 3e^K|W|^2 - K_{,BB^*}\partial B\cdot\partial B^* \\ &-\frac{1}{f}\left(\sqrt{1+2f\,\partial A\cdot\partial A^* + f^2\,(\partial A\cdot\partial A^*)^2 - f^2\,(\partial A)^2(\partial A^*)^2} - 1\right) \\ &+K_{,BB^*}e^{K/3}|F_B|^2 + e^{2K/3}\big(F_B(D_BW) + F_B^*(D_BW)^*\big) \\ &+e^{K/3}|F|^2 + e^{2K/3}\big(F(D_AW) + F^*(D_AW)^*\big) \\ &-32\,e^{K/3}|F|^2\partial A\cdot\partial A^*\,\mathcal{T} + 16e^{2K/3}|F|^4\,\mathcal{T}.\end{aligned} \tag{7.54}$$

In this expression, the auxiliary fields F, F_B of the two chiral multiplets have not yet been eliminated. Their equations of motion are given by

$$F + e^{K/3}(D_AW)^* + 32F\,\mathcal{T}(e^{K/3}|F|^2 - \partial A\cdot\partial A^*) = 0, \tag{7.55}$$

$$K_{,BB^*}F_B + e^{K/3}(D_BW)^* = 0. \tag{7.56}$$

Note that these equations are not coupled and, thus, F can be eliminated as in the previous section. It is also straightforward to substitute for F_B, since its equation of motion is algebraic and linear. In the non-relativistic limit—that is, when f is small—one obtains the usual potential

$$V_{\text{non-rel.,2 superfields}} = e^K\big(|D_AW|^2 + K^{,BB^*}|D_BW|^2 - 3|W|^2\big). \tag{7.57}$$

However, in the relativistic limit the $|D_AW|^2$ term again is subdominant and the potential becomes

$$V_{\text{rel.,2 superfields}} = e^K\big(K^{,BB^*}|D_BW|^2 - 3e^K|W|^2\big). \tag{7.58}$$

Comparing this to expression (7.50), we see that in the two superfield case a new, positive definite terms enters the potential energy! Hence, by choosing the superpotential

[6] One could equally well assume that it also has higher-derivative kinetic terms, but that they are unimportant in the vacuum. For simplicity, we will not pursue this option here.

appropriately, the overall potential can be made positive along the direction(s) of interest in field space—thus enabling inflation to occur.

We will first be interested in the case where one allows the two real scalars in

$$A = \frac{1}{\sqrt{2}}(\phi + i\xi) \tag{7.59}$$

to be dynamically relevant. These scalars both have kinetic terms of the DBI form—as is evident, for example, from (7.49). Our formalism also implies that, after the potential energy has been chosen for the first scalar, the potential of the second scalar is automatically determined. Moreover, when the Kähler potential satisfies certain additional requirements—which we derive below—this second scalar can be stabilized. In this case, our construction allows one to obtain an arbitrary positive potential. Choosing this appropriately leads effectively to a single real component field model of DBI inflation.

We choose for the superpotential W an Ansatz first used in [54] and analyzed, in detail, in [55] within the context of ordinary two-derivative supergravity. This Ansatz is

$$W = Sw(\Phi), \tag{7.60}$$

where $w(\Phi)$ is a "real" holomorphic function of Φ; that is, $w(\Phi) = \sum_n c_n \Phi^n$ with $c_n \in \mathbb{R}$. The coefficients are chosen to be real for simplicity. The lowest component of W is given by $Bw(A)$. On the $B = 0$ plane, we have $W = 0$, $D_B W = w(A)$ and, hence, the potential energy (7.58) becomes

$$V_{B=0} = e^{K(A,A^*)} K^{,BB^*} |w(A)|^2. \tag{7.61}$$

Here, the Kähler potential is also evaluated at $B = 0$. The B field can always be rescaled so that its kinetic term is canonical (when $B = 0$). Correspondingly, we will take $K_{,BB^*}|_{B=0} = 1$. Then the potential further simplifies to

$$V_{B=0} = e^{K(A,A^*)} |w(A)|^2. \tag{7.62}$$

For this expression to be physically relevant, one must ensure that the dynamics is restricted to the $B = 0$ plane. That is, the two real scalar fields b, d, defined by

$$B = \frac{1}{\sqrt{2}}(b + \mathrm{i}d) \tag{7.63}$$

must be stabilized with zero vacuum expectation values. In an inflationary context, this means that around $b = d = 0$ the scalar squared masses m_b^2, m_d^2 must be positive and at least as large as the Hubble expansion scale H^2. A straightforward calculation shows that

$$
\begin{aligned}
m_b^2 &= \frac{\partial^2 V}{\partial b^2}|_{b=d=0} \\
&= \Big(\frac{1}{2}\frac{\partial^2 V}{\partial B^2} + \frac{\partial^2 V}{\partial B \partial B^*} + \frac{1}{2}\frac{\partial^2 V}{\partial B^{*2}}\Big)|_{B=0} \\
&= -e^{K(A,A^*)}|w(A)|^2 K_{,BBB^*B^*}, \qquad (7.64)
\end{aligned}
$$

with a similar expression for m_d^2. One can assume that, during inflation, the dynamics is dominated by the potential and, thus, the Friedmann equation implies that $V \approx 3H^2$. Then the requirement that $m_b^2, m_d^2 \gtrsim H^2$ translates into the stability condition

$$
K_{,BBB^*B^*} \lesssim -\frac{1}{3}. \qquad (7.65)
$$

This condition is analogous to that found in two-derivative supergravity models [55]. It can be satisfied, for example, if the Kähler potential includes a term $\zeta(BB^*)^2$ with $\zeta \lesssim -1/12$.

Now note that for the superpotential (7.60), $D_A W \propto B$ and hence vanishes on the $B = 0$ plane. Thus, the potential term $e^K |D_A W|^2$ that becomes subdominant in the relativistic limit, is actually zero on the inflationary trajectory for models of this type. This can also be seen directly from the equation of motion (7.55) for F–for the Ansatz (7.60) the ordinary branch solution for F is simply the trivial solution $F = 0$ if we restrict to the $B = 0$ plane. In other words, in going from the approximately two-derivative regime to the relativistic DBI regime, the potential does *not* change for the models considered here. This special feature is entirely non-trivial, and arises as a direct consequence of the choice (7.60). It greatly facilitates the analysis of the corresponding inflationary models.

Let us now restrict the theory further, so that only a single real scalar field in (7.59) remains dynamical. For this purpose, choose the Kähler potential to depend on Φ, $\Phi^\dagger$ via the combination $-\frac{1}{2}(\Phi - \Phi^\dagger)^2$ only. Then, the Kähler potential will not depend on ϕ. Correspondingly, if ξ is now stabilized around $\xi = 0$ with a sufficiently high mass, then the dynamics will take place entirely in the ϕ direction with the potential

$$
V_\phi = w\left(\frac{\phi}{\sqrt{2}}\right)^2. \qquad (7.66)
$$

Thus, any smooth positive potential can be engineered in this way, simply by identifying w with the square root of the desired potential and analytically continuing w to the complex plane [55]. However, for consistency, one must check under what conditions ξ is stabilized. Its mass along the putative inflationary trajectory is given by

$$
\begin{aligned}
m_\xi^2 &= \frac{\partial^2 V}{\partial \xi^2}|_{\xi=b=d=0} \\
&= \Big(-\frac{1}{2}\frac{\partial^2 V}{\partial A^2} + \frac{\partial^2 V}{\partial A \partial A^*} - \frac{1}{2}\frac{\partial^2 V}{\partial A^{*2}}\Big)|_{\xi=B=0}
\end{aligned}
$$

$$= -ww'' + w'^2 + 2w^2(1 - K_{,AA^*BB^*}), \tag{7.67}$$

where $w' = w_{,A}|_{\xi=0}$. This mass is identical to that obtained in two-derivative supergravity theories [55]. A working model of single real component field DBI inflation must then satisfy $m_\xi^2 \gtrsim H^2$—otherwise perturbations in the ξ field also become relevant. When w''/w and $(w'/w)^2$ are small (bearing in mind that for DBI inflation they need not be as small as for two-derivative inflation), this translates into the requirement

$$K_{,AA^*BB^*} \lesssim \frac{5}{6}. \tag{7.68}$$

An example of a Kähler potential satisfying all of the above assumptions and stability constraints was discussed in [55]. Here, we will simply repeat it for specificity. It is given by

$$K = -\frac{1}{2}(\Phi - \Phi^\dagger)^2 + SS^\dagger + \zeta(SS^\dagger)^2 + \frac{\gamma}{2}SS^\dagger(\Phi - \Phi^\dagger)^2 \tag{7.69}$$

with $\zeta \lesssim -1/12$ and $\gamma \gtrsim 5/6$.

DBI inflation was inspired by string theory, and is of importance because it has a more direct link to microphysics than most inflationary models. The higher-derivative terms play a crucial role in DBI theories, since they lead to the speed limit (7.45). They also imply the generation of significant equilateral non-Gaussianity [10, 52]. Interestingly, models of single real scalar field DBI inflation are already tightly constrained by current observations—precisely because of the constraints imposed by the underlying microphysics. Such models could be ruled out in the near future [56–59]. However, restricting to a single real scalar field is not necessary within a string theory context. For example, many DBI models that have been considered focus on a D3-brane moving along a warped throat of an internal Calabi-Yau manifold. The radial direction is typically identified with the inflaton. By construction, however, such models naturally have multiple real scalar fields, with the angular directions in the Calabi-Yau space providing the additional degrees of freedom [60]. Hence, it is of interest to also study multi-field models of DBI inflation. For such theories, the constraints arising from the comparison with observational data are typically less severe. An interesting recent example is provided in [61], which is in agreement with all current observations, but where significant non-Gaussianities of both local and equilateral type are predicted.

The models studied in the previous section, if the second real scalar ξ is *not* stabilized, can be regarded as two real scalar field models. This can be achieved by removing restriction (7.68) on the Kähler potential. However, the form of the potential (7.62) is then rather restrictive. We found that an essentially arbitrary positive potential could be obtained in the purely ϕ direction by choosing $w(A)$ appropriately. But, given $w(A)$, the potential for the second field ξ is then determined at the same time. Hence, there is a risk that the second direction spoils the suitability of the potential for inflationary dynamics [62]. It turns out that more flexibility in constructing multi-real-scalar-field potentials can be obtained by coupling our theory to a third chiral superfield $\Psi = C + \Theta^\alpha \Theta_\alpha F_C$, see [19] for details.

7.4 Supersymmetric Galileons

Galileon theories of a real scalar field are special because they have two-derivative equations of motion despite having higher-derivative Lagrangians. They are a subclass of the most general scalar theories with two-derivative equations of motion, known as Horndeski's theories [20] (see also [22]). The "standard" Galileons [21] have the additional property that in the equations of motion there are precisely two derivatives acting on each field. An immediate consequence is that the standard Galileons are invariant under a so-called Galilean shift symmetry $\phi \rightarrow \phi + c + b_\mu x^\mu$ with c, b_μ being constants, whence they derive their name. Many variants of the original model have been constructed, such as conformal Galileons [63], DBI Galileons [64], Galileons with an internal symmetry [65, 66], bi-Galileons [67, 68] and so on. The crucial property of all of these theories is that they have equations of motion with no more than two derivatives acting on a field. This helps to evade Ostrogradsky's theorem [69]—that is, despite the higher-derivative nature of the Lagrangians, for suitable coefficients of the Galileon Lagrangians these theories *do not contain ghosts*.

Galileons have attracted considerable interest due to their rather remarkable properties. For example, they admit de-Sitter-like solutions in the absence of a cosmological constant [70–72] and they lead to a Vainshtein-type screening mechanism so that they can be in agreement with solar system "fifth force" constraints while contributing a fifth force on large scales [73, 74]. Moreover, they allow for solutions that violate the null energy condition without leading to the appearance of ghosts [63, 75]. This last property means that Galileons also have applications to early universe cosmology, allowing the construction of emergent cosmologies (see, for example, the model of Galilean genesis [76]) and non-singular bouncing cosmologies such as new ekpyrotic theory [13–15, 34, 77, 78] or the matter bounce model [79]. Such alternative models to inflation even play a significant role in eternal inflation [80–82].

There exists a suggestive construction of Galileon Lagrangians as the theories describing the dynamics of co-dimension-one branes [64]. This has led people to speculate that Galileons might arise naturally out of string theory and, hence, enjoy a more fundamental status than other higher-derivative terms, in analogy to the Dirac-Born-Infeld action. Brane backgrounds in string theory typically preserve some amount of unbroken supersymmetry. Therefore, if Galileons are to arise from string theory it will be in a supersymmetric context. Hence, it is of importance to study the supersymmetric extensions of Galileon theories. In previous work [18], it was shown that *conformal* Galileons can be made globally $N = 1$ supersymmetric—these theories arising naturally as a way of obtaining correct sign spatial gradients in supersymmetric ghost-condensates (see also [16, 17]). It was found that the new fields required by supersymmetry (a second real scalar, a spin $\frac{1}{2}$ fermion and a complex auxiliary field) admit stable, positive-energy fluctuations around specific backgrounds, namely those where the second scalar field is constant. However, possible ghost instabilities associated with vacua with a *spacetime-dependent* second scalar were not explored. We will do this in the present paper, restricting our discussion for the most part to the cubic Galileons within the context of four-dimensional global $N = 1$ supersymmetry.

7.4.1 Galileons and Complex Fields

In this and the following two sections, we will focus on the simplest non-trivial Galileon Lagrangian given by [21]

$$L_3 = -\frac{1}{2}(\partial\phi)^2\Box\phi. \tag{7.70}$$

By varying with respect to ϕ, one can immediately see that the equation of motion is second order and given by

$$(\Box\phi)^2 - \phi^{,\mu\nu}\phi_{,\mu\nu} = 0. \tag{7.71}$$

Thus, despite the higher-derivative nature of the Lagrangian, the equation of motion is well-behaved and the Cauchy problem is well-posed. In four dimensions, there are two more such Galileon Lagrangians,

$$L_4 = -\frac{1}{2}(\partial\phi)^2\big((\Box\phi)^2 - \phi^{,\mu\nu}\phi_{,\mu\nu}\big), \tag{7.72}$$

$$L_5 = -\frac{1}{2}(\partial\phi)^2\big((\Box\phi)^3 - 3\Box\phi\phi^{,\mu\nu}\phi_{,\mu\nu} + 2\phi^{,\mu\nu}\phi_{,\mu\rho}\phi_{,\nu}{}^{\rho}\big) \tag{7.73}$$

which also lead to second-order equations of motion. In $N = 1$ supersymmetry, the lowest component of a superfield Φ is a *complex* scalar A, which can be decomposed into two real scalars as $A = \frac{1}{\sqrt{2}}(\phi + i\xi)$. One consequence is that supersymmetric scalar-field actions can always be written as hermitian combinations of A and its complex conjugate A^*. Motivated by this, *but before imposing any supersymmetry condition*, it is of interest to consider the possible extensions of the Galileon Lagrangian (7.70) from the real scalar field ϕ to the complex scalar A. Specifically, we are interested in Lagrangians which, when the second real scalar ξ is set to zero, reduce to the Galileon Lagrangian L_3 presented in (7.70). There are, in principle, a large number of such Lagrangians. Let us here just illustrate using two concrete examples that, even though by construction these extended Lagrangians contain the L_3 Lagrangian for ϕ, the properties of the second scalar ξ can vary considerably, and it is in no way guaranteed that the second scalar also shares the desired Galilean symmetries. Having established this, we will then—in Sect. 7.3—move on to supersymmetry (where we will give a completely exhaustive treatment) in order to determine which such complex scalar field generalizations of L_3 supersymmetry allows.

Our first example of a generalization of (7.70) from the real scalar ϕ to a complex scalar field A is straightforward. It is obtained simply by replacing $\phi \to \sqrt{2}A$ and then taking the real part. For L_3 above, this amounts to considering the Lagrangian

$$L_3^{\mathbb{C}} = -\frac{1}{\sqrt{2}}(\partial A)^2 \Box A + h.c., \tag{7.74}$$

where $h.c.$ stands for "hermitian conjugate". It is then evident that the resulting equations of motion are still second order, since they are given by

$$(\Box A)^2 - A^{,\mu\nu}A_{,\mu\nu} = 0, \quad (\Box A^*)^2 - A^{*,\mu\nu}A^*_{,\mu\nu} = 0. \tag{7.75}$$

In terms of the real scalars ϕ and ξ, the Lagrangian and equations of motion are

$$L_3^{\mathbb{C}} = -\frac{1}{2}\big((\partial\phi)^2\Box\phi - (\partial\xi)^2\Box\phi - 2\partial\phi \cdot \partial\xi\Box\xi\big), \tag{7.76}$$

$$0 = (\Box\phi)^2 - \phi^{,\mu\nu}\phi_{,\mu\nu} - (\Box\xi)^2 + \xi^{,\mu\nu}\xi_{,\mu\nu}, \tag{7.77}$$

$$0 = \Box\phi\Box\xi - \phi^{,\mu\nu}\xi_{,\mu\nu}, \tag{7.78}$$

clearly exhibiting that we now have a coupled two-field Galileon system. Not only are the equations of motion of second order, but both fields admit independent Galileon-type shift symmetries $\phi \to \phi + c^{(\phi)} + b_\mu^{(\phi)} x^\mu$ and $\xi \to \xi + c^{(\xi)} + b_\mu^{(\xi)} x^\mu$ respectively.

However, using a second concrete example, we now demonstrate that other extensions of the L_3 Lagrangian to complex scalar field A do *not* necessarily lead to second-order equations of motion. To illustrate this important point, consider the action

$$\tilde{L}_3^{\mathbb{C}} = -\frac{1}{\sqrt{2}}\partial A \cdot \partial A^* \Box A + h.c. \tag{7.79}$$

$$= -\frac{1}{2}\big((\partial\phi)^2\Box\phi + (\partial\xi)^2\Box\phi\big), \tag{7.80}$$

leading to the equations of motion

$$0 = (\Box\phi)^2 - \phi^{,\mu\nu}\phi_{,\mu\nu} - \xi^{,\mu\nu}\xi_{,\mu\nu} - \xi_{,\mu}\xi_{,\nu}{}^{\nu\mu}, \tag{7.81}$$

$$0 = \Box\xi\Box\phi + \xi_{,\mu}\phi_{,\nu}{}^{\nu\mu}. \tag{7.82}$$

Clearly, these are higher-order in time and, thus, by Ostrogradsky's theorem [69], lead to the appearance of ghosts.

Given these two contrasting examples, a crucial question is then: which kinds of complex scalar field generalizations of the Galileon Lagrangian does supersymmetry allow? We now turn to this question.

7.4.2 Supersymmetric Cubic Galileons

In this section, we will construct all possible supersymmetric Lagrangians involving the product of three fields and four space-time derivatives, in order to see if there might exist inequivalent supersymmetric extensions of the L_3 Lagrangian (7.70). This means that we should consider all possible superfield expressions involving the cubic product of a chiral superfield and *two* spacetime derivatives (and linear combinations of all such terms). The superfield Lagrangians of potential interest are straightforward to write down. They are given by the $\theta\theta\bar{\theta}\bar{\theta}$ components of the following expressions (where derivatives act only on the immediately following superfield):

$$\partial^\mu \Phi \partial_\mu \Phi \Phi + h.c. \tag{7.83}$$

$$\partial^\mu \Phi \partial_\mu \Phi^\dagger \Phi + h.c. \tag{7.84}$$

$$\partial^\mu \Phi \partial_\mu \Phi \Phi^\dagger + h.c. \tag{7.85}$$

All other terms of potential interest can be related to these via linear combinations and using integration by parts.

One might be concerned that there could be other allowed terms involving the superspace derivatives D_α and $\bar{D}_{\dot{\alpha}}$. Once again, however, upon integration by parts, using the algebra (7.3) and the chiral superfield constraint, it follows that these are always equivalent to some linear combination of (7.83)–(7.85). As a concrete example, consider the term

$$\int d^4x d^4\theta \bar{D}_{\dot{\alpha}} D^2 \Phi \bar{D}^{\dot{\alpha}} \Phi^\dagger \Phi. \tag{7.86}$$

Using integration by parts, algebra (7.3) and the chiral superfield constraint, this becomes

$$\int d^4x d^4\theta \bar{D}_{\dot{\alpha}} D^2 \Phi \bar{D}^{\dot{\alpha}} \Phi^\dagger \Phi \tag{7.87}$$

$$= \int d^4x d^4\theta (-\bar{D}^2 D^2 \Phi) \Phi^\dagger \Phi \tag{7.88}$$

$$= \int d^4x d^4\theta (-16 \Box \Phi) \Phi^\dagger \Phi \tag{7.89}$$

$$= \int d^4x d^4\theta [16 \partial^\mu \Phi \partial_\mu \Phi^\dagger \Phi + 16 \partial^\mu \Phi \partial_\mu \Phi \Phi^\dagger] \tag{7.90}$$

and, hence, is simply a linear combination of (7.84) and (7.85), as claimed. It is straightforward to show that this is always the case.

Having established this, let us systematically discuss the Lagrangian associated with each of the three supersymmetric terms (7.83)–(7.85). First consider (7.83). Note that this is the only one of the three terms that can possibly lead to the complex Galileon $L_3^{\mathbb{C}}$ given in (7.74) of the previous section. This follows from the fact that it is the sole term containing only Φ's or only $\Phi^\dagger$'s in a single term. Hence, it appears that this might be a suitable supersymmetric extension of the L_3 Lagrangian with purely second order equations of motion. However, the chirality of Φ immediately implies that the supersymmetric Lagrangian associated with (7.83) is, in fact, zero. To see this, instead of integrating over $d^4\theta$, one can make use of the Grassmann nature of the $\theta, \bar{\theta}$ coordinates and replace $d^4\theta$ by a $D^2\bar{D}^2$ derivative of the corresponding superfield expression. Since $\bar{D}$ commutes with partial derivatives, it immediately follows that superfield expressions constructed exclusively out of Φ's and partial derivatives must vanish, since the $\bar{D}$ derivative will necessarily act on a chiral field Φ thus yielding zero. That is, the supersymmetric action associated with (7.83) is

$$\int d^4x d^4\theta \partial^\mu \Phi \partial_\mu \Phi \Phi = 0. \tag{7.91}$$

Note that this argument relies solely on holomorphicity and, thus, also extends to potential supersymmetric extensions of complex Galileons with higher powers of fields, such as $L_4^{\mathbb{C}}$ and $L_5^{\mathbb{C}}$.

It follows that we are left with only two possible supersymmetric extensions of the L_3 Lagrangian—namely, with integrands (7.84) and (7.85). These are

$$\int d^4x d^4\theta \partial^\mu \Phi \partial_\mu \Phi^\dagger \Phi = \int d^4x \left(-A\Box A\Box A^* - \Box A^*(\partial A)^2\right) \tag{7.92}$$

and

$$\int d^4x d^4\theta \partial^\mu \Phi \partial_\mu \Phi \Phi^\dagger = \int d^4x \, \Box A^*(\partial A)^2 \tag{7.93}$$

respectively, plus their hermitian conjugates. Note that we have used integration by parts to simplify these terms as much as possible. Let us first examine the action given in (7.92). We immediately see that this term is *not* an appropriate extension of the L_3 Galileon Lagrangian. This follows from the fact that, when the scalar ξ is set to zero, this Lagrangian does not reduce to L_3 and in fact results in a fourth-order equation of motion for ϕ. Hence, we are left with a single possible supersymmetric extension of the L_3 Galileon Lagrangian, namely the real part of (7.93). We note that this Lagrangian is equivalent to the supersymmetric Galileon Lagrangian used in [18]. Thus, we define the supersymmetric extension of L_3 as

$$\begin{aligned}L_3^{SUSY} &\equiv -\frac{1}{\sqrt{2}}\int d^4\theta\partial^\mu\Phi\partial_\mu\Phi\Phi^\dagger + h.c.\\ &= -\frac{1}{\sqrt{2}}\Box A^*(\partial A)^2 + h.c.\\ &= -\frac{1}{2}\big((\partial\phi)^2\Box\phi - (\partial\xi)^2\Box\phi + 2\partial\phi\cdot\partial\xi\Box\xi\big).\end{aligned} \tag{7.94}$$

Compared to the complex Galileon (7.76), only the sign of the last term has changed! Nevertheless, this has profound consequences, since the resulting equations of motion are now of third order in derivatives. They read

$$0 = (\Box\phi)^2 - \phi^{,\mu\nu}\phi_{,\mu\nu} + (\Box\xi)^2 + \xi^{,\mu\nu}\xi_{,\mu\nu} + 2\xi_{,\mu}\xi_{,\nu}{}^{\nu\mu}, \tag{7.95}$$

$$0 = \xi^{,\mu\nu}\phi_{,\mu\nu} + \xi_{,\mu}\phi_{,\nu}{}^{\nu\mu}. \tag{7.96}$$

As one can clearly see, it is the presence of the second scalar ξ that induces the dangerous higher-derivative terms. That is, L_3^{SUSY} in (7.94), similarly to the second of our concrete examples given in (7.79), has higher-order equations of motion. We will show explicitly in the next section that the presence of these higher derivatives leads to the appearance of a ghost.

7.4.3 Hiding from the Ghost

We would now like to explicitly demonstrate the ghost degree of freedom in L_3^{SUSY}. The presence of a ghost is already implied by Ostrogradsky's theorem [69] and we will, in fact, analyze a supersymmetric version of L_4 from this point of view in the following section. Nevertheless, we prefer to also analyze the Lagrangian L_3^{SUSY} directly, both because it is instructive to see the ghost appearing at the level of the Lagrangian and because such an analysis elucidates in what regime the ghost can be harmless. For this purpose, it suffices to look at the time-derivative terms in the Lagrangian, since it is these that are associated with ghosts. Adding a canonical kinetic term $L_2^{SUSY} = \int d^4\theta\Phi\Phi^\dagger = -\partial^\mu A\partial_\mu A^*$, as well as an overall constant c_3 in front of the L_3^{SUSY} Lagrangian, the Lagrangian of interest becomes

$$L_{2+3}^{SUSY} \equiv L_2^{SUSY} + c_3 L_3^{SUSY} = \frac{1}{2}\dot\phi^2 + \frac{1}{2}\dot\xi^2 + c_3\dot\xi^2\ddot\phi, \tag{7.97}$$

where we have integrated by parts in order to place all double derivatives on ϕ rather than ξ. Note that this is a completely arbitrary choice and does not reduce the generality of our analysis. We consider a time-dependent background and would like to study perturbations around it. Thus, we define

$$\phi = \bar\phi(t) + \delta\phi(x^\mu), \quad \xi = \bar\xi(t) + \delta\xi(x^\mu). \tag{7.98}$$

Even though the perturbations depend on both time and space, we will only be interested in the time dependence here. To quadratic order in fluctuations, the Lagrangian then becomes

$$L^{SUSY}_{2+3\,\mathrm{quad}} = \frac{1}{2}(\delta\dot{\phi})^2 + \frac{1}{2}(1 + 2c_3\ddot{\phi})(\delta\dot{\xi})^2 + 2c_3\dot{\xi}\,\delta\dot{\xi}\delta\ddot{\phi}. \tag{7.99}$$

By defining a new fluctuation variable

$$\delta\dot{b} \equiv \delta\dot{\xi} + \frac{2c_3\dot{\xi}}{1 + 2c_3\ddot{\phi}}\delta\ddot{\phi}, \tag{7.100}$$

the quadratic Lagrangian can then be diagonalized to become

$$L^{SUSY}_{2+3\,\mathrm{quad}} = \frac{1}{2}(\delta\dot{\phi})^2 + \frac{1}{2}(1 + 2c_3\ddot{\phi})\left((\delta\dot{b})^2 - \frac{4c_3^2\dot{\xi}^2}{(1 + 2c_3\ddot{\phi})^2}(\delta\ddot{\phi})^2\right). \tag{7.101}$$

Note that $(\delta\dot{b})^2$ and $(\delta\ddot{\phi})^2$ enter with opposite signs and, hence, one of these two terms is ghost-like.[7] Assuming that the factor $(1 + 2c_3\ddot{\phi})$ is positive, the ghost then resides in $\delta\ddot{\phi}$. As the Lagrangian shows, the significance of the ghost is essentially controlled by the size of $c_3\dot{\xi}$. This can be confirmed by looking at the dispersion relation of $\delta\phi$. If one denotes the four-momentum of $\delta\phi$ by p_μ, then the associated dispersion relation is given by

$$p_0^2\Big(1 - \frac{4c_3^2\dot{\xi}^2}{(1 + 2c_3\ddot{\phi})}p_0^2\Big) = 0, \tag{7.102}$$

where we have assumed that $\dot{\xi}$ and $\ddot{\phi}$ are slowly varying. The mass m is defined via $p^2 = -p_0^2 = -m^2$ and, hence, the dispersion relation implies that $\delta\phi$ consists of two modes. The first is a massless mode which arises from the ordinary correct-sign kinetic term. The second is the ghost, which has a mass

$$m_g^2 = \frac{(1 + 2c_3\ddot{\phi})}{4c_3^2\dot{\xi}^2}. \tag{7.103}$$

Note that, as there is an overall wrong sign for the ghost in the Lagrangian, this mass is formally tachyonic. However, it is important to realize that this mass does not arise from a potential, but rather from the kinetic term $(\delta\dot{\phi})^2$. The implication is that this mass does not indicate a time scale over which the (perturbative) vacuum becomes unstable, but rather an energy scale associated with the ghost. In other words, as

[7] This ghost was not seen in [18] because in that paper the perturbation analysis was performed solely around $\dot{\xi} = constant$ backgrounds.

long as we are considering fluctuations with energy below m_g, the ghost does not get excited. From an effective field theory point of view, we are protected from the catastrophic instabilities associated with the ghost if we take the cut-off Λ of the effective field theory to lie below m_g. At the same time, we must ensure that the background itself, that is, $\dot{\xi}$, remains within the range of validity of the effective theory. Hence, an additional requirement is that $|\dot{\xi}| < \Lambda^2$, and similar inequalities must also hold for higher time derivatives of ξ. Together with the requirement $\Lambda < m_g$, this implies that we must impose (assuming $|c_3\ddot{\phi}| \ll 1$)

$$|\dot{\xi}| < \frac{1}{|c_3|^{2/3}}, \quad |\ddot{\xi}| < \frac{1}{|c_3|}, \cdots \tag{7.104}$$

in order to safely suppress the ghost. Thus, as expected, for general backgrounds one must take the prefactor of the Galileon term to be small for consistency.

7.5 Conclusions and Outlook

This talk was concerned with a new formalism that allows one to obtain an $\mathcal{N} = 1$ supergravity extension of any scalar field theory with higher-derivative kinetic terms. This was accomplished by constructing a superfield—quartic in chiral scalars—which contains the term $(\partial\phi)^4$ and, when the fermions are set to zero, consists entirely of its top component. Thus, when multiplied by any other superfield, the resulting Lagrangian contains only the lowest component of the multiplicative factor. This property enables one to directly construct a supergravity extension any higher-derivative scalar field term of interest. Moreover, the discussed supergravity extension of $(\partial\phi)^4$ is likely to be the unique one that does not modify the gravitational sector of the theory. We discussed the investigation of the properties of the auxiliary fields in this context, which are crucial to the structure of supergravity, and their effect on the expressions for the potentials.

Namely, there is one new, and important, property of our formalism. That is, although the auxiliary fields F satisfy an algebraic equation of motion, that equation is now cubic—as opposed to the linear equation in the usual second-order kinetic theory. Hence, this equation admits up to three distinct solutions. These solutions lead to different theories that cannot dynamically transition from one to another. One solution is directly related to the one ordinarily obtained in the absence of higher-derivative terms. This leads to corrections to both the kinetic and potential terms when substituted into the action. In the limit that the higher-derivative terms become large, the effect of eliminating the auxiliary field is to suppress certain contributions to the potential. The result is that the negative term $-3e^K|W|^2$ becomes the dominant contribution to the potential energy. Thus, in the large higher-derivative limit, supergravity manifests once more its predilection for negative potentials. This feature implies that the supergravity implementation of inflationary and k-essence models—such as DBI inflation—that rely on higher-derivative kinetic terms in an essential way, become more challenging. In addition to this ordinary solution for F, there exist up to two new solutions. We refer to [16] for details.

We hope that our results can eventually be used to bridge the gap between standard model building in cosmology and full-blown string compactifications, leading to well-motivated effective theories of early-universe dynamics. One of the most important problems in cosmology is to find a scenario for the early universe that is not only in agreement with observations, but is also rooted in a sensible microphysical theory. Only in this way can cosmology and particle physics can be united, and a consistent theory of our universe be obtained. While still far from this goal, we discussed a small aspect of the problem—showing how to construct models of DBI inflation in four-dimensional $\mathcal{N} = 1$ supergravity. We saw that if one tries to construct a model of DBI inflation from a single chiral superfield, it is bound to fail—since the potential becomes negative when the higher-derivative terms become important. This obstacle can be circumvented by coupling the theory to one or more additional chiral superfields. In fact, the construction can be generalized to an arbitrary number N of chiral superfields—then, not only can the potential energy be positive but one can construct a wide range of potential functions for the original DBI scalar $\phi = \sqrt{2}\Re(\phi^1)$ and $N - 1$ additional real scalars. Our analysis shows that models of multi-real-scalar-field DBI inflation in fact can be constructed in $\mathcal{N} = 1$ supergravity. Of course this leaves open the question of whether there exist other ways of realizing DBI inflation within the context of supergravity. More importantly, however, is the question of whether or not such constructions can be obtained from a full-fledged string compactification, or from some other fundamental theory of particle physics. These are pertinent questions for future research.

We also discussed the $N = 1$ supersymmetric extension of Galileons theories. Those containing the product of three chiral fields necessarily admit higher-derivative equations of motion, which implies that these theories contain ghosts. This means that when supersymmetry is included, cubic Galileons, both of the standard and the conformal variety, lose their special status among higher-derivative scalar theories and should be treated in much the same way as other higher-derivative terms. That is to say, they should be regarded as correction terms in a perturbative, effective field theory framework. By extension, our results are also likely to apply to the relevant parts of Horndeski's most general scalar-tensor theory [20]. We stress that our work has been done in the context of minimal $N = 1$ supersymmetry. It would be interesting to carry out a similar analysis for extended supersymmetries.

Acknowledgments M. Koehn would like to thank Stefano Bellucci for the invitation to the workshop "Breaking of supersymmetry and ultraviolet divergences in extended supergravity" at the INFN in Frascati in March 2013, where the content of these proceedings was presented. M. Koehn and J. L. Lehners gratefully acknowledge the support of the European Research Council via the Starting Grant numbered 256994. B. A. Ovrut is supported in part by the DOE under contract No. DE-AC02-76-ER-03071 and the NSF under grant No. 1001296.

References

1. Y.A. Golfand, E.P. Likhtman, Extension of algebra of Poincare Group generators and violation of P invariance. JETP Lett. **13**(8), 323 (1971)
2. D.V. Volkov, V.P. Akulov, Is the neutrino a goldstone particle? Phys. Lett. B **46**(1), 109–110 (1973)
3. J. Wess, B. Zumino, Supergauge transformations in four dimensions. Nucl. Phys. B **70**(1), 39–50 (1974)
4. R. Haag, J.T. Lopuszanski, M. Sohnius, All possible generators of supersymmetries of the S-matrix. Nucl. Phys. B **88**(2), 257–274 (1975)
5. P. Langacker, Proton decay, grand unification, and supersymmetry, in Proceedings of AIP Conference (University of Pennsylvania, Philadelphia, Pennsylvania 19104, 1984). pp. 251–276
6. S. Dimopoulos, H. Georgi, Softly broken supersymmetry and SU(5). Nucl. Phys. B **193**(1), 150–162 (1981)
7. S. Dimopoulos, S. Raby, F. Wilczek, Supersymmetry and the scale of unification. Phys. Rev. D **24**(6), 1681–1683 (1981)
8. V. Braun, Y.H. He, B.A. Ovrut, T. Pantev, The exact MSSM spectrum from string theory. JHEP 05, 043 (2006)
9. A. Lukas, B.A. Ovrut, K.S. Stelle, D. Waldram, Universe as a domain wall. Phys. Rev. D **59**(8), 86001 (1999)
10. E. Silverstein, D. Tong, Scalar speed limits and cosmology: acceleration from D-cceleration. Phys. Rev. D **70**(1), 103505 (2004)
11. J. Khoury, B.A. Ovrut, P.J. Steinhardt, N. Turok, Ekpyrotic universe: colliding branes and the origin of the hot big bang. Phys. Rev. D **64**(1), 123522 (2001)
12. J. Khoury, B.A. Ovrut, N. Seiberg, P.J. Steinhardt, N. Turok, From big crunch to big bang. Phys. Rev. D **65**(8), 86007 (2002)
13. E.I. Buchbinder, J. Khoury, B.A. Ovrut, New Ekpyrotic cosmology. Phys. Rev. D **76**, 123503 (2007)
14. E.I. Buchbinder, J. Khoury, B.A. Ovrut, On the initial conditions in new ekpyrotic cosmology. JHEP **11**(1), 076 (2007)
15. E.I. Buchbinder, J. Khoury, B.A. Ovrut, Non-gaussianities in new ekpyrotic cosmology. Phys. Rev. Lett. **100**(1), 171302 (2008)
16. M. Koehn, J.L. Lehners, B.A. Ovrut, Higher-derivative chiral superfield actions coupled to N=1 supergravity. Phys. Rev. D **86**(8), 85019 (2012)
17. J. Khoury, J.L. Lehners, B. Ovrut, Supersymmetric $P(X, \phi)$ and the ghost condensate. Phys. Rev. D **83**(1), 125031 (2011)
18. J. Khoury, J.L. Lehners, B.A. Ovrut, Supersymmetric galileons. Phys. Rev. D **84**(4), 43521 (2011)
19. M. Koehn, J.L. Lehners, B.A. Ovrut, Dbi inflation in $\mathcal{N} = 1$ supergravity. Phys. Rev. D **86**, 123510 (2012)
20. G.W. Horndeski, Second-order scalar-tensor field equations in a four-dimensional space. Int. J. Theor. Phys. **10**(6), 363–384 (1974)
21. A. Nicolis, R. Rattazzi, E. Trincherini, Galileon as a local modification of gravity. Phys. Rev. D **79**(6), 064036 (2009)
22. C. Deffayet, X. Gao, D.A. Steer, G. Zahariade, From k-essence to generalized Galileons. Phys. Rev. D **84**(6), 064039 (2011)
23. M. Koehn, J.L. Lehners, B.A. Ovrut, Supersymmetric cubic Galileons have ghosts. Phys. Rev. D **88**(2), 023528 (2013)
24. M. Koehn, J.L. Lehners, B.A. Ovrut, Ghost condensate in N = 1 supergravity. Phys. Rev. D **87**(6), 65022 (2013)
25. R. Donagi, B.A. Ovrut, D. Waldram, Moduli spaces of fivebranes on elliptic Calabi-Yau threefolds. J. High Energ. Phys **11**(1), 030 (1999)

26. E. Buchbinder, R. Donagi, B. Ovrut, Vector bundle moduli and small instanton transitions. J. High Energ. Phys. **06**(0), 054 (2002)
27. E.I. Buchbinder, R. Donagi, B.A. Ovrut, Vector bundle moduli superpotentials in heterotic superstrings and M-theory. J. High Energ. Phys. **07**(0), 066 (2002)
28. C. Armendáriz-Picón, T. Damour, V. Mukhanov, k-Inflation. Phys. Lett. B **458**(2), 209–218 (1999)
29. P. Steinhardt, N. Turok, Cosmic evolution in a cyclic universe. Phys. Rev. D **65**(12) (2002)
30. J.L. Lehners, Ekpyrotic and cyclic cosmology. Phys. Rep. **465**(6), 223–263 (2008)
31. M. Trodden, K. Hinterbichler, Generalizing Galileons. Class. Quant. Grav. **28**(2), 204003 (2011)
32. N. Arkani-Hamed, H.S. Cheng, M.A. Luty, S. Mukohyama, Ghost condensation and a consistent infrared modification of gravity. JHEP **05**, 074 (2004)
33. P. Creminelli, M.A. Luty, A. Nicolis, L. Senatore, Starting the Universe: stable violation of the null energy condition and non-standard cosmologies. J. High Energ. Phys. **12**(1), 080 (2006)
34. J.L. Lehners, Cosmic bounces and cyclic universes. Class. Quant. Grav. **28**(20), 204004 (2011)
35. J. Wess, J. Bagger, *Supersymmetry and Supergravity* (Princeton University Press, Princeton, NJ, 1992)
36. D. Baumann, D. Green, Signature of supersymmetry from the early universe. Phys. Rev. D **85**(1), 103520 (2012)
37. D. Baumann, D. Green, Supergravity for effective theories. J. High Energ. Phys. **03**, 001 (2012)
38. F. Farakos, A. Kehagias, Emerging potentials in higher-derivative gauged chiral models coupled to N=1 supergravity. J. High Energ. Phys. **11**, 077 (2012)
39. I.L. Buchbinder, S.M. Kuzenko, Nonlocal action for supertrace anomalies in superspace of N=1 supergravity. Phys. Lett. B **202**(2), 233–237 (1988)
40. I. Buchbinder, S. Kuzenko, J. Yarevskaya, Supersymmetric effective potential: superfield approach. Nucl. Phys. B **411**(2), 665–692 (1994)
41. A.T. Banin, I.L. Buchbinder, N.G. Pletnev, Quantum properties of the four-dimensional generic chiral superfield model. Phys. Rev. D **74**(4), 045010 (2006)
42. F. Brandt, Anomaly candidates and invariants of D = 4, N = 1 supergravity theories. Class. Quant. Grav. **11**(4), 849–863 (1994)
43. F. Brandt, Local BRST cohomology in minimal D = 4, N = 1 supergravity. Ann. Phys. **259**, 253–312 (1997)
44. I. Antoniadis, E. Dudas, D.M. Ghilencea, Supersymmetric models with higher dimensional operators. J. High Energ. Phys. **03**, 045 (2008)
45. R.G. Leigh, Dirac-Born action from Dirichlet Sigma model. Mod. Phys. Lett. A **4**(2), 2767–2772 (1989)
46. J. Khoury, B.A. Ovrut, J. Stokes, The worldvolume action of kink solitons in AdS spacetime. JHEP **08**, 015 (2012)
47. B.A. Ovrut, J. Stokes, Heterotic kink solitons and their worldvolume action. JHEP **09**, 065 (2012)
48. G. Dvali, S.H.H. Tye, Brane inflation. Phys. Lett. B **450**(1), 72–82 (1999)
49. E. Lima, B. Ovrut, J. Park, Five-brane superpotentials in heterotic M-theory. Nucl. Phys. B **626**(1), 113–164 (2002)
50. S. Kachru, R. Kallosh, A. Linde, J. Maldacena, L. McAllister, S.P. Trivedi, Towards inflation in string theory. J. Cosmol. Astropart. Phys. **10**(1), 013 (2003)
51. J. Gray, A. Lukas, B. Ovrut, Perturbative antibrane potentials in heterotic M theory. Phys. Rev. D **76**(6), 066007 (2007)
52. M. Alishahiha, E. Silverstein, D. Tong, DBI in the sky: non-Gaussianity from inflation with a speed limit. Phys. Rev. D **70**(1), 123505 (2004)
53. M. Rocek, A.A. Tseytlin, Partial breaking of global D = 4 supersymmetry, constrained superfields, and 3-brane actions. Phys. Rev. D **59**(1), 106001 (1999)
54. M. Kawasaki, M. Yamaguchi, T. Yanagida, Natural chaotic inflation in supergravity. Phys. Rev. Lett. **85**(1), 3572–3575 (2000)

55. R. Kallosh, A. Linde, T. Rube, General inflaton potentials in supergravity. Phys. Rev. D **83**(4), 043507 (2011)
56. D. Baumann, L. McAllister, A microscopic limit on gravitational waves from D-brane inflation. Phys. Rev. D **75**(1), 123508 (2007)
57. J.E. Lidsey, I. Huston, Gravitational wave constraints on Dirac Born Infeld inflation. J. Cosmol. Astropart. Phys. **07**, 002 (2007)
58. T. Kobayashi, S. Mukohyama, S. Kinoshita, Constraints on wrapped Dirac-Born-Infeld inflation in a warped throat. J. Cosmol. Astropart. Phys. **01**, 028 (2008)
59. R. Bean, X. Chen, H. Peiris, Comparing infrared Dirac-Born-Infeld brane inflation to observations. Phys. Rev. D **77**(2), 023527 (2008)
60. D. Langlois, S. Renaux-Petel, D.A. Steer, T. Tanaka, Primordial perturbations and non-Gaussianities in DBI and general multifield inflation. Phys. Rev. D **78**(6), 63523 (2008)
61. T. Kidani, K. Koyama, S. Mizuno, Non-Gaussianities in multifield DBI inflation with a waterfall phase transition. Phys. Rev. D **86**(8), 083503 (2012)
62. B.A. Ovrut, P.J. Steinhardt, Supersymmetry and inflation: a new approach. Phys. Lett. B **133**(3), 161–168 (1983)
63. A. Nicolis, R. Rattazzi, E. Trincherini, Energy's and amplitudes' positivity. J. High Energ. Phys. **05**, 095 (2010)
64. C. de Rham, A.J. Tolley, DBI and the Galileon reunited. J. Cosmol. Astropart. Phys. **05**, 015 (2010)
65. K. Hinterbichler, M. Trodden, D. Wesley, Multifield Galileons and higher codimension branes. Phys. Rev. D **82**(1), 124018 (2010)
66. G. Goon, K. Hinterbichler, M. Trodden, Symmetries for Galileons and DBI scalars on curved space. J. Cosmol. Astropart. Phys. **07**, 017 (2011)
67. C. Deffayet, S. Deser, G. Esposito-Farese, Arbitrary p-form Galileons. Phys. Rev. D **82**(6), 61501 (2010)
68. A. Padilla, P.M. Saffi, S.Y. Zhou, Bi-galileon theory I: motivation and formulation. J. High Energ. Phys. **12**, 031 (2010)
69. M. Ostrogradsky, Sur les racines égales des polynomes entiers. Mem. Ac. St. Petersbourg, IV **4**, 385 (1850)
70. F.P. Silva, K. Koyama, Self-accelerating universe in Galileon cosmology. Phys. Rev. D **80**(1), 121301(R) (2009)
71. A. De Felice, S. Tsujikawa, Cosmology of a Covariant Galileon field. Phys. Rev. Lett. **105**(1), 111301 (2010)
72. N. Chow, J. Khoury, Galileon cosmology. Phys. Rev. D **80**(2), 024037 (2009)
73. A.I. Vainshtein, To the problem of nonvanishing gravitation mass. Phys. Lett. B **39**(3), 393–394 (1972)
74. C. Deffayet, G. Dvali, G. Gabadadze, A. Vainshtein, Nonperturbative continuity in graviton mass versus perturbative discontinuity. Phys. Rev. D **65**(4), 044026 (2002)
75. K. Hinterbichler, A. Joyce, J. Khoury, G.E.J. Miller, Dirac-Born-Infeld Genesis: an improved violation of the null energy condition. Phys. Rev. Lett. **110**(2), 241303 (2013)
76. P. Creminelli, A. Nicolis, E. Trincherini, Galilean genesis: an alternative to inflation. J. Cosmol. Astropart. Phys. **11**(1), 021 (2010)
77. J.L. Lehners, S. Renaux-Petel, Multifield cosmological perturbations at third order and the ekpyrotic trispectrum. Phys. Rev. D **80**(6), 063503 (2009)
78. J.L. Lehners, Ekpyrotic nongaussianity: a review. Adv. Astron. **2010**, 1–20 (2010)
79. Y.F. Cai, D.A. Easso, R. Brandenberger, Towards a nonsingular bouncing cosmology. J. Cosmol. Astropart. Phys. **08**, 020 (2012)
80. J.L. Lehners, Eternal inflation with noninflationary pocket universes. Phys. Rev. D **86**(4), 43518 (2012)
81. M.C. Johnson, J.L. Lehners, Cycles in the multiverse. Phys. Rev. D **85**(1), 103509 (2012)
82. Z.G. Liu, Y.S. Piao, Galilean Islands in eternally inflating background (2013). arXiv.org

Chapter 8
The Leading Eikonal Operator in String-Brane Scattering at High Energy

G. D'Appollonio, P. Di Vecchia, R. Russo and G. Veneziano

In this paper we present two (a priori independent) derivations of the eikonal operator in string-brane scattering. The first one is obtained by summing surfaces with any number of boundaries, while in the second one the eikonal operator is derived from the three-string vertex in a suitable light-cone gauge. This second derivation shows that the bosonic oscillators present in the leading eikonal operator are to be identified with the string bosonic oscillators in a suitable light-cone gauge, while the first one shows that it exponentiates recovering unitarity. This paper is a review of results obtained in [1, 2].

G. D'Appollonio
Dipartimento di Fisica, Università di Cagliari and INFN 5 Cittadella Universitaria, 09042 Monserrato, Italy

P. Di Vecchia (✉)
The Niels Bohr Institute, University of Copenhagen, Blegdamsvej 17, 2100 Copenhagen, Denmark
e-mail: divecchi@nbi.dk

P. Di Vecchia
Nordita, KTH Royal Institute of Technology and Stockholm University, Roslagstullsbacken 23, SE-10691 Stockholm, Sweden

R. Russo
Queen Mary University of London, Mile End Road, London E1 4NS, UK

R. Russo
Laboratoire de Physique Thèorique de L'Ecole Normale Supèrieure, 24 rue Lhomond, 75231 Paris cedex, France

G. Veneziano
Collège de France, 11 place M. Berthelot, 75005 Paris, France

G. Veneziano
Theory Division, CERN, 1211 Geneva 23, Switzerland

S. Bellucci (ed.), *Breaking of Supersymmetry and Ultraviolet Divergences in Extended Supergravity*, Springer Proceedings in Physics 153,
DOI: 10.1007/978-3-319-03774-5_8,

8.1 Introduction

High energy scattering in the Regge limit in superstring theory has been investigated since more than 25 years. It was originally studied in elastic string-string collisions [1] and has more recently been extended to the elastic scattering of a closed string on a Dp-brane [2]. Due to the fact that, in the Regge limit, the amplitude is dominated by the exchange of the leading Regge trajectory that has the graviton as the lowest state, one gets a lowest order (sphere or disc) amplitude that diverges with the energy violating unitarity at high energy. Unitarity is restored by adding higher order corrections (torus or annulus etc.) and summing them up. In this way, while in field theory one gets an exponential with a phase divergent at high energy that is consistent with unitarity,what one obtains in string theory can be written in terms of an infinite set of bosonic oscillators, introduced to write the amplitude in a simple and compact form, and is called the leading eikonal operator.

This construction poses, however, various problems. What are these bosonic oscillators? Are they connected to the bosonic oscillators of superstring theory? Since we are studying superstring theory, why don't we get also fermionic oscillators? Although the connection of these oscillators with the string oscillators was unclear, it was believed that they were somehow directly related to the string bosonic oscillators. Evidence of this connection came from a paper by Black and Monni [3] where the disk amplitude for the production of massive states, lying on the leading Regge trajectory, from the scattering of a massless state on a Dp-brane was computed and found to agree with what one gets from the eikonal operator. It turns out, however, that this comparison is more subtle because one has to take into account that the longitudinal polarization of the massive state gets enhanced at high energy pretty much as the longitudinal component of the gauge boson $W^{\pm}$ in the Standard Model without the Higgs boson.

In a recent paper [1] the problems raised above were clarified showing that the bosonic oscillators appearing in the eikonal operator are the bosonic oscillators of superstring in a suitable light-cone gauge and that the fermionic oscillators are not relevant at high energy. Furthermore, it was shown how to correctly treat the longitudinal polarization of the massive state. This means that, if we scatter a massless state on a Dp-brane, we produce, at high energy, only massive states involving an arbitrary number of bosonic oscillators together with only the fermionic oscillators already present in the massless state. Actually, the analysis of [1] is more general because it provides the production amplitude in the Regge high energy limit of an arbitrary state of superstring theory from the scattering of an arbitrary state on the Dp-brane. In particular, it has been shown [1] that the leading eikonal operator can be directly derived starting from the three-string light-cone vertex (either in the form of Green-Schwarz or in that of Ramond-Neveu-Schwarz) and then inserting in one of the three legs the string propagator and by closing it with the boundary state that takes care of the presence of the Dp-branes. This provides a direct construction of

[1] For a complete list of references see [1].

the leading eikonal operator from the string operator formalism. The aim of this talk is to present these recent results. In [1] the leading eikonal operator has been also constructed by using a covariant formalism in terms of the Reggeon vertex operator, but this will not be reviewed here.

The content of this paper is the following. In Sect. 8.2 we derive the eikonal operator as it was originally constructed in [2] starting from the scattering amplitudes. In Sect. 8.3 we give a description of the physical spectrum of the first massive level in the two light-cone formalisms (GS and RNS) and in the covariant formalism. Then, interpreting the bosonic oscillators of the eikonal operator as the light-cone bosonic oscillators of string theory, we show that, at high energy, the states that can be produced by the scattering of a graviton on a Dp-brane, are only those of the type $A_{-1;j}|i,0\rangle$, while those of the type $Q_{-1;a}|\dot{a},0\rangle$ are not. This is consistent with what one gets from the eikonal operator that does not contain any fermionic oscillator. In Sect. 8.4 we show how to derive the eikonal operator from the light-cone three-string vertex and the boundary state. Finally, an Appendix with a discussion of the kinematics of the scattering process is added at the end of the paper.

8.2 The Eikonal Operator I

In this section we derive the leading eikonal operator from the elastic scattering of a massless state of superstring theory on a Dp-brane, following [2]. The starting point is the disk amplitude given by:

$$\begin{aligned} \mathcal{A}_1(E,t) &\sim \langle 0| \int \frac{d^2z_1 d^2z_2}{dV_{abc}} W_1(z_1,\bar{z}_1) W_2(z_2,\bar{z}_2)|B\rangle \\ &= -\frac{\pi^{\frac{9-p}{2}} R_p^{7-p}}{\Gamma(\frac{7-p}{2})} \mathcal{K}(p_1,\epsilon_1;p_2,\epsilon_2) \frac{\Gamma(-\alpha' E^2)\Gamma(-\frac{\alpha'}{4}t)}{\Gamma(1-\alpha' E^2 - \frac{\alpha'}{4}t)} \end{aligned} \tag{8.1}$$

where

$$R_p^{7-p} = gN \frac{(2\pi\sqrt{\alpha'})^{7-p}}{(7-p)V_{S^{8-p}}}, \qquad V_{S^n} = \frac{2\pi^{\frac{n+1}{2}}}{\Gamma(\frac{n+1}{2})}, \tag{8.2}$$

W_1 and W_2 are the vertex operators of a massless state and $|B\rangle$ is the boundary state that identifies the right with the left oscillators and imposes Dirichlet (Neumann) boundary conditions along the directions transverse (longitudinal) to the world-volume of the stack of N parallel Dp-branes. The scattering is described by two Mandelstam-like variables:

$$t = -(p_{1\perp} + p_{2\perp})^2 = -4E^2 \sin^2\frac{\Theta}{2}; \quad s = E^2 = |p_{1\perp}|^2 = |p_{2\perp}|^2 \tag{8.3}$$

Θ is the angle between the $(9-p)$-dim vectors $p_{1\perp}$ and $-p_{2\perp}$.

Along the directions of the world-volume of the Dp-branes, there is conservation of energy and momentum:

$$(p_1+p_2)_{\parallel}=0;\quad p_1^2=p_2^2=0 \tag{8.4}$$

The amplitude has simultaneously poles for E^2 such that $1+\alpha' E^2=n\ (n=1,2\ldots)$ corresponding to open strings exchanged in the s-channel and poles for t such that $2+\frac{\alpha'}{2}t=2m(m=1,2\ldots)$ corresponding to closed strings exchanged in the t-channel. At high energy:

$$\mathcal{K}(p_1,\epsilon_1;p_2,\epsilon_2)=(\alpha' E^2)^2 Tr(\epsilon_1\epsilon_2^t) \tag{8.5}$$

and the amplitude has Regge behaviour for $\alpha' s \gg \alpha' t \sim 0\ (s\equiv E^2)$:

$$T_1(E,t)\equiv\frac{\mathcal{A}_1(E,t)}{2E}=\frac{R_p^{7-p}\pi^{\frac{9-p}{2}}}{\Gamma(\frac{7-p}{2})}\,\frac{\pi e^{-i\frac{\alpha'}{4}t}(\alpha' s)^{1+\frac{\alpha'}{4}t}}{2E\ \sin(\pi\frac{\alpha'}{4}(-t))\ \Gamma(1+\frac{\alpha' t}{4})} \tag{8.6}$$

T_1 has a real and an imaginary part. The real part describes the scattering of the closed string on the Dp-brane, while the imaginary part describes the absorption of the closed string by the Dp-brane. When $\alpha'\to 0$ the real part reduces to the field theoretical result (graviton exchange), while for $\alpha'\neq 0$ we have the graviton exchange dressed with string corrections. Notice that the imaginary part is a pure string correction that, however, is not relevant at very large impact parameter because it is not divergent at $t=0$ as the real part. The disk amplitude in (8.6) diverges at high energy and violates unitarity. In order to restore unitarity we have to include higher order corrections and sum them up.

Before we proceed further it is instructive to write the corresponding amplitude that one gets in the bosonic string for the elastic scattering of a closed string tachyon on a Dp-brane:

$$\mathcal{A}_1\sim\frac{\Gamma\left(-1-\alpha' s\right)\Gamma\left(-\frac{\alpha' t}{4}-1\right)}{\Gamma\left(-\alpha' s-\frac{\alpha' t}{4}-2\right)}=\frac{\Gamma\left(-\alpha_{open}(s)\right)\Gamma\left(-\frac{\alpha_{closed}(t)}{2}\right)}{\Gamma\left(-\alpha_{open}(s)-\frac{\alpha_{closed}(t)}{2}\right)} \tag{8.7}$$

where $\alpha_{open}(s)=1+\alpha' s$ and $\alpha_{closed}(t)=2+\frac{\alpha'}{2}t$. It has the same form as the original Veneziano model except having two different trajectories in the two channels: one corresponding to the open string and the other to the closed string.

The next diagram is the annulus diagram that is given by:

$$\mathcal{A}_2=\mathcal{N}\int d^2z_1 d^2z_2\sum_{\alpha,\beta}{}_{\alpha\beta}\langle B|W_1^{(0)}(z_1,\bar z_1)W_2^{(0)}(z_2,\bar z_2)D|B\rangle_{\alpha,\beta}$$

$\mathcal{N}$ is a normalization factor and $\sum_{\alpha,\beta}$ is the sum over the spin structures.

The sum over the spin structures can be explicitly performed obtaining in practice only the contribution of the bosonic degrees of freedom without the bosonic partition function.

The final result is rather explicit. In the closed string channel the coefficient of the term with $\mathrm{Tr}(\epsilon_1 \epsilon_2^T)$ (relevant at high energy) of the annulus is equal to:

$$\mathcal{A}_2(s,t) = \frac{\pi^3 (\alpha' s)^2}{\Gamma^2\left(\frac{7-p}{2}\right)} \frac{R_p^{14-2p}}{(2\alpha')^{\frac{7-p}{2}}} \times \left[2\int_0^\infty \frac{d\lambda}{\lambda^{\frac{5-p}{2}}} \int_0^{\frac{1}{2}} d\rho_1 \int_0^{\frac{1}{2}} d\rho_2 \int_0^1 d\omega_1 \int_0^1 d\omega_2\, \mathcal{I} \right] \tag{8.8}$$

where

$$\mathcal{I} \equiv \mathrm{e}^{-\alpha' s V_s - \frac{\alpha'}{4} t V_t}; \quad z_{1,2} \equiv \mathrm{e}^{2\pi(-\lambda\rho_{1,2} + i\omega_{1,2})} \tag{8.9}$$

and

$$V_s = -2\pi\lambda\rho^2 + \log \frac{\Theta_1(i\lambda(\zeta+\rho)|i\lambda)\Theta_1(i\lambda((\zeta-\rho|)i\lambda)}{\Theta_1(i\lambda\zeta+\omega)|i\lambda)\Theta_1(i\lambda\zeta-\omega)|i\lambda)} \tag{8.10}$$

and

$$V_t = 8\pi\lambda\rho_1\rho_2 + \log \frac{\Theta_1(i\lambda\rho+\omega)|i\lambda)\Theta_1(i\lambda\rho-\omega)|i\lambda)}{\Theta_1(i\lambda\zeta+\omega)|i\lambda)\Theta_1(i\lambda\zeta-\omega)|i\lambda)} \tag{8.11}$$

with $\rho \equiv \rho_1 - \rho_2;\ \zeta = \rho_1 + \rho_2;\ \omega \equiv \omega_1 - \omega_2$.

The high energy behaviour ($E \to \infty$) of the annulus diagram can be studied, by the saddle point technique, looking for points where V_s vanishes. This happens for $\lambda \to \infty$ and $\rho \to 0$.

Performing the calculation one gets the leading term for $E \to \infty$:

$$\frac{\mathcal{A}_2^{(3)}(E,t)}{2E} \to \frac{i}{2} \prod_{i=1}^{2} \left[\int \frac{d^{8-p}\mathbf{k_i}}{(2\pi)^{8-p}} \frac{\mathcal{A}_1(E,t_i)}{2E} \right] \times \delta^{(8-p)}\left(\sum_{i=1}^{2} k_i - q\right) V_2(t_1,t_2,t); \quad t_i \equiv -\mathbf{k}_i^2;\ t = -\mathbf{q}^2 \tag{8.12}$$

where

$$V_2(t_1,t_2,t) = \frac{\Gamma\left(1 + \frac{\alpha'}{2}(t_1+t_2-t)\right)}{\Gamma^2\left(1 + \frac{\alpha'}{4}(t_1+t_2-t)\right)} \tag{8.13}$$

In order to find the complete leading eikonal operator we write it in a more suggestive way, in terms of an infinite set of $(8-p)$-dim bosonic oscillators:

$$V_2(t_1,t_2,t) = \langle 0| \prod_{i=1}^{2} \left[\int_0^{2\pi} \frac{d\sigma_i}{2\pi} : \mathrm{e}^{i\mathbf{k}_i \cdot X(\sigma_i)} : \right] |0\rangle \tag{8.14}$$

where

$$\hat{X}(\sigma) = i\sqrt{\frac{\alpha'}{2}} \sum_{n\neq 0} \left(\frac{\alpha_n}{n} e^{in\sigma} + \frac{\tilde{\alpha}_n}{n} e^{-in\sigma} \right) \tag{8.15}$$

The two vacuum states correspond to the two external massless states (states with no bosonic excitations: $(\epsilon_{\mu\nu}\psi^{\mu}_{-\frac{1}{2}}\tilde{\psi}^{\nu}_{-\frac{1}{2}}|0\rangle)$.

Then the leading order from the annulus can be written as follows:

$$\frac{\mathcal{A}_2^{(3)}(E,t)}{2E} \to \frac{i}{2} \prod_{i=1}^{2} \left[\int \frac{d^{8-p}\mathbf{k}_i}{(2\pi)^{8-p}} \frac{\mathcal{A}_1(E,-\mathbf{k}_i^2)}{2E} \right] \delta^{(8-p)}\Big(\sum_{i=1}^{2} \mathbf{k}_i - \mathbf{q}\Big) \\ \times \langle 0| \prod_{i=1}^{2} \left[\int_0^{2\pi} \frac{d\sigma_i}{2\pi} : \mathrm{e}^{i\mathbf{k}_i \cdot X(\sigma_i)} : \right] |0\rangle \tag{8.16}$$

where the two vertex operators correspond to the two leading Reggeons exchanged in the two t-channels: t_1 and t_2.

It can be naturally generalized to the leading term coming from a surface with h boundaries:

$$\frac{\mathcal{A}_h^{(h+1)}(s,t)}{2E} \sim \frac{i^{h-1}}{h!} \prod_{i=1}^{h} \left[\int \frac{d^{8-p}\mathbf{k}_i}{(2\pi)^{8-p}} \frac{\mathcal{A}_1(s,-\mathbf{k}_i^2)}{2E} \right] \\ \times \delta^{(8-p)}\Big(\sum_{i=1}^{h} \mathbf{k}_i - \mathbf{q}\Big) \langle 0| \prod_{i=1}^{h} \left[\int_0^{2\pi} \frac{d\sigma_i}{2\pi} : \mathrm{e}^{i\mathbf{k}_i \cdot X(\sigma_i)} : \right] |0\rangle \tag{8.17}$$

Going to impact parameter space

$$i\frac{\mathcal{A}_h^{(h+1)}(s,\mathbf{b})}{2E} = \int \frac{d^{8-p}\mathbf{q}}{(2\pi)^{8-p}} e^{i\mathbf{b}\mathbf{q}}\, i\frac{\mathcal{A}_h^{(h+1)}(s,t)}{2E} \\ = \frac{i^h}{h!} \prod_{i=1}^{h} \left[\int \frac{d^{8-p}\mathbf{k}_i}{(2\pi)^{8-p}} \frac{\mathcal{A}_1(s,-\mathbf{k}_i^2)}{2E} \right] \\ \langle 0| \prod_{i=1}^{h} \left[\int_0^{2\pi} \frac{d\sigma_i}{2\pi} : \mathrm{e}^{i\mathbf{k}_i(\mathbf{b}+\hat{X}(\sigma_i))} : \right] |0\rangle \tag{8.18}$$

and summing all contributions:

$$\sum_{h=1}^{\infty} \frac{\mathcal{A}_h^{(h+1)}(s, \mathbf{b})}{2E} \sim \langle 0| \frac{1}{i} \left[e^{2i\hat{\delta}(s,b)} - 1 \right] |0\rangle, \tag{8.19}$$

we get the leading eikonal operator

$$\begin{aligned} 2\hat{\delta}(s, b) &= \int_0^{2\pi} \frac{d\sigma}{2\pi} \int \frac{d^{8-p}\mathbf{k}}{(2\pi)^{8-p}} \frac{\mathcal{A}_1(s, -\mathbf{k}^2)}{2E} : e^{i\mathbf{k}(\mathbf{b}+\hat{\mathbf{X}}(\sigma))} : \\ &= \int_0^{2\pi} \frac{d\sigma}{2\pi} \frac{: \mathcal{A}_1\left(s, \mathbf{b} + \hat{\mathbf{X}}(\sigma)\right) :}{2E} \end{aligned} \tag{8.20}$$

The final result that includes all string corrections is obtained from the field theoretical one with the substitution:

$$\mathbf{b} \Longrightarrow \mathbf{b} + \hat{\mathbf{X}}; \quad \hat{\mathbf{X}}(\sigma) = i\sqrt{\frac{\alpha'}{2}} \sum_{n \neq 0} \left(\frac{\alpha_n}{n} e^{in\sigma} + \frac{\tilde{\alpha}_n}{n} e^{-in\sigma} \right) \tag{8.21}$$

and normal ordering.

This is the way that the leading eikonal operator was originally constructed both in string-string and string-brane scattering. From this derivation it is not clear what the bosonic oscillators represent. It was, however, somehow believed that, when the eikonal operator is saturated with a couple of physical states, it will reproduce the high energy behaviour of their scattering amplitude.

For the states of the leading Regge trajectory it has been shown [3] that the quantity

$$\int \frac{d^{8-p}\mathbf{k}}{(2\pi)^{8-p}} \frac{\mathcal{A}_1(E, -\mathbf{k}^2)}{2E} \delta^{(8-p)}(\mathbf{k} - \mathbf{q}) \; \langle 0| \int_0^{2\pi} \frac{d\sigma}{2\pi} : \mathrm{e}^{i\mathbf{k}\cdot X(\sigma)} : |\lambda\rangle \tag{8.22}$$

reproduces the high energy behaviour of the disk amplitude involving a massless state ($\langle 0|$) and a state of the leading Regge trajectory ($|\lambda\rangle$). It turned out, however, that this computation is more subtle because the longitudinal polarization of the massive state gets enhanced at high energy. The annulus diagram for a massless state and an excited state of the leading Regge trajectory has also been computed [4].

In any case, the problem of the nature of the bosonic oscillators present in the eikonal operator remains. Given the fact that in string-string collisions they are along the eight directions orthogonal to both the time and the direction of the fast moving string and similarly in string-brane collisions they are along the $8 - p$ transverse directions again orthogonal to the time and to the direction of the fast moving string, strongly suggests that they should be interpreted as the string bosonic oscillators in the light-cone gauge. But even so, why does the eikonal operator not contain the fermionic oscillators?

Putting this problem for a moment aside, in the next section we compute the amplitude for the production of a massive state belonging to the first excited level of superstring theory from the scattering of a graviton on a Dp-brane and we compare with what one gets from the eikonal operator. We will show that, in agreement with the eikonal operator, we produce, at high energy, only excited states of the graviton ($|i\rangle|\tilde{i}\rangle$) of the type $A_{-1,j}|i\rangle\tilde{A}_{-1;\tilde{j}}|\tilde{i}\rangle$. The remaining massive states of the type $Q_{-1,b}|a\rangle\tilde{Q}_{-1;\tilde{b}}|\tilde{a}\rangle$, $A_{-1,j}|i\rangle\tilde{Q}_{-1;\tilde{b}}|\tilde{a}\rangle$ and $Q_{-1,b}|a\rangle\tilde{A}_{-1;\tilde{j}}|\tilde{i}\rangle$ are not produced at high energy.

8.3 States of the First Massive Level Produced at High Energy

In order to understand the problems listed at the end of the last section, in this section we consider the production of a massive state, belonging to the first massive level, from the scattering of a massless state on a Dp-brane and we study which of the 128×128 bosonic states are produced at high energy in the Regge limit. This section is divided in three subsections. In the first one we compare the spectrum of physical states at the first excited level in the Green-Schwarz light-cone formalism, in the RNS light-cone formalism and in the covariant formalism. We introduce also the DDF operators that connect the states in the light-cone RNS with those in the covariant formalism. In the second short subsection we compute the three-point amplitudes involving two gravitons and a bosonic state of the first excited level. Finally, in the third subsection, we compute the inelastic amplitude for the production of the states of the first excited level and we check which of them are produced at high energy.

8.3.1 Spectrum of the First Excited Level

In this subsection we discuss the spectrum of physical states of the first massive level in closed superstring theory in the two light-cone gauges (Green-Schwarz (GS) and Ramond-Neveu-Schwarz (RNS)) and in the covariant formalism. Any closed string state is a product of a state with left moving oscillators times a state with right moving oscillators. In the following we discuss only the states with one type of oscillators. Those with the other type of oscillators can be obtained exactly in the same way.

1. **GS light-cone**
   In the GS light-cone the bosonic physical states at the first massive level are the following:
$$\begin{array}{c} \alpha^i_{-1}|j\rangle \Longrightarrow 64 \ states \\ Q^a_{-1}|b\rangle \Longrightarrow 64 \ states \end{array} \tag{8.23}$$
   where $i, j = 1\ldots 8$ are vector indices and $a, b = 1\ldots 8$ are spinor indices of $SO(8)$.

2. **RNS light-cone**
In the RNS light-cone the bosonic physical states are the following:

$$\begin{aligned} &A^i_{-1}B^j_{-\frac{1}{2}}|0\rangle &&\Longrightarrow 64\ states\\ &B^i_{-\frac{3}{2}}|0\rangle &&\Longrightarrow 8\ states \\ &B^i_{-\frac{1}{2}}B^j_{-\frac{1}{2}}B^k_{-\frac{1}{2}}|0\rangle &&\Longrightarrow 56\ states \end{aligned} \tag{8.24}$$

where $i, j, k = 1\ldots 8$ are vector indices of $SO(8)$. The states in the first line of (8.23) correspond to those in the first line of (8.24), while the states in the second line of (8.23) correspond to those in the second and third line of (8.24).

3. **Covariant formalism**
In the covariant formalism the physical states in the center of mass frame ($p = (M, \mathbf{0})$) are:

$$\begin{aligned} T^{IJ} &= \left(\alpha^I_{-1}\psi^J_{-\frac{1}{2}} + \alpha^J_{-1}\psi^I_{-\frac{1}{2}} - \frac{2}{9}\eta^{IJ}\eta^{HK}\alpha^H_{-1}\psi^K_{-\frac{1}{2}}\right)|0, p\rangle \Longrightarrow 44\ states\\ V^{IJK} &= \psi^I_{-\frac{1}{2}}\psi^J_{-\frac{1}{2}}\psi^K_{-\frac{1}{2}}|0, p\rangle \Longrightarrow 84\ states \end{aligned} \tag{8.25}$$

where $I, J, K, H = 1\ldots 9$ are vector indices of $SO(9)$. We can decompose the 9-dim indices $I = i, v$; $J = j, v$ in 8-dim indices and a longitudinal one that we call v:

$$\begin{aligned} T^{ij} &\Longrightarrow 36\ states; \quad T^{iv} \Longrightarrow 8\ states\\ V^{ijk} &\Longrightarrow 56\ states; \quad V^{ijv} \Longrightarrow 28\ states \end{aligned} \tag{8.26}$$

T^{ij} and V^{ijv} correspond to the 64 states in the first line of (8.24), while the others correspond to those in the second and third line of (8.24). The two states in (8.25) can be given a covariant $SO(1, 9)$ form by a boost, In this way one gets the following states:

$$|\phi_1\rangle = T^{\alpha\rho}_{\alpha'\rho'}\alpha^{\rho'}_{-1}\psi^{\alpha'}_{-\frac{1}{2}}|0, p\rangle \tag{8.27}$$

where

$$\begin{aligned} T^{\alpha\rho}_{\alpha'\rho'} &= (\eta_\perp)^\rho_{\rho'}(\eta_\perp)^\alpha_{\alpha'} + (\eta_\perp)^\rho_{\alpha'}(\eta_\perp)^\alpha_{\rho'} - \frac{2}{9}\eta^{\rho\alpha}_\perp\eta_{\perp\alpha'\rho'}\\ \eta^{\mu\nu}_\perp &= \eta^{\mu\nu} - \frac{p^\mu p^\nu}{p^2} \end{aligned} \tag{8.28}$$

and

$$|\phi_2\rangle = \eta^\rho_{\perp\rho'}\eta^\sigma_{\perp\sigma'}\eta^\tau_{\perp\tau'}\psi^{\rho'}_{-\frac{1}{2}}\psi^{\sigma'}_{-\frac{1}{2}}\psi^{\tau'}_{-\frac{1}{2}}|0, p\rangle \tag{8.29}$$

It can be shown that the two states in (8.27) and (8.29) are physical states:

$$G_{\frac{1}{2}}|\phi_{1,2}\rangle = G_{\frac{3}{2}}|\phi_{1,2}\rangle = 0 \tag{8.30}$$

The connection between the RNS oscillators in the light-cone gauge and those in the covariant formalism is provided by the DDF operators [5]. In the case of superstring they can be found in [6] and they are reviewed in [1]. In particular, as discussed in [1], one gets for the states at the first massive level made with one A and one B oscillators:

$$\begin{aligned} &A_{-1,j}B_{-\frac{1}{2},k}|p_T,0\rangle \\ &= \left\{\frac{1}{2}\left[\alpha^j_{-1}\psi^k_{-\frac{1}{2}} + \alpha^k_{-1}\psi^j_{-\frac{1}{2}} - \frac{\delta^{jk}}{3}\left(\sum_{i=1}^{8}\alpha^i_{-1}\psi^i_{-\frac{1}{2}} - 2\alpha^v_{-1}\psi^v_{-\frac{1}{2}}\right)\right]\right. \\ &\left.+ \frac{1}{\sqrt{2}}(v\psi_{-\frac{1}{2}})\psi^j_{-\frac{1}{2}}\psi^k_{-\frac{1}{2}}\right\}|p;0\rangle; \quad j,k = 1\dots 8. \end{aligned} \tag{8.31}$$

where $(\epsilon_j)_\mu\psi^\mu_{-\frac{1}{2}} \equiv \psi^j_{-\frac{1}{2}}$, $(\epsilon_j)_\mu\alpha^\mu_{-1}$, p_T is the momentum of the tachyon present in the DDF state and v is the longitudinal polarization of the massive state that is orthogonal to the momentum p. Analogously, one can also compute the connection with the covariant states of the other two DDF states: $B_{-\frac{1}{2},i}B_{-\frac{1}{2},j}B_{-\frac{1}{2},k}|0,p_T\rangle$ and $B_{-\frac{3}{2},i}|0,p_T\rangle$.

8.3.2 Three-Point Amplitudes

In this subsection we provide the three-point amplitude, in the covariant formalism, involving two gravitons and one of the states of the first massive level. In a closed string theory the amplitude is the product of two amplitudes of open string theory, one for the left movers and the other for the right movers. Here, we quote only the result for the left movers.

For the massive state in (8.28) one gets:

$$A^{\mu;IJ}_\nu(\phi_1) \sim \epsilon^{IJ}_{\alpha\rho}\frac{\alpha'}{2}\left[\eta^{\mu\alpha}p_3^\rho p_1^\nu - \eta^{\nu\alpha}p_3^\rho p_3^\mu + \eta^{\mu\nu}p_3^\alpha p_3^\rho + \eta^{\mu\alpha}\eta^{\nu\rho}\right] \tag{8.32}$$

where p_1 and p_3 are the momenta of the two gravitons and we have assumed that the polarization matrix is symmetric, traceless and orthogonal to the four-momentum p_2 of the massive state:

$$p_2^\alpha\epsilon^{IJ}_{\alpha\rho} = \eta^{\alpha\rho}\epsilon^{IJ}_{\alpha\rho} = 0 \tag{8.33}$$

For the state in (8.29) one gets:

$$A_{\nu}^{\mu;I,J,K}(\phi_2) \sim \epsilon_{\rho\sigma\tau}^{IJK}\sqrt{\frac{\alpha'}{2}}\left[\eta^{\nu\rho}\left(p_3^{\sigma}\eta^{\mu\tau} - p_3^{\tau}\eta^{\mu\sigma}\right)\right.$$
$$\left. - p_3^{\rho}\left(\eta^{\nu\sigma}\eta^{\mu\tau} - \eta^{\mu\sigma}\eta^{\nu\tau}\right) + \eta^{\mu\rho}\left(\eta^{\nu\sigma}p_3^{\tau} - \eta^{\nu\tau}k_3^{\sigma}\right)\right] \quad (8.34)$$

In this case the polarization is completely antisymmetric and orthogonal to the four-momentum of the massive state p_2. The indices μ and ν must be saturated with the left moving part of the polarization of the two gravitons. We have assumed that all three states are incoming: $p_1 + p_2 + p_3 = 0$.

8.3.3 Inelastic Amplitudes

In this subsection we use the three-point amplitudes of the previous section to compute the inelastic amplitude where the graviton with momentum p_1 scatters on a Dp-brane producing a massive state with momentum p_2. This can be done by considering the product of any of the two amplitudes (one for the right movers and the other for the left movers) constructed above and by saturating the indices ν and $\bar{\nu}$ of the graviton with momentum p_3 first with the graviton propagator in the De Donder gauge:

$$D^{\nu\lambda;\bar{\nu}\bar{\lambda}} = \frac{\eta^{\nu\lambda}\eta^{\bar{\nu}\bar{\lambda}} + \eta^{\nu\bar{\lambda}}\eta^{\bar{\nu}\lambda} - \frac{1}{4}\eta^{\nu\bar{\nu}}\eta^{\lambda\bar{\lambda}}}{2p_3^2} \quad (8.35)$$

and then with the coupling of the graviton to the Dp-brane given by

$$\frac{1}{2}T_p\frac{\eta^{\lambda\bar{\lambda}} + R^{\lambda\bar{\lambda}}}{2}; \quad T_p = \sqrt{\pi}(2\pi\sqrt{\alpha'})^{3-p} \quad (8.36)$$

where R is the reflection matrix:

$$R^{\mu}_{\ \nu} = \delta^{\mu}_{\ \nu},\ \mu,\nu = 0,\ldots,p; \quad R^{\mu}_{\ \nu} = -\delta^{\mu}_{\ \nu},\ \mu,\nu = p+1,\ldots,9. \quad (8.37)$$

In this way one obtains:

$$\frac{1}{2}T_p\kappa_{10}\ A_{\nu}\frac{\left(R^{\nu\bar{\nu}} + \frac{3-p}{4}\eta^{\nu\bar{\nu}}\right)}{(-t)}\bar{A}_{\bar{\nu}} \quad (8.38)$$

where $2\kappa_{10}^2 = (2\pi)^7 g^2(\alpha')^4$, A and $\bar{A}$ stand for one of the two amplitudes of the previous subsection and $t = -p_3^2 = -(p_1+p_2)^2$ is the momentum transfer in the inelastic process. It is easy to check that

$$\frac{1}{2}T_p\kappa_{10} = \frac{\pi^{\frac{9-p}{2}} R_p^{7-p}}{\Gamma(\frac{7-p}{2})} \tag{8.39}$$

appearing in (8.1). Let us consider the case where both the right and left three-point amplitudes are as in (8.32). We get:

$$\frac{1}{2}T_p\kappa_{10}\ \epsilon^{IJ}_{\rho\alpha}\left\{\frac{\alpha'}{2}\left[\eta^{\mu\alpha}k_1^{\nu}+\eta^{\nu\alpha}q^{\mu}-\eta^{\mu\nu}q^{\alpha}\right]q^{\rho}-\eta^{\mu\alpha}\eta^{\nu\rho}\right\}\left(R_{\nu\bar{\nu}}+\frac{3-p}{4}\eta_{\nu\bar{\nu}}\right)$$
$$\epsilon^{\bar{I}\bar{J}}_{\bar{\rho}\bar{\alpha}}\left\{\frac{\alpha'}{2}\left[\eta^{\bar{\mu}\bar{\alpha}}k_1^{\bar{\nu}}+\eta^{\bar{\nu}\bar{\alpha}}q^{\bar{\mu}}-\eta^{\bar{\mu}\bar{\nu}}q^{\bar{\alpha}}\right]q^{\bar{\rho}}-\eta^{\bar{\mu}\bar{\alpha}}\eta^{\bar{\nu}\bar{\rho}}\right\} \tag{8.40}$$

The term $\frac{\alpha'}{2}k_1Rk_1 = (-\alpha' s)$ gives a divergent term at high energy. Furthermore, we have to remember that in the case of a massive state the longitudinal polarization is also enhanced at high energy. Taking this into account we get the following amplitude:

$$\frac{1}{2}T_p\kappa_{10}\ \epsilon^{IJ}_{\rho\alpha}\epsilon^{\bar{I}\bar{J}}_{\bar{\rho}\bar{\alpha}}(-\alpha' s)\frac{\alpha'}{2}\left[\eta^{\mu\alpha}\left(q^{\rho}-\frac{v^{\rho}}{\sqrt{\alpha'}}\right)+\frac{\alpha'}{2}q^{\rho}q^{\mu}\frac{v^{\alpha}}{\sqrt{\alpha'}}\right]$$
$$\times\left[\eta^{\bar{\mu}\bar{\alpha}}\left(q^{\bar{\rho}}-\frac{v^{\bar{\rho}}}{\sqrt{\alpha'}}\right)+\frac{\alpha'}{2}q^{\bar{\rho}}q^{\bar{\mu}}\frac{v^{\bar{\alpha}}}{\sqrt{\alpha'}}\right] \tag{8.41}$$

If we use the two amplitudes as those in (8.34), we get

$$\frac{1}{2}T_p\kappa_{10}\ \epsilon^{IJK}_{\rho\sigma\tau}\left[q^{\rho}\left(\eta^{\nu\sigma}\eta^{\mu\tau}-\eta^{\mu\sigma}\eta^{\nu\tau}\right)+q^{\sigma}\left(\eta^{\mu\rho}\eta^{\nu\tau}-\eta^{\nu\rho}\eta^{\mu\tau}\right)\right.$$
$$\left.+\,q^{\tau}\left(\eta^{\nu\rho}\eta^{\mu\sigma}-\eta^{\mu\rho}\eta^{\nu\sigma}\right)\right]\left(R_{\nu\bar{\nu}}+\frac{3-p}{4}\eta_{\nu\bar{\nu}}\right)\frac{\alpha'}{2}\epsilon^{\bar{I}\bar{J}\bar{K}}_{\bar{\rho}\bar{\sigma}\bar{\tau}}$$
$$\times\left[q^{\bar{\rho}}\left(\eta^{\bar{\nu}\bar{\sigma}}\eta^{\bar{\mu}\bar{\tau}}-\eta^{\bar{\mu}\bar{\sigma}}\eta^{\bar{\nu}\bar{\tau}}\right)+q^{\bar{\sigma}}\left(\eta^{\bar{\mu}\bar{\rho}}\eta^{\bar{\nu}\bar{\tau}}-\eta^{\bar{\nu}\bar{\rho}}\eta^{\bar{\mu}\bar{\tau}}\right)+q^{\bar{\tau}}\left(\eta^{\bar{\nu}\bar{\rho}}\eta^{\bar{\mu}\bar{\sigma}}-\eta^{\bar{\mu}\bar{\rho}}\eta^{\bar{\nu}\bar{\sigma}}\right)\right] \tag{8.42}$$

Taking again into account the enhancement at high energy due to the longitudinal polarization one gets:

$$\frac{1}{2}T_p\kappa_{10}\ \frac{(-\alpha' s)}{2}\frac{\alpha'}{2}\epsilon^{IJK}_{\rho\sigma\tau}$$
$$\times\left[q^{\rho}\left(\eta^{\mu\sigma}v^{\tau}-\eta^{\mu\tau}v^{\sigma}\right)+q^{\sigma}\left(\eta^{\mu\rho}v^{\tau}-v^{\rho}\eta^{\mu\tau}\right)+q^{\tau}\left(\eta^{\mu\rho}v^{\sigma}-v^{\rho}\eta^{\mu\sigma}\right)\right]$$
$$\epsilon^{\bar{I}\bar{J}\bar{K}}_{\bar{\rho}\bar{\sigma}\bar{\tau}}\left[q^{\bar{\rho}}\left(\eta^{\bar{\mu}\bar{\sigma}}v^{\bar{\tau}}-\eta^{\bar{\mu}\bar{\tau}}v^{\bar{\sigma}}\right)+q^{\bar{\sigma}}\left(\eta^{\bar{\mu}\bar{\rho}}v^{\bar{\tau}}-v^{\bar{\rho}}\eta^{\bar{\mu}\bar{\tau}}\right)+q^{\bar{\tau}}\left(\eta^{\bar{\mu}\bar{\rho}}v^{\bar{\sigma}}-v^{\bar{\rho}}\eta^{\bar{\mu}\bar{\sigma}}\right)\right] \tag{8.43}$$

Using the kinematics of the Appendix one can write the quantity in one of the two squared brackets in (8.41) as follows:

$$
\begin{aligned}
A_k^{IJ} &= \epsilon_\mu^k \epsilon_{\rho\alpha}^{IJ}\left[\eta^{\mu\alpha}\left(q^\rho - \frac{v^\rho}{\sqrt{\alpha'}}\right) + \frac{\alpha'}{2} q^\rho q^\mu \frac{v^\alpha}{\sqrt{\alpha'}}\right] \\
&= \frac{1}{2}\left[\bar{p}_1^I \delta^{kJ} + \bar{p}_1^J \delta^{kI} - \frac{1}{3}\bar{p}_1^k \delta^{IJ}\right] + \frac{1}{2}\bar{p}_1^k \delta^{Iv}\delta^{Jv}
\end{aligned}
\tag{8.44}
$$

where $k = 1 \dots 8$; $I, J = 1 \dots 8, v$. If we divide the 9-dim indices $I = (i, v)$ and $J = (j, v)$ in an 8-dim part and a part along v, from the previous expression we get:

$$
\begin{aligned}
A_k^{ij} &= \tfrac{1}{2}\left[\bar{p}_1^i \delta^{kj} + \bar{p}_1^j \delta^{ki}\right]; \quad i \neq j \\
A_k^{ii} &= \bar{p}_1^k \left(\delta^{ik} - \tfrac{1}{6}\right); \quad i = 1 \dots 8 \\
A_k^{vv} &= -\textstyle\sum_{i=1}^8 A_k^{ii} = \tfrac{1}{3}\bar{p}_1^k \\
A^{iv} &= A^{vi} = 0
\end{aligned}
\tag{8.45}
$$

Performing the same analysis with the antisymmetric amplitude in (8.43), we get:

$$
\begin{aligned}
A_k^{IJH} = \frac{1}{2}\epsilon_\mu^k \epsilon_{\rho\sigma\tau}^{IJH}\, &\left[\bar{p}_1^\rho \left(\eta^{\mu\sigma} v^\tau - \eta^{\mu\tau} v^\sigma\right) - \bar{p}_1^\sigma \left(\eta^{\mu\rho} v^\tau - v^\rho \eta^{\mu\tau}\right)\right. \\
&\left. + \bar{p}_1^\tau \left(\eta^{\mu\rho} v^\sigma - v^\rho \eta^{\mu\sigma}\right)\right]
\end{aligned}
\tag{8.46}
$$

that implies

$$
\begin{aligned}
A_k^{ijh} &= A^{ivv} = 0 \\
A^{ijv} &= \frac{1}{2}\left(\bar{p}_1^i \delta^{Kj} - \bar{p}_1^j \delta^{Ki}\right)
\end{aligned}
\tag{8.47}
$$

Remembering the connection between covariant and light-cone states, from the previous expressions we see that the scattering of a graviton on a Dp-brane will produce only closed string states with left or right movers of the type $A_{-1;j} B_{-\frac{1}{2};k}|0\rangle$ in the RNS case corresponding to the states $A_{-1;j}|k\rangle$ in the GS case, while the states with left or right movers of the type $B_{-\frac{3}{2};j}|0\rangle$ and to $B_{-\frac{1}{2};i} B_{-\frac{1}{2};j} B_{-\frac{1}{2};k}|0\rangle$, corresponding to $Q_{-1;i}|j\rangle$ in the GS case, are not produced at high energy. This is in agreement with what one gets from the eikonal operator interpreting the bosonic oscillators as the string bosonic oscillators in the light-cone gauge. In the next section we will derive the eikonal operator directly from string theory without needing to go through the scattering amplitude and require unitarity as it was done in Sect. 8.2.

8.4 The Eikonal Operator II

In this section we sketch the construction of the eikonal operator that was done in [1]. The first ingredient is the GS three-string vertex given by:

$$|V_{GS}\rangle = \left(P_i - \alpha_1\alpha_2\alpha_3 \frac{n}{\alpha_q} N_n^q A_{-n,i}^q\right) V_b V_f |V_i\rangle |V_0\rangle, \tag{8.48}$$

where

$$\begin{aligned}
V_b &= \exp\left(\frac{1}{2} A_{-n,i}^p N_{mn}^{pq} A_{-m,i}^q + P_i N_n^q A_{-n,i}^q\right),\\
V_f &= \exp\left(\frac{1}{2} Q_{-n,a}^p X_{mn}^{pq} Q_{-m,a}^q - S_a \frac{n}{\alpha_q} N_n^q Q_{-n,a}^q\right),\\
|V_i\rangle &= \frac{1}{\alpha_1}|ijj\rangle + \frac{1}{\alpha_2}|jij\rangle + \frac{1}{\alpha_3}|jji\rangle + \frac{\alpha_1-\alpha_2}{4\alpha_3}|aai\rangle + \frac{\alpha_1-\alpha_3}{4\alpha_2}|aia\rangle\\
&\quad + \frac{\alpha_2-\alpha_3}{4\alpha_1}|iaa\rangle + \frac{1}{4}\gamma_{ab}^{ij}\left(|baj\rangle + |bja\rangle + |jba\rangle\right).
\end{aligned} \tag{8.49}$$

To insure momentum conservation we have included in the vertex a part with the bosonic zero modes given by:

$$\begin{aligned}
|V_0\rangle &= \int d^{10}x\; |x\rangle_1\; |x\rangle_2\; |x\rangle_3\\
&= (2\pi)^{10}\delta^{(10)}(\hat{p}_1 + \hat{p}_2 + \hat{p}_3)|x=0\rangle_1\; |x=0\rangle_2\; |x=0\rangle_3
\end{aligned} \tag{8.50}$$

where the state $|x\rangle$ is an eigenstate of the position operator: $\hat{q}|x\rangle = x|x\rangle$. The operators P_i and S_a stand for the following combinations of the bosonic and fermionic zero-modes

$$P_i \equiv \left(\alpha_r \bar{p}_{iL}^{(r+1)} - \alpha_{r+1}\bar{p}_{iL}^{(r)}\right), \quad S_a \equiv \alpha_r Q_{0a}^{(r+1)} - \alpha_{r+1} Q_{0a}^{(r)}. \tag{8.51}$$

which, with the cyclic identification between $r = 4$ and $r = 1$, are independent of the choice of $r = 1, 2, 3$. Finally, the 'Neumann' coefficients encoding the actual value of the various couplings are

$$N_{nm}^{rs} = -\frac{nm\alpha_1\alpha_2\alpha_3}{n\alpha_s + m\alpha_r} N_n^r N_m^s; \quad X_{nm}^{rs} = \frac{n\alpha_s - m\alpha_r}{2\alpha_r\alpha_s} N_{nm}^{rs}, \tag{8.52}$$

$$N_n^r = -\frac{1}{n\alpha_{r+1}}\binom{-n\frac{\alpha_{r+1}}{\alpha_r}}{n} = \frac{1}{\alpha_r n!}\frac{\Gamma\left(-n\frac{\alpha_{r+1}}{\alpha_r}\right)}{\Gamma\left(-n\frac{\alpha_{r+1}}{\alpha_r} + 1 - n\right)}. \tag{8.53}$$

Remember that the light-cone three-string vertex depends on a light-like vector k that in general can be chosen as we want. It turns out, however, that, if we choose it to be along the direction of the two energetic strings, at high energy the vertex gets enormously simplified. Since we have chosen the momentum of incoming graviton and massive state as in (8.68) and (8.71), this means that we have to choose $k = \frac{1}{\sqrt{2}}(-1, 0_p; 0_{8-p}, 1)$. Momentum conservation implies that the momentum of the third string is given by $p_2 = (0, 0_p; -\bar{p}_1, -q_9)$.[2] Proceeding in this way, at high energy, we get the following GS vertex:

$$|V_{GS}\rangle \sim \frac{P_i}{\alpha_2} \exp\left\{-\sqrt{\frac{\alpha'}{2}} \frac{\bar{p}_{1\ell}}{n}\left(A^3_{-n\ell} + (-1)^n A^1_{-n\ell}\right)\right\}\left[|jij\rangle + \frac{\alpha_1 - \alpha_3}{4}|aia\rangle\right]. \tag{8.54}$$

The second ingredient is the boundary state in the light-cone gauge that was constructed in [7]. We use a slightly modified version of it where we impose Neumann (Dirichlet) boundary conditions along the longitudinal (transverse) directions to the world volume of the Dp-branes. It is given by:

$$|B, \eta, y\rangle \sim \exp\left\{-\sum_{n=1}^{\infty}\left[\frac{1}{n}\alpha^i_{-n} D_{ij}\tilde{\alpha}^j_{-n} + i\eta S^a_{-n} M_{a\dot{b}} \tilde{S}^{\dot{b}}_{-n}\right]\right\}|B_0, \eta, y\rangle \tag{8.55}$$

where R is the reflection matrix given in (8.37),

$$|B_0, \eta, y\rangle = \left(R_{ij}|i\rangle|\tilde{j}\rangle + i\eta M_{\dot{a}b}|\dot{a}\rangle|\tilde{b}\rangle\right)\delta^{(9-p)}(\hat{q} - y)|0_\alpha, p = 0\rangle \tag{8.56}$$

and

$$M_{\dot{a}b} = i\left(\gamma^1\gamma^2 \ldots \gamma^{p+1}\right)_{\dot{a}b}; \quad M_{a\dot{b}} = i\left(\gamma^1\gamma^2 \ldots \gamma^{p+1}\right)_{a\dot{b}}. \tag{8.57}$$

The third ingredient is the light-cone propagator:

$$P = \frac{\pi\alpha'}{2}\int_0^\infty dt \; \mathrm{e}^{-\pi t\left(\frac{\alpha'}{2}\hat{p}_i^2 + N + \tilde{N}\right)}; \quad i = 1 \ldots 8 \tag{8.58}$$

where N and $\tilde{N}$ are the bosonic and fermionic number operators.

Using the three previous ingredients, we compute the quantity:

$$\frac{T_p}{2}\, {}_2\langle B|P\left(\kappa_{10}|V_{GS}\rangle|\widetilde{V_{GS}}\rangle\right) \sim \frac{R_p^{7-p}\pi^{\frac{9-p}{2}}}{\Gamma\left(\frac{7-p}{2}\right)}\, {}_2\langle B_0|\frac{1}{-t}\left(|V_{GS}\rangle|\widetilde{V_{GS}}\rangle\right). \tag{8.59}$$

In particular, in the previous equation we limit ourselves only to the pole of the graviton, as we have done in the previous section. Then we can neglect all oscillators in the boundary state and in the propagator and we need only to consider the contribution of the bosonic zero modes:

[2] Notice that the state labelled here by $r = 3$ has momentum p_2 in (8.68).

$$
{}_2\langle p=0|\,\delta^{9-p}(\hat{q})\,\frac{1}{\hat{p}_i^2}\,|x\rangle_2 = {}_2\langle p=0|\int\frac{d^{9-p}k}{(2\pi)^{9-p}}\mathrm{e}^{ik\cdot\hat{q}}\,\frac{1}{\hat{p}_i^2}\,|x\rangle_2
$$

$$
=\int\frac{d^{9-p}k}{(2\pi)^{9-p}}\frac{\mathrm{e}^{ik\cdot x}}{k_i^2} \tag{8.60}
$$

Then, assuming that the strings 1 and 3 have momentum p_1 and p_3, we get ($\langle x|p\rangle = \mathrm{e}^{-ipx}$):

$$
\int d^{10}x\,\langle p_1|x\rangle_1\langle p_3|x\rangle_3\int\frac{d^{9-p}k}{(2\pi)^{9-p}}\frac{\mathrm{e}^{ik\cdot x}}{k_i^2} = (2\pi)^{p+1}\delta^{(p+1)}(p_1+p_3)\frac{1}{(-t)} \tag{8.61}
$$

where $t=-(p_1+p_3)^2$ is the momentum transfer. Using the following equation [1]:

$$
\frac{2}{\alpha'}\frac{P_h R_{hk} P_k}{\alpha_2^2(-t)} = \frac{\alpha_3^2}{\alpha_2^2}\frac{(\bar{p}_1)^2}{t} = -\frac{\alpha_3^2}{\alpha_2^2}\left(1+\frac{q_9^2}{t}\right) \sim -\frac{4E^2}{q_9^2}-\frac{4E^2}{t}, \tag{8.62}
$$

and neglecting the term without the pole at $t=0$ we arrive at

$$
|W\rangle \sim \frac{R_p^{7-p}\pi^{\frac{9-p}{2}}}{\Gamma\left(\frac{7-p}{2}\right)}\frac{4E^2}{-t}\exp\left\{-\sqrt{\frac{\alpha'}{2}}\frac{\bar{p}_{1\ell}}{n}(A^3_{-n\ell}+(-1)^n A^1_{-n\ell})\right\}\left[|j\rangle_1|j\rangle_3+\frac{\alpha_1}{2}|a\rangle_1|a\rangle_3\right]
$$

$$
\times\exp\left\{-\sqrt{\frac{\alpha'}{2}}\frac{\bar{p}_{1\ell}}{n}(\tilde{A}^3_{-n\ell}+(-1)^n\tilde{A}^1_{-n\ell})\right\}\left[|\tilde{j}\rangle_1|\tilde{j}\rangle_3+\frac{\alpha_1}{2}|\tilde{a}\rangle_1|\tilde{a}\rangle_3\right]. \tag{8.63}
$$

Following [1] we can finally write it in a single Hilbert space getting:

$$
W \sim \frac{R_p^{7-p}\pi^{\frac{9-p}{2}}}{\Gamma\left(\frac{7-p}{2}\right)}\frac{4E^2}{-t}:\exp\left\{-\sqrt{\frac{\alpha'}{2}}\frac{\bar{p}_{1\ell}}{n}(A_{-n\ell}-A_{n\ell})\right\}:
$$

$$
\times\ :\exp\left\{-\sqrt{\frac{\alpha'}{2}}\frac{\bar{p}_{1\ell}}{n}(\tilde{A}_{-n\ell}-\tilde{A}_{n\ell})\right\}: \tag{8.64}
$$

Introducing an auxiliary string coordinate (without zero modes):

$$
\hat{X}^i(\sigma) = \mathrm{i}\sqrt{\frac{\alpha'}{2}}\sum_{n\neq 0}\left(\frac{A_{ni}}{n}\mathrm{e}{in\sigma}+\frac{\tilde{A}_{ni}}{n}\mathrm{e}{-in\sigma}\right). \tag{8.65}
$$

we can write (8.64) in an operator form as follows

$$W(\bar{p}_1) = \int_0^{2\pi} \frac{d\sigma}{2\pi} : \mathrm{e}i\bar{p}_1 \hat{X}(\sigma) : \left(\frac{R_p^{7-p} \pi^{\frac{9-p}{2}}}{\Gamma\left(\frac{7-p}{2}\right)} \frac{4E^2}{-t} \right), \tag{8.66}$$

that provides the same amplitude as in (8.64) when we saturate them with physical states satisfying the level matching condition. This operator is identical to the eikonal operator in (8.20) if we take the limit $\alpha' \to 0$ in the amplitude $\mathcal{A}_1$ given in (8.6). The α' corrections are recovered if one does not include just the contribution of the graviton as we have done above, but add also the contribution of the other string states.

In conclusion, we have provided two independent derivations of the eikonal operator. The one in this section shows that the bosonic oscillators are the bosonic oscillators of superstring theory in a suitably chosen light-cone gauge. This means that when we sandwich the eikonal operator between two arbitrary string states, we obtain the production amplitude of one of them from the scattering of the other on a Dp-brane at high energy and small transverse momentum.

8.5 Kinematics

The scattering amplitude for the production of a massive string with momentum p_2 from the scattering of a graviton with momentum p_1 on a Dp-brane is described by the two (Mandelstam like) variables:

$$t = -q^2 = -(p_1 + p_2)^2, \quad s = -\frac{1}{4}(p_1 + Rp_1)^2 = -\frac{1}{4}(p_2 + Rp_2)^2 \equiv E^2, \tag{8.67}$$

where in the second equation we used the momentum conservation along the Neumann directions and $E > 0$ will denote, hereafter, the common energy of the incoming and outgoing closed strings. It is convenient to choose the massive string to move along the 9-th space direction:

$$p_2^\mu = \left(-E, 0_p; 0_{8-p}, -\sqrt{E^2 - M^2}\right), \tag{8.68}$$

where the first $p + 1$ directions are parallel to the (Neumann directions of the) Dp-branes and the entries after the semicolon are along the Dirichlet directions. The most direct way to describe the physical polarization of massive particles is to introduce 9 vectors perpendicular to their momentum. For instance, in the case of the outgoing state (8.68) we have the unit vectors $\hat{w}^i$

$$\hat{w}_1 = \left(0, 1, 0_{p-1}; 0_{8-p}, 0\right), \ldots, \quad \hat{w}_8 = \left(0, 0_p; 0_{7-p}1, 0\right) \tag{8.69}$$

and, as the ninth one, v^μ corresponding to the longitudinal polarization:

$$v_2^\mu = \left(\frac{\sqrt{E^2 - M^2}}{M}, 0_p; 0_{8-p}, \frac{E}{M} \right). \tag{8.70}$$

The possible momenta of the ingoing massless string take the following form

$$p_1^\mu = \left(E, 0_p; \bar{p}_1, \sqrt{E^2 - M^2} + q_9 \right), \tag{8.71}$$

$$q^{\mu=9} = \frac{t + M^2}{2\sqrt{E^2 - M^2}}, \quad (\bar{p}_1)^2 + (q^{\mu=9})^2 = -t \equiv (p_1 + p_2)^2, \tag{8.72}$$

where $\bar{p}_1$ is a $(8-p)$-dim vector orthogonal to the direction of motion of the massive string. It is convenient to choose the eight polarizations of the massless string as follows:

$$\epsilon_k^\mu = \left(\frac{\bar{p}_1^k}{E + \sqrt{E^2 - M^2} + q^9}, \delta_k^i, -\frac{\bar{p}_1^k}{E + \sqrt{E^2 - M^2} + q^9} \right) \tag{8.73}$$

It is easy to check that $\epsilon_k^\mu p_{1\mu} = 0$ for any $k = 1 \dots 8$. Using this we can compute

$$\epsilon_k q \equiv \epsilon_k (p_1 + p_2) = \epsilon_k p_2 = \bar{p}_1^k \tag{8.74}$$

where we have kept only the leading term at high energy.

References

1. G. D'Appollonio, P. Di Vecchia, R. Russo and G. Veneziano, Microscopic unitary description of tidal excitations in high-energy string-brane scattering. J. High Energy Phys. (2010). arXiv: 1310.1254 [hep-th].
2. G. D'Appollonio, P. Di Vecchia, R. Russo, G. Veneziano, High-energy string-brane scattering: Leading eikonal and beyond. J. High Energy Phys. **1011**, 100 (2010). arXiv:1008.4773 [hep-th]
3. W. Black, C. Monni, High-energy string-brane scattering for massive states. Nucl. Phys. B **859**, 299–320 (2012). arXiv:1107.4321 [hep-th]
4. M. Bianchi, P. Teresi, Scattering higher spins off D-branes. J. High Energy Phys. **1201**, 161 (2012). arXiv:1108.1071 [hep-th]
5. E. Del Giudice, P. Di Vecchia, S. Fubini, General properties of the dual resonance model. Ann. Phys. **70**, 378–398 (1972)
6. K. Hornfeck, Three reggeon light cone vertex of the Neveu-Schwarz string. Nucl. Phys. **B293**, 189 (1987)
7. M.B. Green, M. Gutperle, Light cone supersymmetry and d-branes. Nucl. Phys. **B476**, 484–514 (1996). arXiv:hep-th/9604091

Chapter 9
Ghost Condensation in $N = 1$ Supergravity

Michael Koehn, Jean-Luc Lehners and Burt Ovrut

We present the theory of an $N = 1$ supersymmetric ghost condensate coupled to supergravity using a general formalism for constructing locally supersymmetric higher-derivative chiral superfield actions. The theory admits a ghost condensate vacuum in de Sitter spacetime. Expanded around this vacuum, the scalar sector is shown to be ghost-free with no spatial gradient instabilities. The fermion sector is found to consist of a massless chiral fermion and a massless gravitino. The ghost condensate vacuum spontaneously breaks local supersymmetry with the chiral field as the Goldstone fermion. Although potentially able to get a mass through the super-Higgs effect, the vanishing superpotential in the ghost condensate theory renders the gravitino massless.

9.1 Motivation

Higher-derivative scalar field theories coupled to gravitation appear in

- DBI theories [1]

Talk presented at "Breaking of Supersymmetry and Ultraviolet Divergences in Extended Supergravity", INFN, Frascati, Italy, March 2013

M. Koehn · J.-L. Lehners
Max Planck Institute for Gravitational Physics (Albert Einstein Institute), 14476 Potsdam, Germany
e-mail: michael.koehn@aei.mpg.de

J.-L. Lehners
e-mail: jlehners@aei.mpg.de

B. Ovrut (✉)
Department of Physics, University of Pennsylvania, 209 South 33rd Street, Philadelphia, PA 19104-6395, USA
e-mail: ovrut@elcapitan.hep.upenn.edu

S. Bellucci (ed.), *Breaking of Supersymmetry and Ultraviolet Divergences in Extended Supergravity*, Springer Proceedings in Physics 153,
DOI: 10.1007/978-3-319-03774-5_9,

- ghost-condensate theories of NEC violation [2–5]
- Galileon theories of cosmology [6, 7]
- worldvolume actions of solitonic branes [8, 9].

Using a general formalism for constructing global $N = 1$ supersymmetric higher-derivative chiral superfield Lagrangians [10], these scalar theories have been supersymmetrized in [10–12] respectively. Can these be extended to $N = 1$ local supersymmetry? Yes! We have

- given a general formalism for coupling higher-derivative chiral superfield Lagrangians to $N = 1$ supergravity [13] (also see [14, 15])
- applied this to DBI [16], ghost-condensates [17] and Galileons [18].

9.2 Scalar Ghost Condensation

Consider a real scalar field ϕ. Denote the standard kinetic term as $X = -\frac{1}{2}(\partial\phi)^2$. A ghost condensate arises from higher-derivative theories of the form

$$\mathcal{L} = \sqrt{-g}P(X) \tag{9.2.1}$$

where $P(X)$ is an arbitrary differentiable function of X. In a flat spacetime with $ds^2 = -dt^2 + a(t)^2\delta_{ij}dx^i dx^j$ and assuming $\phi = \phi(t)$, the scalar equation of motion is

$$\frac{d}{dt}\left(a^3 P_{,X}\dot{\phi}\right) = 0. \tag{9.2.2}$$

The trivial solution is $\phi =$ constant. More interesting is the solution

$$X = \frac{1}{2}\dot{\phi}^2 = \text{constant}, \quad P_{,X} = 0. \tag{9.2.3}$$

Denoting by X_{ext} a constant extremum of $P(X)$, the equation of motion admits the "ghost condensate" solution

$$\phi = ct, \quad c^2 = 2X_{\text{ext}}. \tag{9.2.4}$$

This vacuum spontaneously breaks Loretz invariance. It can also lead to violations of the "null energy condition" (NEC). To see this, evaluating the energy and pressure densities $\Rightarrow$

$$\rho = 2XP_{,X} - P, \quad p = P \;\Rightarrow\; \rho + p = 2XP_{,X}. \tag{9.2.5}$$

The NEC corresponds to the requirement that

$$\rho + p \geq 0. \tag{9.2.6}$$

Since $X > 0$, $\Rightarrow$ the NEC can be violated if

$$P_{,X} < 0. \tag{9.2.7}$$

That is, if we are close to an extremum of $P(X)$, then on one side the NEC is violated while on the other side it is not. Since Einstein's equations ⇒

$$\dot{H} = -\frac{1}{2}(\rho + p) \tag{9.2.8}$$

it is now possible to obtain a non-singular "bouncing" universe where H increases from negative to positive values. However, is this NEC violating vacuum "stable"?

Expanding the Lagrangian around the ghost condensate

$$\phi = ct + \delta\phi(x^m) \tag{9.2.9}$$

gives to quadratic order

$$\frac{\mathcal{L}}{\sqrt{-g}} = \frac{1}{2}\Big((2XP_{,XX} + P_{,X})(\dot{\delta\phi})^2 - P_{,X}\delta\phi^{,i}\delta\phi_{,i}\Big). \tag{9.2.10}$$

Note that Lorentz violation ⇒ that the coefficients of the time- and space-derivatives are different. The vacuum will be ghost-free iff

$$2XP_{,XX} + P_{,X} > 0. \tag{9.2.11}$$

This can be achieved by choosing the condensate to be at a minimum

$$P_{,XX} > 0. \tag{9.2.12}$$

Note that the theory can remain ghost-free even in the NEC violating region where $P_{,X} < 0$. However, in the NEC violating region the coefficient $-P_{,X}$ in front of the spatial derivative term has the wrong sign. This ⇒ the theory suffers from "gradient instabilities"! These can be softened by adding small higher-derivative terms—not of the $P(X)$ type—such as

$$-(\Box\phi)^2\,. \tag{9.2.13}$$

These modify the dispersion relation for $\delta\phi$ at high momenta and suppress instabilities for a short—but sufficient—period of time.

Finally, a prototypical choice for $P(X)$ that shows all interesting properties is

$$P(X) = -X + X^2 \quad (\Rightarrow \; c = 1). \tag{9.2.14}$$

9.3 Review of Globally $N = 1$ Supersymmetric Ghost Condensation

9.3.1 Higher-Derivative Chiral Superfield Lagrangian

Consider the chiral superfield

$$\Phi = A + i\theta\sigma^m\bar{\theta}A_{,m} + \frac{1}{4}\theta\theta\bar{\theta}\bar{\theta}\Box A + \theta\theta F + \sqrt{2}\theta\chi - \frac{i}{\sqrt{2}}\theta\theta\chi_{,m}\sigma^m\bar{\theta}. \tag{9.3.1}$$

The ordinary kinetic Lagrangian is

$$\mathcal{L}_{\Phi^\dagger\Phi} = \int d^4\theta\, \Phi^\dagger\Phi \;=\; \Phi^\dagger\Phi\,|_{\theta\theta\bar{\theta}\bar{\theta}} = -\partial A \cdot \partial A^* + F^*F + \frac{i}{2}\left(\chi_{,m}\sigma^m\bar{\chi} - \chi\sigma^m\bar{\chi}_{,m}\right). \tag{9.3.2}$$

Defining $A = \frac{1}{\sqrt{2}}(\phi + i\xi)$, the Lagrangian becomes

$$\mathcal{L}_{\Phi^\dagger\Phi} = -\frac{1}{2}(\partial\phi)^2 - \frac{1}{2}(\partial\xi)^2 + F^*F + \frac{i}{2}\left(\chi_{,m}\sigma^m\bar{\chi} - \chi\sigma^m\bar{\chi}_{,m}\right). \tag{9.3.3}$$

This is the global $N = 1$ supersymmetric generalization of X.

What is the supersymmetric generalization of X^2? Consider

$$\mathcal{L}_{D\Phi D\Phi\bar{D}\Phi^\dagger\bar{D}\Phi^\dagger} = \frac{1}{16}D\Phi D\Phi\bar{D}\Phi^\dagger\bar{D}\Phi^\dagger\big|_{\theta\theta\bar{\theta}\bar{\theta}}. \tag{9.3.4}$$

To quadratic order in the spinor component field

$$\begin{aligned}\mathcal{L}_{D\Phi D\Phi\bar{D}\Phi^\dagger\bar{D}\Phi^\dagger} =& (\partial A)^2(\partial A^*)^2 - 2F^*F\partial A\cdot\partial A^* + F^{*2}F^2\\ &-\frac{i}{2}\left(\chi\sigma^m\bar{\sigma}^l\sigma^n\bar{\chi}_{,n}\right)A_{,m}A^*_{,l} + \frac{i}{2}\left(\chi_{,n}\sigma^n\bar{\sigma}^m\sigma^l\bar{\chi}\right)A_{,m}A^*_{,l}\\ &+ i\chi\sigma^m\bar{\chi}^{,n}A_{,m}A^*_{,n} - i\chi^{,m}\sigma^n\bar{\chi}A_{,m}A^*_{,n}\\ &+\frac{i}{2}\chi\sigma^m\bar{\chi}\left(A^*_{,m}\Box A - A_{,m}\Box A^*\right)\\ &+\frac{1}{2}\left(F\Box A - \partial F\partial A\right)\bar{\chi}\bar{\chi}\\ &+\frac{1}{2}\left(F^*\Box A^* - \partial F^*\partial A^*\right)\chi\chi + \frac{1}{2}FA_{,m}\left(\bar{\chi}\bar{\sigma}^m\sigma^n\bar{\chi}_{,n} - \bar{\chi}_{,n}\bar{\sigma}^m\sigma^n\bar{\chi}\right)\\ &+\frac{1}{2}F^*A^*_{,m}\left(\chi_{,n}\sigma^n\bar{\sigma}^m\chi - \chi\sigma^n\bar{\sigma}^m\chi_{,n}\right) + \frac{3i}{2}F^*F\left(\chi_{,m}\sigma^m\bar{\chi} - \chi\sigma^m\bar{\chi}_{,m}\right)\\ &+\frac{i}{2}\chi\sigma^m\bar{\chi}\left(FF^*_{,m} - F^*F_{,m}\right).\end{aligned} \tag{9.3.5}$$

Written in terms of ϕ, ξ the pure A term in this Lagrangian is

$$(\partial A)^2(\partial A^*)^2 = \frac{1}{4}(\partial\phi)^4 + \frac{1}{4}(\partial\xi)^4 - \frac{1}{2}(\partial\phi)^2(\partial\xi)^2 + (\partial\phi\cdot\partial\xi)^2. \tag{9.3.6}$$

This is the global $N = 1$ supersymmetric generalization of X^2. It is the unique generalization with the properties:

(a) When the spinor is set to zero, the only non-vanishing term in $\frac{1}{16}D\Phi D\Phi\bar{D}\Phi^\dagger\bar{D}\Phi^\dagger$ is the top $\theta^2\bar{\theta}^2$ component.
This is very helpful in producing higher-derivative terms that include X^2.
(b) When coupled to supergravity, $\frac{1}{16}D\Phi D\Phi\bar{D}\Phi^\dagger\bar{D}\Phi^\dagger$ leads to minimal coupling of ϕ, ξ to gravity.

For example, an alternative generalization of X^2

$$-\frac{1}{16}\left(\Phi - \Phi^\dagger\right)^2\bar{D}D\Phi D\bar{D}\Phi^\dagger \quad\Rightarrow\quad \phi^2(\partial\xi)^2\mathcal{R}\,. \tag{9.3.7}$$

9.3.2 Globally Supersymmetric Ghost Condensate

Choose the scalar function $P(X)$ to be

$$P(X) = -X + X^2. \tag{9.3.8}$$

For a pure ghost condensate can take the superpotential

$$W = 0 \Rightarrow F = 0. \tag{9.3.9}$$

The associated globally supersymmetric Lagrangian, to quadratic order in the spinor, is

$$\begin{aligned}
\mathcal{L}^{\text{SUSY}} &= \left(-\Phi^\dagger\Phi + \frac{1}{16}D\Phi D\Phi\bar{D}\Phi^\dagger\bar{D}\Phi^\dagger\right)\Big|_{\theta\theta\bar{\theta}\bar{\theta}} \\
&= +\frac{1}{2}(\partial\phi)^2 + \frac{1}{4}(\partial\phi)^4 + \frac{1}{2}(\partial\xi)^2 + \frac{1}{4}(\partial\xi)^4 - \frac{1}{2}(\partial\phi)^2(\partial\xi)^2 + (\partial\phi\cdot\partial\xi)^2 \\
&\quad -\frac{\mathrm{i}}{2}\left(\chi_{,m}\sigma^m\bar{\chi} - \chi\sigma^m\bar{\chi}_{,m}\right) - \frac{1}{2}(\partial\phi)^2\frac{\mathrm{i}}{2}\left(\chi_{,m}\sigma^m\bar{\chi} - \chi\sigma^m\bar{\chi}_{,m}\right) \\
&\quad -\phi_m\phi_{,n}\frac{\mathrm{i}}{2}\left(\chi^{,n}\sigma^m\bar{\chi} - \chi\sigma^m\bar{\chi}^{,n}\right).
\end{aligned} \tag{9.3.10}$$

The equations of motion admit a ghost condensate vacuum

$$\phi = ct, \quad \xi = 0, \quad \chi = 0. \tag{9.3.11}$$

To assess stability, expand in the small fluctuations

$$\phi = t + \delta\phi(t,\vec{x}), \quad \xi = \delta\xi(t,\vec{x}), \quad \chi = \delta\chi(t,\vec{x}). \tag{9.3.12}$$

To quadratic order, the result is

$$\begin{aligned}\mathcal{L}^{\mathrm{SUSY}} &= \left(\dot{\delta\phi}\right)^2 + 0\cdot\delta\phi^{,i}\delta\phi_{,i} \\ &\quad + 0\cdot\left(\dot{\delta\xi}\right)^2 + \delta\xi^{,i}\delta\xi_{,i} \\ &\quad + \frac{1}{2}\frac{\mathrm{i}}{2}\left(\delta\chi_{,0}\sigma^0\delta\bar{\chi} - \delta\chi\sigma^0\delta\bar{\chi}_{,0}\right) - \frac{1}{2}\frac{\mathrm{i}}{2}\left(\delta\chi_{,i}\sigma^i\delta\bar{\chi} - \delta\chi\sigma^i\delta\bar{\chi}_{,i}\right).\end{aligned} \tag{9.3.13}$$

1. $\delta\phi$ kinetic term: As previously, has a gradient instability in the NEC violating region. $\Rightarrow$ In the pure boson case, added a $-(\Box\phi)^2$ term. The appropriate SUSY extension is

$$\begin{aligned}&-\frac{1}{2^{11}}D\Phi D\Phi\bar{D}\Phi^\dagger\bar{D}\Phi^\dagger\left(\{D,\bar{D}\}\{D,\bar{D}\}(\Phi+\Phi^\dagger)\right)^2\Big|_{\theta\theta\bar{\theta}\bar{\theta}} \\ &= -(\Box\phi)^2\left(\frac{1}{4}(\partial\phi)^4 + \frac{1}{4}(\partial\xi)^4 + (\partial\phi\cdot\partial\xi)^2 - \frac{1}{2}(\partial\phi)^2(\partial\xi)^2\right).\end{aligned} \tag{9.3.14}$$

Expanding around the ghost condensate using $(\partial\phi)^2 = -1$

$$\mathcal{L}^{\mathrm{SUSY}} = \left(\dot{\delta\phi}\right)^2 + 0\cdot\delta\phi^{,i}\delta\phi_{,i} - \frac{1}{4}(\Box\delta\phi)^2 + \cdots \tag{9.3.15}$$

which softens gradient instabilities.

2. $\delta\xi$ kinetic term: New to SUSY. Has vanishing time and wrong sign spatial kinetic terms. Cured by adding supersymmetric higher-derivative terms. The appropriate terms are

$$\begin{aligned}&+\frac{8}{16^2}D\Phi D\Phi\bar{D}\Phi^\dagger\bar{D}\Phi^\dagger\left(\{D,\bar{D}\}(\Phi-\Phi^\dagger)\{D,\bar{D}\}(\Phi^\dagger-\Phi)\right)\Big|_{\theta\theta\bar{\theta}\bar{\theta}} \\ &-\frac{4}{16^3}D\Phi D\Phi\bar{D}\Phi^\dagger\bar{D}\Phi^\dagger\left(\{D,\bar{D}\}(\Phi+\Phi^\dagger)\{D,\bar{D}\}(\Phi-\Phi^\dagger)\right) \\ &\left(\{D,\bar{D}\}(\Phi+\Phi^\dagger)\{D,\bar{D}\}(\Phi^\dagger-\Phi)\right)\Big|_{\theta\theta\bar{\theta}\bar{\theta}} = -2(\partial\phi)^4(\partial\xi)^2 - (\partial\phi)^4\left(\partial\phi\cdot\partial\xi\right)^2.\end{aligned} \tag{9.3.16}$$

Expanding around the ghost condensate $\Rightarrow$

$$\mathcal{L}^{\mathrm{SUSY}} = \cdots + (\dot{\delta\xi})^2 - \delta\xi^{,i}\delta\xi_{,i} + \cdots \tag{9.3.17}$$

which is Lorentz covariant with the correct sign.

3. $\delta\chi$ kinetic term: Ghost free with gradient "instability". Can be cured within the context of supersymmetric Galileons but re-grow a ghost! Won't discuss here. To summarize: The entire supersymmetric ghost condensate Lagrangian is

$$\begin{aligned}\mathcal{L}^{\text{SUSY}} &= -\,\Phi^\dagger\Phi\mid_{\theta\theta\bar\theta\bar\theta} + \frac{1}{16} D\Phi D\Phi\bar D\Phi^\dagger\bar D\Phi^\dagger\mid_{\theta\theta\bar\theta\bar\theta}\\ &+ D\Phi D\Phi\bar D\Phi^\dagger\bar D\Phi^\dagger\Big[-\frac{1}{2^{11}}\Big(\{D,\bar D\}\{D,\bar D\}(\Phi+\Phi^\dagger)\Big)^2\\ &\qquad + \frac{1}{2^5}\{D,\bar D\}(\Phi-\Phi^\dagger)\{D,\bar D\}(\Phi^\dagger-\Phi)\\ &\qquad - \frac{1}{2^{10}}\Big(\{D,\bar D\}(\Phi+\Phi^\dagger)\{D,\bar D\}(\Phi-\Phi^\dagger)\Big)^2\Big]\Big|_{\theta\theta\bar\theta\bar\theta}.\end{aligned} \tag{9.3.18}$$

In components, writing out all terms that are relevant for a stability analysis in a ghost condensate background, this corresponds to

$$\begin{aligned}\mathcal{L}^{\text{SUSY}} &= +\frac{1}{2}(\partial\phi)^2 + \frac{1}{4}(\partial\phi)^4 - \frac{1}{4}(\partial\phi)^4(\Box\phi)^2\\ &+ \frac{1}{2}(\partial\xi)^2 - \frac{1}{2}(\partial\phi)^2(\partial\xi)^2 - 2(\partial\phi)^4(\partial\xi)^2\\ &+ (\partial\phi\cdot\partial\xi)^2 - (\partial\phi)^4(\partial\phi\cdot\partial\xi)^2\\ &+ \frac{\mathrm{i}}{2}(\chi_{,m}\sigma^m\bar\chi - \chi\sigma^m\bar\chi_{,m})\Big(-1-\frac{1}{2}(\partial\phi)^2\Big)\\ &- \phi_m\phi_{,n}\frac{\mathrm{i}}{2}(\chi^{,n}\sigma^m\bar\chi - \chi\sigma^m\bar\chi^{,n}).\end{aligned} \tag{9.3.19}$$

The ghost condensate vacuum of this theory breaks $N = 1$ supersymmetry spontaneously in a new form. Consider the SUSY transformation

$$\delta\chi = \mathrm{i}\sqrt{2}\sigma^m\bar\zeta\partial_m A + \sqrt{2}\zeta F. \tag{9.3.20}$$

Usually supersymmetry is broken by a non-vanishing VEV $\langle F\rangle \neq 0$ of the auxiliary field. However, since in the ghost condensate Lagrangian $W = 0 \Rightarrow F = 0$. Recall that for the ghost condensate $\langle\phi\rangle = ct \Rightarrow$

$$\langle\dot A\rangle = \langle\dot\phi\rangle/\sqrt{2} = c/\sqrt{2}. \tag{9.3.21}$$

Therefore,

$$\delta\chi = \mathrm{i}\sqrt{2}\sigma^m\bar\zeta\partial_m A = \mathrm{i}\sigma^0\bar\zeta c \tag{9.3.22}$$

and the spinor trasnsforms inhomogeneously. $\Rightarrow$ SUSY is broken by the time-dependent condensate.

9.4 The Ghost Condensate in $N = 1$ Supergravity

In previous work, we showed that a global $N = 1$ supersymmetric Lagrangian of the general form

$$\mathcal{L}^{\text{SUSY}} = K(\Phi, \Phi^\dagger)\,|_{\theta\theta\bar{\theta}\bar{\theta}} + \frac{1}{16} D\Phi D\Phi \bar{D}\Phi^\dagger \bar{D}\Phi^\dagger T\left(\Phi, \Phi^\dagger, \partial_m\Phi, \partial_n\Phi^\dagger\right)|_{\theta\theta\bar{\theta}\bar{\theta}} + \left(W(\Phi)\,|_{\theta\theta} + W^\dagger(\Phi^\dagger)\,|_{\bar{\theta}\bar{\theta}}\right) \tag{9.4.1}$$

where K is any real function, T is an arbitrary hermitian function (with all derivative indices constracted) and W is a holomorphic superpotential, can be consistently coupled to $N = 1$ supergravity.

Notation: Curved $N = 1$ superspace

$$\left(x^m, \Theta^\alpha, \bar{\Theta}_{\dot{\alpha}}\right), \quad \mathcal{D}_A = \left(\mathcal{D}_a, \mathcal{D}_\alpha, \bar{\mathcal{D}}_{\dot{\alpha}}\right) \tag{9.4.2}$$

Gravity supermultiplet

$$\left(e_m^a . \psi_m, M, b_m\right) \tag{9.4.3}$$

Two superfield expansions we will need are the chiral curvature superfield

$$\begin{aligned} R = &-\frac{1}{6}M - \frac{1}{6}\Theta^\alpha\left(\sigma_{\alpha\dot{\alpha}}{}^a\bar{\sigma}^{b\dot{\alpha}\beta}\psi_{ab\beta} - \mathrm{i}\sigma_{\alpha\dot{\alpha}}{}^a\bar{\psi}_a^{\dot{\alpha}}M + \mathrm{i}\psi_{a\alpha}b^a\right) \\ &+ \Theta^\alpha\Theta_\alpha\Big(\frac{1}{12}\mathcal{R} - \frac{1}{6}\mathrm{i}\bar{\psi}_{\dot{\alpha}}^a\bar{\sigma}^{b\dot{\alpha}\beta}\psi_{ab\beta} - \frac{1}{9}MM^* - \frac{1}{18}b^a b_a + \frac{1}{6}\mathrm{i}e_a{}^m\mathcal{D}_m b^a \\ &- \frac{1}{12}\bar{\psi}_{\dot{\alpha}}\bar{\psi}^{\dot{\alpha}}M + \frac{1}{12}\psi_a{}^\alpha\sigma_{\alpha\dot{\alpha}}{}^a\bar{\psi}_c^{\dot{\alpha}}b^c \\ &- \frac{1}{48}\varepsilon^{abcd}\left[\bar{\psi}_{a\dot{\alpha}}\bar{\sigma}_b^{\dot{\alpha}\beta}\psi_{cd\beta} + \psi_a{}^\alpha\sigma_{\alpha\dot{\alpha}b}\bar{\psi}_{cd}^{\dot{\alpha}}\right]\Big) \end{aligned} \tag{9.4.4}$$

and the chiral density superfield

$$2\mathcal{E} = e\left(1 + \mathrm{i}\Theta^\alpha\sigma_{\alpha\dot{\alpha}}{}^a\bar{\psi}_a^{\dot{\alpha}} - \Theta^\alpha\Theta_\alpha\left[M^* + \bar{\psi}_{a\dot{\alpha}}\bar{\sigma}^{ab\dot{\alpha}}{}_{\dot{\beta}}\bar{\psi}_b^{\dot{\beta}}\right]\right). \tag{9.4.5}$$

In terms of these quantities, the supergravity extension of global $\mathcal{L}^{SUGRA}$ is

$$\mathcal{L}^{SUGRA} = \int d^2\Theta 2\mathcal{E}\Big[\frac{3}{8}(\bar{\mathcal{D}}^2 - 8R)e^{-K/3} - \frac{1}{8}(\bar{\mathcal{D}}^2 - 8R)(\mathcal{D}\Phi\mathcal{D}\Phi\bar{\mathcal{D}}\Phi^\dagger\bar{\mathcal{D}}\Phi^\dagger T) + W(\Phi)\Big] + \text{h.c.} \tag{9.4.6}$$

Since we are interested in the pure ghost condensate, we can take

$$W = 0 \Rightarrow F = M = 0. \tag{9.4.7}$$

The component expansion of $\mathcal{L}^{SUGRA}$ then becomes

$$\begin{aligned}\mathcal{L}^{SUGRA} =& \Big[-\frac{3}{32}e\left(\mathcal{D}^2\bar{\mathcal{D}}^2\mathrm{e}^{-K/3}\right) + \mathrm{i}\frac{3}{16}e\bar{\psi}_{a\dot{\alpha}}\bar{\sigma}^{a\dot{\alpha}\alpha}\left(\mathcal{D}_\alpha\bar{\mathcal{D}}^2\mathrm{e}^{-K/3}\right)\\ &-\frac{3}{8}e\bar{\psi}_a\bar{\sigma}^{ab}\bar{\psi}_b\left(\bar{\mathcal{D}}^2\mathrm{e}^{-K/3}\right) + \mathrm{i}\frac{1}{4}e\left(\bar{\psi}_a\bar{\sigma}^a\right)^\alpha\left(\mathcal{D}_\alpha\mathrm{e}^{-K/3}\right)\\ &-\frac{1}{4}e\left(\psi_{ab}\sigma^b\bar{\psi}^a + \mathrm{i}\psi_a b^a\right)^\alpha\left(\mathcal{D}_\alpha\mathrm{e}^{-K/3}\right) + \frac{1}{32}e\mathcal{D}^2\bar{\mathcal{D}}^2(\mathcal{D}\Phi\mathcal{D}\Phi\bar{\mathcal{D}}\Phi^\dagger\bar{\mathcal{D}}\Phi^\dagger T)\\ &-\frac{1}{16}e\mathrm{i}\left(\bar{\psi}_a\bar{\sigma}^a\right)^\alpha\mathcal{D}_\alpha\bar{\mathcal{D}}^2(\mathcal{D}\Phi\mathcal{D}\Phi\bar{\mathcal{D}}\Phi^\dagger\bar{\mathcal{D}}\Phi^\dagger T)\\ &+\frac{1}{8}e\bar{\psi}_a\bar{\sigma}^{ab}\bar{\psi}_b\bar{\mathcal{D}}^2(\mathcal{D}\Phi\mathcal{D}\Phi\bar{\mathcal{D}}\Phi^\dagger\bar{\mathcal{D}}\Phi^\dagger T)\Big]\Big| + \mathrm{h.c.}\\ &+e\Big(-\frac{1}{2}\mathcal{R} + \frac{1}{3}b^a b_a + \frac{1}{4}\varepsilon^{abcd}\left(\bar{\psi}_a\bar{\sigma}_b\psi_{cd} - \psi_a\sigma_b\bar{\psi}_{cd}\right)\Big)\mathrm{e}^{-K(A,A^*)/3}\end{aligned} \tag{9.4.8}$$

where $|$ specifies taking the lowest component of the superfield and

$$\psi_{mn}{}^\alpha = \tilde{\mathcal{D}}_m\psi_n^\alpha - \tilde{\mathcal{D}}_n\psi_m^\alpha, \quad \tilde{\mathcal{D}}_m\psi_n^\alpha = \partial_m\psi_n^\alpha + \psi_n^\beta\omega_{m\beta}^{\alpha}. \tag{9.4.9}$$

Note that the auxiliary field b_m remains undetermined. We must evaluate the lowest component of the superfield term. Evaluating the first part of the Lagrangian ⇒

$$\begin{aligned}\frac{1}{e}\mathcal{L}^{SUGRA}_{K(\Phi,\Phi^\dagger)} =& \frac{1}{e}\Big[\int \mathrm{d}^2\Theta 2\mathcal{E}\frac{3}{8}(\bar{\mathcal{D}}^2 - 8R)\mathrm{e}^{-K/3}\Big] + \mathrm{h.c.}\\ =& \Big(-\frac{1}{2}\mathcal{R} + \frac{1}{3}b^a b_a + \frac{1}{4}\varepsilon^{abcd}(\bar{\psi}_a\bar{\sigma}_b\psi_{cd} - \psi_a\sigma_b\bar{\psi}_{cd})\Big)\mathrm{e}^{-K(A,A^*)/3}\\ &+3|\partial A|^2(\mathrm{e}^{-K/3})_{,AA^*} + \mathrm{i}b^a\left(A_{,a}(\mathrm{e}^{-K/3})_{,A} - A^*{}_{,a}(\mathrm{e}^{-K/3})_{,A^*}\right)\\ &-\mathrm{i}\frac{1}{\sqrt{2}}b^a\left(\psi_a\chi(\mathrm{e}^{-K/3})_{,A} - \bar{\psi}_a\bar{\chi}(\mathrm{e}^{-K/3})_{,A^*}\right)\\ &-\sqrt{2}\chi\sigma^{mn}\psi_{mn}(\mathrm{e}^{-K/3})_{,A} - \sqrt{2}\bar{\chi}\bar{\sigma}^{mn}\bar{\psi}_{mn}(\mathrm{e}^{-K/3})_{,A^*}\\ &-\mathrm{i}\frac{3}{2}\psi_a\sigma^{ab}\sigma^c\bar{\psi}_b A_{,c}(\mathrm{e}^{-K/3})_{,A} - \mathrm{i}\frac{3}{2}\bar{\psi}_a\bar{\sigma}^{ab}\bar{\sigma}^c\psi_b A^*_{,c}(\mathrm{e}^{-K/3})_{,A^*}\\ &+\frac{1}{2}\chi\sigma^a\bar{\chi}b_a(\mathrm{e}^{-K/3})_{,AA^*} + \mathrm{i}\frac{3}{2}\left(\chi\sigma^a e_a{}^m\mathcal{D}_m\bar{\chi} + \bar{\chi}\bar{\sigma}^a e_a{}^m\mathcal{D}_m\chi\right)(\mathrm{e}^{-K/3})_{,AA^*}\\ &+\frac{3}{2}\sqrt{2}A^*_{,b}\psi_a\sigma^b\bar{\sigma}^a\chi(\mathrm{e}^{-K/3})_{,AA^*} + \frac{3}{2}\sqrt{2}A_{,b}\bar{\psi}_a\bar{\sigma}^b\sigma^a\bar{\chi}(\mathrm{e}^{-K/3})_{,AA^*}\\ &-\frac{3}{2}(\partial A)^2(\mathrm{e}^{-K/3})_{,AA} - \frac{3}{2}(\partial A^*)^2(\mathrm{e}^{-K/3})_{,A^*A^*}\\ &+\mathrm{i}\frac{3}{2}\chi\sigma^a\bar{\chi}\Big(A^*_{,a}(\mathrm{e}^{-K/3})_{,AA^*A^*} - A_{,a}(\mathrm{e}^{-K/3})_{,AAA^*}\Big).\end{aligned} \tag{9.4.10}$$

This is the supergravitaty extension of the $-X$ scalar term if one takes

$$K(\Phi, \Phi^\dagger) = -\Phi\Phi^\dagger. \tag{9.4.11}$$

Evaluating the second part of the Lagrangian taking

$$T = \frac{\mathcal{T}}{16} \quad \Rightarrow \tag{9.4.12}$$

$$\begin{aligned}
\frac{1}{e}\mathcal{L}^{SUGRA}_{\mathcal{D}\Phi\mathcal{D}\Phi\bar{\mathcal{D}}\Phi^\dagger\bar{\mathcal{D}}\Phi^\dagger,\mathcal{T}} &= \frac{1}{e}\Big(-\frac{\mathcal{T}}{2^7}\int \mathrm{d}^2\Theta 2\mathcal{E}(\bar{\mathcal{D}}^2 - 8R)(\mathcal{D}\Phi\mathcal{D}\Phi\bar{\mathcal{D}}\Phi^\dagger\bar{\mathcal{D}}\Phi^\dagger)\Big) + \text{h.c.} \\
&= \Big(+\frac{\mathcal{T}}{2^9}\mathcal{D}^2\bar{\mathcal{D}}^2(\mathcal{D}\Phi\mathcal{D}\Phi\bar{\mathcal{D}}\Phi^\dagger\bar{\mathcal{D}}\Phi^\dagger) \\
&\quad -\frac{\mathcal{T}}{2^8}\mathrm{i}(\bar{\psi}_a\bar{\sigma}^a)^\alpha\mathcal{D}_\alpha\bar{\mathcal{D}}^2(\mathcal{D}\Phi\mathcal{D}\Phi\bar{\mathcal{D}}\Phi^\dagger\bar{\mathcal{D}}\Phi^\dagger) \\
&\quad +\frac{\mathcal{T}}{2^7}\bar{\psi}_a\bar{\sigma}^{ab}\bar{\psi}_b\bar{\mathcal{D}}^2(\mathcal{D}\Phi\mathcal{D}\Phi\bar{\mathcal{D}}\Phi^\dagger\bar{\mathcal{D}}\Phi^\dagger)\Big)\Big| + \text{h.c.} \\
&= +\mathcal{T}(\partial A)^2(\partial A^*)^2 - \frac{1}{2}\sqrt{2}\mathcal{T}\bar{\psi}_a\bar{\sigma}^a\sigma^c\bar{\chi}A^*_{,c}(\partial A)^2 \\
&\quad -\frac{1}{2}\sqrt{2}\mathcal{T}\chi\sigma^c\bar{\sigma}^a\psi_a A_{,c}(\partial A^*)^2 - \sqrt{2}\mathcal{T}(\partial A^*)^2 A_{,m}\chi\psi^m \\
&\quad -\sqrt{2}\mathcal{T}(\partial A)^2 A^*_{,m}\bar{\psi}^m\bar{\chi} - \frac{\mathrm{i}}{2}\mathcal{T}\chi\sigma^a\bar{\chi}A_{,a}e_b{}^m\mathcal{D}_m A^*_{,b} \\
&\quad +\frac{5}{6}\mathcal{T}\chi\sigma^a\bar{\chi}A_{,a}A^*_{,b}b^b + \frac{\mathrm{i}}{2}\mathcal{T}\chi\sigma^a\bar{\chi}A^*_{,a}e_b{}^m\mathcal{D}_m A_{,b} \\
&\quad +\frac{5}{6}\mathcal{T}\chi\sigma^a\bar{\chi}A^*_{,a}A_{,b}b^b - \mathrm{i}\mathcal{T}(\mathcal{D}_m\chi)\sigma^b\bar{\chi}A^{,m}A^*_{,b} \\
&\quad +\sqrt{2}\mathcal{T}\bar{\psi}_a\bar{\sigma}^c\sigma^b\bar{\chi}A^{,a}A^*_{,b}A_{,c} + \frac{1}{3}\mathcal{T}\bar{\chi}\bar{\sigma}^b\sigma_c\bar{\sigma}_a\chi b^c A^{,a}A^*{}_{,b} \\
&\quad +\mathrm{i}\mathcal{T}\chi\sigma^b(\mathcal{D}_m\bar{\chi})A^{*,m}A_{,b} + \sqrt{2}\mathcal{T}\chi\sigma^b\bar{\sigma}^c\psi_a A^{*,a}A_{,b}A^*{}_{,c} \\
&\quad -\frac{\mathrm{i}}{2}\mathcal{T}\chi\sigma^a\bar{\sigma}^b\sigma^m(\mathcal{D}_m\bar{\chi})A_{,a}A^*{}_{,b} - \frac{1}{12}\mathcal{T}\chi\sigma^a\bar{\sigma}^b\sigma^c\bar{\chi}b_c A_{,a}A^*{}_{,b} \\
&\quad +\frac{\mathrm{i}}{2}\mathcal{T}(\mathcal{D}_m\chi)\sigma^m\bar{\sigma}^b\sigma^a\bar{\chi}A^*{}_{,a}A_{,b} - \frac{1}{12}\mathcal{T}\chi\sigma^c\bar{\sigma}^b\sigma^a\bar{\chi}b_c A^*{}_{,a}A_{,b}.
\end{aligned} \tag{9.4.13}$$

This is the supergravity extension of the X^2 scalar term if one takes

$$\mathcal{T} = 1. \tag{9.4.14}$$

The equation of motion of b_m is given by

$$\begin{aligned}
b_m =& -\frac{3}{2}\mathrm{i}\left(A_{,m}(\mathrm{e}^{-K/3})_{,A} - A^*_{,m}(\mathrm{e}^{-K/3})_{,A^*}\right)\mathrm{e}^{K/3} - \frac{3}{4}\chi\sigma_m\bar{\chi}(\mathrm{e}^{-K/3})_{,AA^*}\mathrm{e}^{K/3} \\
&+\frac{3}{4}\sqrt{2}\mathrm{i}\left(\psi_m\chi(\mathrm{e}^{-K/3})_{,A} - \bar{\psi}_m\bar{\chi}(\mathrm{e}^{-K/3})_{,A^*}\right)\mathrm{e}^{K/3}
\end{aligned}$$

$$
\begin{aligned}
&-\frac{5}{4}\tau\chi\sigma^a\bar{\chi}(A_{,a}A^*_{,m} + A^*_{,a}A_{,m})e^{K/3} \\
&+\frac{1}{2}\tau\chi\sigma^a\bar{\sigma}_m\sigma^b\bar{\chi}A_{,a}A^*_{,b}e^{K/3} \\
&+\frac{1}{8}\tau(\chi\sigma^a\bar{\sigma}^b\sigma_m\bar{\chi} + \chi\sigma_m\bar{\sigma}^a\sigma^b\bar{\chi})A_{,a}A^*{}_{,b}e^{K/3}.
\end{aligned}
\tag{9.4.15}
$$

Inserting this back into the Lagrangian, Weyl rescaling as

$$
\begin{aligned}
e_n{}^a &\xrightarrow{\text{WEYL}} e^{K/6}e_n{}^a \\
\chi &\xrightarrow{\text{WEYL}} e^{-K/12}\chi \\
\psi_m &\xrightarrow{\text{WEYL}} e^{K/12}\psi_m
\end{aligned}
\tag{9.4.16}
$$

and shifting

$$
\psi_m \xrightarrow{\text{SHIFT}} \psi_m + i\frac{\sqrt{2}}{6}\sigma_m\bar{\chi}K_{,A^*} \tag{9.4.17}
$$

$\Rightarrow$ keeping terms with at most two fermions

$$
\begin{aligned}
\frac{1}{e}\mathcal{L}^{SUGRA}_{K(\Phi,\Phi^\dagger),\text{Weyl}} &= \frac{1}{e}\Big[\int d^2\Theta 2\mathcal{E}\frac{3}{8}(\bar{\mathcal{D}}^2 - 8R)e^{-K/3}\Big]_{\text{Weyl}} + \text{h.c.} \\
&= -\frac{1}{2}\mathcal{R} - K_{,AA^*}|\partial A|^2 \\
&\quad - iK_{,AA^*}\bar{\chi}\bar{\sigma}^m\mathcal{D}_m\chi + \varepsilon^{klmn}\bar{\psi}_k\bar{\sigma}_l\tilde{\mathcal{D}}_m\psi_n \\
&\quad - \frac{1}{2}\sqrt{2}K_{,AA^*}A^*{}_{,n}\chi\sigma^m\bar{\sigma}^n\psi_m - \frac{1}{2}\sqrt{2}K_{,AA^*}A_{,n}\bar{\chi}\bar{\sigma}^m\sigma^n\bar{\psi}_m
\end{aligned}
\tag{9.4.18}
$$

and

$$
\begin{aligned}
\frac{1}{e}\mathcal{L}^{SUGRA}_{\mathcal{D}\Phi\mathcal{D}\Phi\bar{\mathcal{D}}\Phi^\dagger\bar{\mathcal{D}}\Phi^\dagger,\tau,\text{Weyl}} &= \frac{1}{e}\Big[\int d^2\Theta 2\mathcal{E}(-\frac{\tau}{2^7})(\bar{\mathcal{D}}^2 - 8R)(\mathcal{D}\Phi\mathcal{D}\Phi\bar{\mathcal{D}}\Phi^\dagger\bar{\mathcal{D}}\Phi^\dagger)\Big]_{\text{Weyl}} + \text{h.c.} \\
&= +\tau(\partial A)^2(\partial A^*)^2 - \frac{1}{2}\sqrt{2}\tau\bar{\psi}_a\bar{\sigma}^a\sigma^c\bar{\chi}A^*{}_{,c}(\partial A)^2 \\
&\quad - \frac{1}{2}\sqrt{2}\tau\chi\sigma^c\bar{\sigma}^a\psi_a A_{,c}(\partial A^*)^2 \\
&\quad - \sqrt{2}\tau(\partial A^*)^2 A_{,m}\chi\psi^m - \sqrt{2}\tau(\partial A)^2 A^*{}_{,m}\bar{\psi}^m\bar{\chi} \\
&\quad - \frac{i}{2}\tau\chi\sigma^a\bar{\chi}A_{,a}e^{bm}(\mathcal{D}_m A^*{}_{,b}) + \frac{i}{2}\tau\chi\sigma^a\bar{\chi}A^*{}_{,a}e^{bm}(\mathcal{D}_m A_{,b}) \\
&\quad - \frac{i}{6}\tau\chi\sigma^a\bar{\chi}A_{,a}A^*{}_{,b}K^{,b} + \frac{i}{6}\tau\chi\sigma^a\bar{\chi}A^*{}_{,a}A_{,b}K^{,b} \\
&\quad - i\tau(\mathcal{D}_m\chi)\sigma_n\bar{\chi}A^{,m}A^{*,n} + \sqrt{2}\tau\bar{\psi}_a\bar{\sigma}^c\sigma^b\bar{\chi}A^{,a}A^*{}_{,b}A_{,c} \\
&\quad + \frac{i}{12}\tau\chi\sigma^a\bar{\chi}A_{,b}A^*{}_{,a}K^{,b} + \frac{i}{6}\tau\chi\sigma^{cb}\sigma^a\bar{\chi}A_{,c}A^*{}_{,a}K_{,b}
\end{aligned}
$$

$$
\begin{aligned}
&+\mathrm{i}\tau\chi\sigma^b(\mathcal{D}_m\bar{\chi})A^{*,m}A_{,b}+\sqrt{2}\tau\chi\sigma^b\bar{\sigma}^c\psi_aA^{*,a}A_{,b}A^*{}_{,c}\\
&-\frac{\mathrm{i}}{12}\tau\chi\sigma^a\bar{\chi}A^*{}_{,b}A_{,a}K^{,b}-\frac{\mathrm{i}}{6}\tau\chi\sigma^a\bar{\sigma}^{bc}\bar{\chi}A^*{}_{,c}A_{,a}K_{,b}\\
&-\frac{\mathrm{i}}{2}\tau\chi\sigma^p\bar{\sigma}^q\sigma^m(\mathcal{D}_m\bar{\chi})A_{,p}A^*{}_{,q}+\frac{\mathrm{i}}{2}\tau(\mathcal{D}_m\chi)\sigma^m\bar{\sigma}^p\sigma^q\bar{\chi}A_{,p}A^*{}_{,q}\\
&+\frac{\mathrm{i}}{6}\tau\chi\sigma^c\bar{\sigma}^b\sigma^a\bar{\chi}K_{,a}A^*{}_{,b}A_{,c}-\frac{\mathrm{i}}{6}\tau\chi\sigma^a\bar{\sigma}^b\sigma^c\bar{\chi}K_{,a}A_{,b}A^*{}_{,c}\\
&-\frac{7}{4}\mathrm{i}\tau\chi\sigma^a\bar{\chi}\big(A^*{}_{,a}(\partial A)^2(\mathrm{e}^{-K/3})_{,A}-A_{,a}(\partial A^*)^2(\mathrm{e}^{-K/3})_{,A^*}\big)\mathrm{e}^{K/3}\\
&-\frac{3}{2}\mathrm{i}\tau\chi\sigma^a\bar{\chi}\big(A_{,a}(\mathrm{e}^{-K/3})_{,A}-A^*{}_{,a}(\mathrm{e}^{-K/3})_{,A^*}\big)|\partial A|^2\mathrm{e}^{K/3}.
\end{aligned}
\tag{9.4.19}
$$

9.4.1 The $N = 1$ Supergravity Ghost Condensate

Taking $K(\Phi,\Phi^\dagger) = -\Phi\Phi^\dagger$ and $\tau = 1$, the sum of these two terms is the $N = 1$ supergravity extension of the prototype scalar ghost condensate $P(X) = -X + X^2$ given by

$$
\mathcal{L}^{SUGRA}_{T=1/16,\mathrm{Weyl}}=\frac{1}{8}\Big[\int\mathrm{d}^2\Theta 2\mathcal{E}(\bar{\mathcal{D}}^2-8R)\Big(3\mathrm{e}^{\Phi\Phi^\dagger/3}-\frac{1}{2^4}(\mathcal{D}\Phi\mathcal{D}\Phi\bar{\mathcal{D}}\Phi^\dagger\bar{\mathcal{D}}\Phi^\dagger)\Big)\Big]_{\mathrm{Weyl}}+\mathrm{h.c.}
\tag{9.4.20}
$$

The purely scalar part of this supergravity Lagrangian is simply

$$
\frac{1}{e}\mathcal{L}^{SUGRA}_{T=1/16,\mathrm{Weyl}}=-\frac{1}{2}\mathcal{R}+|\partial A|^2+(\partial A)^2(\partial A^*)^2+\cdots
\tag{9.4.21}
$$

For $A = \frac{1}{\sqrt{2}}(\phi + \mathrm{i}\xi)$ this becomes

$$
\begin{aligned}
\frac{1}{e}\mathcal{L}^{SUGRA}_{T=1/16,\mathrm{Weyl}}=&-\frac{1}{2}\mathcal{R}+\frac{1}{2}(\partial\phi)^2+\frac{1}{4}(\partial\phi)^4\\
&+\frac{1}{2}(\partial\xi)^2+\frac{1}{4}(\partial\xi)^4-\frac{1}{2}(\partial\phi)^2(\partial\xi)^2+(\partial\phi\cdot\partial\xi)^2+\cdots
\end{aligned}
\tag{9.4.22}
$$

The Einstein and gravitino equations can be solved in an FRW spacetime $ds^2 = -dt^2 + a(t)^2\delta_{ij}dx^idx^j$ with

$$
a(t)=e^{\pm\frac{1}{\sqrt{12}}t},\quad \psi_m=0.
\tag{9.4.23}
$$

The ϕ, ξ and χ equations continue to admit the ghost condensate vacuum of the form

$$\phi = ct, \quad \xi = 0, \quad \chi = 0. \tag{9.4.24}$$

To assess stability, expand in the small fluctuations

$$\phi = t + \delta\phi(t,\vec{x}), \quad \xi = \delta\xi(t,\vec{x}), \quad \chi = \delta\chi(t,\vec{x}). \tag{9.4.25}$$

To quadratic order, the result is

$$\begin{aligned}\frac{1}{e}\mathcal{L}^{SUGRA}_{T=1/16,\mathrm{Weyl}} &= (\dot{\delta\phi})^2 + 0\cdot\delta\phi^{,i}\delta\phi_{,i}\\ &\quad + 0\cdot(\dot{\delta\xi})^2 + \delta\xi^{,i}\delta\xi_{,i}\\ &\quad + \cdots\end{aligned} \tag{9.4.26}$$

1. $\delta\phi$ kinetic term: As previously, has a gradient instability in the NEC violating region. $\Rightarrow$ In the global SUSY case, this was solved by adding the term

$$-\frac{1}{2^{11}}D\Phi D\Phi\bar{D}\Phi^\dagger\bar{D}\Phi^\dagger\left(\{D,\bar{D}\}\{D,\bar{D}\}(\Phi+\Phi^\dagger)\right)^2\Big|_{\theta\theta\bar{\theta}\bar{\theta}} \tag{9.4.27}$$

to the Lagrangian. In the supergravity case, this is easily generalized to

$$-\frac{1}{8}\int d^2\Theta 2\mathcal{E}(\bar{\mathcal{D}}^2 - 8R)(\mathcal{D}\Phi\mathcal{D}\Phi\bar{\mathcal{D}}\Phi^\dagger\bar{\mathcal{D}}\Phi^\dagger T_\phi) + \text{h.c.} \tag{9.4.28}$$

where

$$T_\phi = \frac{\kappa}{2^9}\left(\{\mathcal{D}^\alpha,\bar{\mathcal{D}}_{\dot{\alpha}}\}\{\mathcal{D}_\alpha,\bar{\mathcal{D}}^{\dot{\alpha}}\}(\Phi+\Phi^\dagger)\right)^2 \tag{9.4.29}$$

and κ is any real number (chosen arbitrarily to be $\kappa = 1/4$ in the global SUSY case). Setting $F = M = 0$, its bosonic contribution to the Lagrangian is

$$\begin{aligned}&-\frac{1}{8e}\Big[\int d^2\Theta 2\mathcal{E}(\bar{\mathcal{D}}^2 - 8R)\mathcal{D}\Phi\mathcal{D}\Phi\bar{\mathcal{D}}\Phi^\dagger\bar{\mathcal{D}}\Phi^\dagger T_\phi\Big]_{\mathrm{Weyl}} + \text{h.c.}\\ &\quad = \kappa(\Box\phi)^2\left((\partial\phi)^4 + (\partial\xi)^4 - 2(\partial\phi)^2(\partial\xi)^2 + 4(\partial\phi\cdot\partial\xi)^2\right).\end{aligned} \tag{9.4.30}$$

Adding this to the original scalar Lagrangian $\frac{1}{e}\mathcal{L}^{SUGRA}_{T=1/16,\mathrm{Weyl}}$, the metric and ϕ solutions of their equations of motion change—unlike in the global SUSY case. Expanded perturbatively in small κ, they become

$$\langle\dot{\phi}\rangle^2 = 1 - 3\kappa + \mathcal{O}(\kappa^2), \tag{9.4.31}$$

$$\langle H\rangle^2 = \frac{1}{12} + \frac{1}{4}\kappa + \mathcal{O}(\kappa^2). \tag{9.4.32}$$

That is, there is a shift in the condensate/FRW solution without altering its fundamental features. However, expanded around this new vacuum ⇒

$$\mathcal{L}^{\text{SUGRA}} = \frac{1}{2}\left(3\langle\dot{\phi}\rangle^2 - 1\right)\left(\dot{\delta\phi}\right)^2 + \frac{1}{2a^2}\left(1 - \langle\dot{\phi}\rangle^2\right)\delta\phi^{,i}\delta\phi_{,i} + \kappa\left(\Box\delta\phi\right)^2 + \cdots . \tag{9.4.33}$$

which, for $\kappa < 0$, softens the gradient instability—as anticipated.

2. $\delta\xi$ kinetic term: Has vanishing time and wrong sign spatial kinetic terms. In global SUSY, this is cured by adding the higher-derivative terms

$$\begin{aligned}&+\frac{8}{16^2}D\Phi D\Phi\bar{D}\Phi^\dagger\bar{D}\Phi^\dagger\left(\{D,\bar{D}\}(\Phi-\Phi^\dagger)\{D,\bar{D}\}(\Phi^\dagger-\Phi)\right)\Big|_{\theta\theta\bar{\theta}\bar{\theta}}\\&-\frac{4}{16^3}D\Phi D\Phi\bar{D}\Phi^\dagger\bar{D}\Phi^\dagger\left(\{D,\bar{D}\}(\Phi+\Phi^\dagger)\{D,\bar{D}\}(\Phi-\Phi^\dagger)\right)\\&\qquad\times\left(\{D,\bar{D}\}(\Phi+\Phi^\dagger)\{D,\bar{D}\}(\Phi^\dagger-\Phi)\right)\Big|_{\theta\theta\bar{\theta}\bar{\theta}}\end{aligned} \tag{9.4.34}$$

to the Lagrangian. In the supergravity case, this is easily generalized to

$$-\frac{1}{8}\int d^2\Theta 2\mathcal{E}(\bar{\mathcal{D}}^2 - 8R)(\mathcal{D}\Phi\mathcal{D}\Phi\bar{\mathcal{D}}\Phi^\dagger\bar{\mathcal{D}}\Phi^\dagger T_\xi) + \text{h.c.} \tag{9.4.35}$$

where

$$\begin{aligned}T_\xi =& +2^{-5}\{\mathcal{D}^\alpha,\bar{\mathcal{D}}_{\dot{\alpha}}\}(\Phi-\Phi^\dagger)\{\mathcal{D}_\alpha,\bar{\mathcal{D}}^{\dot{\alpha}}\}(\Phi^\dagger-\Phi)\\&-2^{-10}\left(\{\mathcal{D}^\alpha,\bar{\mathcal{D}}_{\dot{\alpha}}\}(\Phi+\Phi^\dagger)\{\mathcal{D}_\alpha,\bar{\mathcal{D}}^{\dot{\alpha}}\}(\Phi-\Phi^\dagger)\right)^2.\end{aligned} \tag{9.4.36}$$

Setting $F = M = 0$, its bosonic contribution is

$$\begin{aligned}&-\frac{1}{8e}\left[\int d^2\Theta 2\mathcal{E}(\bar{\mathcal{D}}^2 - 8R)\mathcal{D}\Phi\mathcal{D}\Phi\bar{\mathcal{D}}\Phi^\dagger\bar{\mathcal{D}}\Phi^\dagger T_\xi\right]_{\text{Weyl}} + \text{h.c.}\\&\quad= -2(\partial\phi)^4(\partial\xi)^2 - (\partial\phi)^4(\partial\phi\cdot\partial\xi)^2 .\end{aligned} \tag{9.4.37}$$

The addition of these terms does not alter the supergravity ghost condensate vacuum given above. Expanding around this vacuum, the ξ fluctuations are

$$\begin{aligned}\frac{1}{e}\mathcal{L}^{\text{SUGRA}} &= \cdots + \left(-\frac{1}{2} + \frac{1}{2}\langle\dot{\phi}\rangle^2 + 2\langle\dot{\phi}\rangle^4 - \langle\dot{\phi}\rangle^6\right)(\dot{\delta\xi})^2\\&\quad+\left(\frac{1}{2} + \frac{1}{2}\langle\dot{\phi}\rangle^2 - 2\langle\dot{\phi}\rangle^4\right)\delta\xi^{,i}\delta\xi_{,i} + \cdots\\&= \cdots + \left(1 - \frac{9}{2}\kappa + \mathcal{O}(\kappa^2)\right)\left((\dot{\delta\xi})^2 - \delta\xi^{,i}\delta\xi_{,i}\right) + \cdots\end{aligned} \tag{9.4.38}$$

$\Rightarrow$ the scalar $\delta\xi$ kinetic energy is rendered Lorentz covariant and stable by the addition of these terms. By suitably choosing the coefficients, this kinetic energy can be made canonical.

3. $\delta\chi$ kinetic term: Ghost free with gradient "instability". Can be cured within the context of supergravitational Galileons—but re-grow a ghost! Won't discuss here.

The ghost condensate vacuum of this theory breaks $N = 1$ supersymmetry spontaneously in a specific way. The SUSY transformations of the fermions in the ghost condensate vacuum are

$$\delta\chi = \mathrm{i}\sqrt{2}\sigma^m\bar{\zeta}\partial_m A = \mathrm{i}\sigma^0\bar{\zeta}c, \quad \delta\psi_m = 2\mathcal{D}_m\zeta. \tag{9.4.39}$$

Redefining

$$\psi_{m\alpha} = \tilde{\psi}_{m\alpha} - \frac{2\mathrm{i}}{(\partial\phi)^2}\mathcal{D}_m(\phi_{,n}\sigma^n_{\alpha\dot{\alpha}}\bar{\chi}^{\dot{\alpha}}) \tag{9.4.40}$$

$\Rightarrow$

$$\delta\tilde{\psi}_m = 0. \tag{9.4.41}$$

This identifies χ as the Goldstone fermion and $\tilde{\psi}_{m\alpha}$ as the physical gravitino. Since $m_{3/2} = e^{K/2}|W|$, then

$$W = 0 \quad \Rightarrow \quad m_{3/2} = 0 \tag{9.4.42}$$

consistent with an explicit calculation. Specifically–using various identities, redefining the gravitino as above and evaluating on the ghost condensate FRW background, we find that

$$\begin{aligned} \frac{1}{e}\mathcal{L}^{SUGRA}_{T=1/16,\text{Weyl}} = & \cdots + \frac{1}{2}\varepsilon^{klmn}\left(\bar{\tilde{\psi}}_k\bar{\sigma}_l\tilde{\mathcal{D}}_m\tilde{\psi}_n - \tilde{\psi}_k\sigma_l\tilde{\mathcal{D}}_m\bar{\tilde{\psi}}_n\right) \\ & + \frac{\mathrm{i}}{2}\left(\chi\sigma^m\mathcal{D}_m\bar{\chi} + \bar{\chi}\bar{\sigma}^m\mathcal{D}_m\chi\right) \\ & + \mathrm{i}\phi^{,m}\phi_{,n}\left(\bar{\chi}\bar{\sigma}^n(\mathcal{D}_m\chi) + \chi\sigma^n(\mathcal{D}_m\bar{\chi})\right) + \cdots \end{aligned} \tag{9.4.43}$$

$\Rightarrow$ canonical gravitino kinetic term, Lorentz violating ghost-free/gradient unstable χ kinetic term, and vanishing masses for both $\tilde{\psi}_m$ and χ.

Acknowledgments B.A.O. would like to thank Stefano Bellucci for the invitation to the workshop "Breaking of Supersymmetry and Ultrtaviolet Divergences in Extended Supergravity" at the INFN in Frascati, Italy in March 2013, where the content of these proceedings were presented. B.A.O. is supported in part by the DOE under contract No. DE-AC02-76-ER-03071 and the NSF under grant No. 1001296. M.K. and J.L.L. acknowledge the support of the European Research Council via the Starting Grant No. 256994.

References

1. See, for example, R.G. Leigh, Dirac-Born-Infeld action from Dirichlet Sigma model. Mod. Phys. Lett. **A4**, 2767 (1989)
2. N. Arkani-Hamed, H.C. Cheng, M.A. Luty, S. Mukohyama, Ghost condensation and a consistent infrared modification of gravity. JHEP **0405**, 074 (2004). arXiv:hep-th/0312099
3. E.I. Buchbinder, J. Khoury, B.A. Ovrut, New Ekpyrotic cosmology. Phys. Rev. D **76**, 123503 (2007). arXiv:hep-th/0702154
4. E.I. Buchbinder, J. Khoury, B.A. Ovrut, On the initial conditions in new ekpyrotic cosmology. JHEP **0711**, 076 (2007). arXiv:0706.3903 [hep-th]
5. E.I. Buchbinder, J. Khoury, B.A. Ovrut, Non-Gaussianities in new ekpyrotic cosmology. Phys. Rev. Lett. **100**, 171302 (2008). arXiv:0710.5172 [hep-th]
6. A. Nicolis, R. Rattazzi, E. Trincherini, The galileon as a local modification of gravity. Phys. Rev. D **79**, 064036 (2009). arXiv:0811.2197 [hep-th]
7. N. Chow, J. Khoury, Galileon cosmology. Phys. Rev. D **80**, 024037 (2009). arXiv:0905.1325 [hep-th]
8. R. Gregory, Effective actions for bosonic topological defects. Phys. Rev. D **43**, 520 (1991)
9. J. Khoury, B.A. Ovrut, J. Stokes, The worldvolume action of Kink Solitons in AdS Spacetime. JHEP **1208**, 015 (2012). arXiv:1203.4562 [hep-th]
10. J. Khoury, J.-L. Lehners, B. Ovrut, Supersymmetric P(X,ϕ) and the ghost condensate. Phys. Rev. D **83**, 125031 (2011). arXiv:1012.3748 [hep-th]
11. J. Khoury, J.-L. Lehners, B.A. Ovrut, Supersymmetric galileons. Phys. Rev. D **84**, 043521 (2011). arXiv:1103.0003 [hep-th]
12. B.A. Ovrut, J. Stokes, Heterotic Kink Solitons and their worldvolume action. JHEP **1209**, 065 (2012). arXiv:1205.4236 [hep-th]
13. M. Koehn, J.-L. Lehners, B.A. Ovrut, Higher-derivative Chiral superfield actions coupled to N=1 supergravity. Phys. Rev. D **86**, 085019 (2012). arXiv:1207.3798 [hep-th]
14. D. Baumann, D. Green, Supergravity for effective theories. J. High Energy Phys. **03**, 001 (2012)
15. F. Farakos, A. Kehagias, Emerging potentials in higher-derivative Gauged Chiral models coupled to $N = 1$ supergravity. J. High Energy Phys. **11**, 077 (2012)
16. M. Koehn, J.-L. Lehners, B.A. Ovrut, DBI inflation in N=1 supergravity. Phys. Rev. D **86**, 123510 (2012). arXiv:1208.0752 [hep-th]
17. M. Koehn, J.-L. Lehners, B. Ovrut, The ghost condensate in N=1 supergravity. arXiv:1212.2185 [hep-th]
18. M. Koehn, J.-L. Lehners, B. Ovrut, Supersymmetric galileons have ghosts. arXiv:1302.0840 [hep-th]

Chapter 10
Thermodynamic Curvature and Black Holes

George Ruppeiner

In my talk, I will discuss black hole thermodynamics, particularly what happens when you add thermodynamic curvature to the mix. Although black hole thermodynamics is a little off the main theme of this workshop, I hope nevertheless that my message will be of some interest to researchers in supersymmetry and supergravity. Black hole thermodynamics would appear very much in need of some microscopic foundation. We might ask: what are black holes made out of? I will give no answer, but I would like to suggest that what I present here might offer some guidance about the microscopic character of black holes.

Thermodynamic curvature is an element of thermodynamic metric geometry. A pioneering paper on this was by Weinhold [1] who introduced a thermodynamic energy inner product. This led to the work of Ruppeiner [2] who wrote a Riemannian thermodynamic entropy metric to represent thermodynamic fluctuation theory, and was the first to systematically calculate the thermodynamic Ricci curvature scalar R. A parallel effort was by Andresen et al. [3] who began the systematic application of the thermodynamic entropy metric to characterize finite time thermodynamic processes.

This talk presents a review of thermodynamic curvature R broad in scope, though far from complete in its coverage. I extend the themes discussed in a previous talk [4]. My main focus is on achieving some understanding of thermodynamic curvature in the black hole setting. To accomplish this, my working assumption is that for black holes, R follows the same physical interpretation as for ordinary thermodynamic systems, where R gives the size of organized microscopic structures. I present a review of what is known about ordinary thermodynamics, and what this might tell us about black holes.

G. Ruppeiner (✉)
Division of Natural Sciences, New College of Florida, Sarasota, FL 34243-2109, USA
e-mail: ruppeiner@ncf.edu

S. Bellucci (ed.), *Breaking of Supersymmetry and Ultraviolet Divergences in Extended Supergravity*, Springer Proceedings in Physics 153,
DOI: 10.1007/978-3-319-03774-5_10,

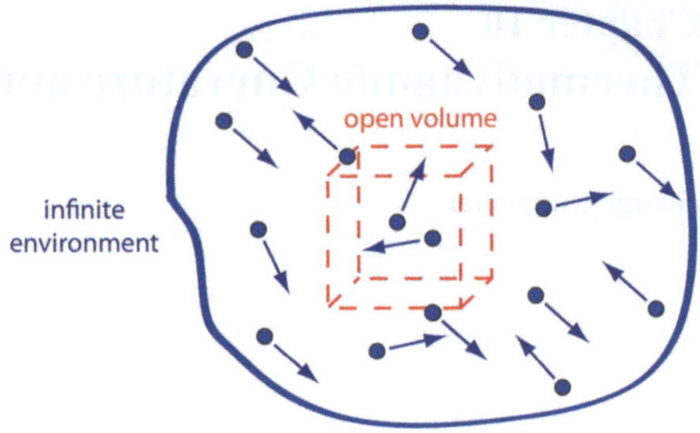

Fig. 10.1 An infinite environment of particles and an open volume, with fixed volume V, into which particles fluctuate in and out

10.1 What is Thermodynamic Curvature R?

Thermodynamic curvature comes from thermodynamic fluctuation theory. This classical theory is described in every book on statistical mechanics; it is chapter twelve in Landau and Lifshitz [5]. For a fluid system, the basic set-up is shown in Fig. 10.1. There is a infinite universe of particles and some imaginary open volume with fixed volume V, into which the particles can travel freely in and out. What is the probability of finding some energy U and some number of particles N in the open volume? Thermodynamic fluctuation theory gives the answer.

Let the particles in the open volume, and the environment consisting of the rest of the particles, be two thermodynamic systems. Denote the fixed thermodynamic state of the environment by "0". The thermodynamic state of the open volume fluctuates about an equilibrium characterized by maximum total entropy. The probability of a fluctuation away from equilibrium is given by Einstein's famous Gaussian thermodynamic fluctuation formula [5–8]:

$$\text{probability} \propto \exp\left[-\frac{V}{2}(\Delta\ell)^2\right], \tag{10.1}$$

where

$$(\Delta\ell)^2 = g_{\mu\nu}\Delta x^\mu \Delta x^\nu, \tag{10.2}$$

x^1 and x^2 denote a pair of independent fluctuating thermodynamic variables of the open volume, $\Delta x^\alpha = \left(x^\alpha - x_0^\alpha\right)$ denotes the difference between x^α and its equilibrium value x_0^α, where the total entropy is maximized, and $g_{\mu\nu}$ denotes the elements of the thermodynamic entropy metric discussed below.

Let S, X^1, and X^2 be the entropy, internal energy U, and particle number N, respectively, of the open volume. Regard $S = S(X^1, X^2, V)$, with V fixed. X^1 and X^2 correspond to conserved quantities, and S is additive between the open volume and its environment. If $x^\alpha = X^\alpha$, then

$$g_{\alpha\beta} = -\frac{1}{k_B V}\frac{\partial^2 S}{\partial X^\alpha \partial X^\beta}, \tag{10.3}$$

where k_B is Boltzmann's constant [5, 9, 10]. Since probability depends only on the thermodynamic state, the metric elements $g_{\alpha\beta}$ constitute a second-rank tensor. $g_{\alpha\beta}$ is a positive definite matrix, since the entropy has a maximum value in equilibrium. This is the condition of thermodynamic stability.

This is all found in Landau and Lifshitz [5]. Let me now get into some things Landau and Lifshitz did not say. The quadratic form $(\Delta\ell)^2$ in (10.2) has the look of a distance between thermodynamic states, a distance in the form of a Riemannian metric. The physical interpretation is that: the less the probability of a fluctuation between two states, the further apart they are.

A Riemannian metric in any context leads directly to a Ricci curvature scalar R [11], and this is certainly the case here. R is the only geometric scalar invariant function in thermodynamics, and so it must be very fundamental. The units of the thermodynamic curvature are those of volume per particle, and this limits its possible physical interpretation greatly. Units alone suggest that R is a measure of the characteristic size of some sort of organized fluctuating structures within the system.

R is readily calculable from the thermodynamic metric elements $g_{\alpha\beta}$. For example, in $(x^1, x^2) = (T, \rho)$ coordinates, where T is the temperature and ρ is the particle number density, we have the Helmholtz free energy per volume $f = f(T, \rho)$, the entropy per volume $s = -f_{,T}$ (where the comma notation indicates partial differentiation), and the chemical potential $\mu = f_{,\rho}$. The diagonal metric elements ($g_{12} = 0$) are [12]

$$g_{11} = \frac{1}{k_B T}\left(\frac{\partial s}{\partial T}\right)_\rho, \tag{10.4}$$

and

$$g_{22} = \frac{1}{k_B T}\left(\frac{\partial \mu}{\partial \rho}\right)_T. \tag{10.5}$$

For a diagonal metric [11]

$$R = \frac{1}{\sqrt{g}}\left[\frac{\partial}{\partial x^1}\left(\frac{1}{\sqrt{g}}\frac{\partial g_{22}}{\partial x^1}\right) + \frac{\partial}{\partial x^2}\left(\frac{1}{\sqrt{g}}\frac{\partial g_{11}}{\partial x^2}\right)\right], \tag{10.6}$$

where

$$g = g_{11}\, g_{22}. \tag{10.7}$$

A simple example is the ideal gas, in which there is no interaction between the particles. Here

$$f(T, \rho) = \rho k_B T \ln \ln \rho + \rho k_B h(T), \tag{10.8}$$

Fig. 10.2 Three surfaces with constant Ricci curvature scalar R the *sphere*, the *plane*, and the *pseudosphere*. For pure fluids, in Weinberg's sign convention, $R < 0$ if attractive interparticle interactions dominate, and $R > 0$ if repulsive interactions dominate. Regardless the sign convention for R, attractive interactions correspond to the geometry of the sphere, and repulsive interactions to the geometry of the pseudosphere

where $h(T)$ is some function of the temperature with negative second derivative. Equation (10.6) now yields $R = 0$ [2]. This suggests that R is some type of measure of interactions between particles.

Calculations in critical point models show that $|R|$ diverges as the correlation volume ξ^d, where d is the spatial dimension of the system [2, 9, 13]. The connection of $|R|$ to fluctuating structure size has also been established directly by means of a covariant thermodynamic fluctuation theory [9, 12, 14–16].

R is a signed quantity, as shown in Fig. 10.2. I use the sign convention of Weinberg [17]. (Sign conventions differ among authors. I express all results reported here in Weinberg's sign convention). For fluid and solid systems, an overall pattern is that R is negative for systems where attractive interparticle interactions dominate, and positive where repulsive interactions dominate. The sign of R alone thus offers direct information about the character of the interactions among the particles.

10.2 *R* for Ordinary Thermodynamics

R has been worked out in a number of cases in ordinary thermodynamics. On systematic tabulation, patterns readily become evident. Such patterns might lend insight into the nature of black hole microscopic properties.

In this section I attempt a classification of the "basic food groups" of R for ordinary thermodynamics. Thermodynamics divides neatly into atomic and molecular systems, like fluids and solids, and discrete lattice systems, like magnetic spin systems. I will treat them separately.

10.2.1 R for Fluid and Solid Systems, Basic Models

In this section, I tabulate results for fluid and solid systems, including the quantum gasses. I pay special attention to Lennard-Jones type interacting systems, for which there are a number of interesting recent results.

Table 10.1 The thermodynamic curvature R for a number of simple models for which R has only one sign

System	n	d	R sign	$\|R\|$ divergence
Ideal Bose gas [18, 19]	2	3	−	$T \to 0$
q-Deformed bosons [20]	2	3	−	Critical line
Critical regime [2, 9, 21]	2	...	−	Critical point
Mean-field theory [22]	2	...	−	Critical point
van der Waals (critical regime) [9, 21, 23]	2	3	−	Critical point
Spherical model [13, 24]	2	3	−	Critical point
Tonks gas [25]	2	1	−	$\|R\|$ small
Ideal gas [2, 26]	2	3	0	$\|R\|$ small
Multicomponent ideal gas [27]	>2	3	+	$\|R\|$ small
Ideal gas paramagnet [28]	3	3	+	$\|R\|$ small
q-Deformed fermions [20]	2	3	+	$T \to 0$
Ideal Fermi gas [18, 19, 29]	2	2,3	+	$T \to 0$
Ideal gas Fermi paramagnet [28]	3	3	+	$T \to 0$

Tabulated are the number of independent thermodynamic parameters n, the spatial dimension d, the sign of R, and the possible divergences of R. For some models, there is no particular spatial dimension d, and this is denoted by .... "$|R|$ small" means that the value of $|R|$ is on the order of the volume of an interparticle spacing or less

Table 10.1 shows R for a number of simple models for which R has only one sign. These models were worked out by a number of authors over a period of years. In these models interactions between particles may take place by virtue of a potential between the particles, or through quantum statistics. In either case, particles tend to either bunch together (attract) or to push apart (repel) compared with the ideal gas. The results in Table 10.1 clearly show the relation between the character of the interparticle interactions and the sign of R. If interactions between particles are attractive, R is negative. Prominent here is the ideal Bose gas,[1] and the typical critical point models. If interactions are repulsive, R is positive. Prominent examples are ideal Fermi gasses. In systems with weak interactions, $|R|$ is zero or "small," where "small" means on the order of the molecular volume v, $|R| \sim v$ or smaller. Cases with $|R| \sim v$ are typical also of systems dominated by strong short-range repulsive interactions, such as dense liquids and solids. Table 10.1 also shows where $|R|$ diverges, typically either at absolute zero or at critical points.

Table 10.2 shows four additional models, each having R with both signs. The Takahashi gas has negative R for the gas-like phase, where attractive interactions dominate, and small $|R|$ in the liquid-like phase. Increasing the density at constant low temperature yields a pseudophase transition from a gas-like phase to a liquid-like phase. This pseudophase transition is accompanied by a sharp *positive* spike in R. Conceptually simpler than the Takahashi gas are the remaining three models in Table 10.2, which are all quantum gasses intermediate between Fermions and Bosons, and

[1] The calculation of R for the ideal Bose gas was done with a continuous density of states, and so a possible divergence of R at a Bose-Einstein phase transition with $T > 0$ would not have been revealed.

Table 10.2 The thermodynamic curvature R for models where R has both signs

System	n	d	R sign	$R = 0$	$\|R\|$ divergence
Takahashi gas [25]	2	1	$\pm$	Yes	$T \to 0$
Gentile's statistics [19]	2	3	$\pm$	Yes	$T \to 0$
M-statistics [30]	2	2, 3	$\pm$	Yes	$T \to 0$
Anyons [31]	2	2	$\pm$	Yes	$T \to 0$

In each case, R changes sign through $R = 0$

with sign of R switching from positive to negative through $R = 0$ on transitioning from Fermionic to Bosonic behavior. Gentile statistics have an integer parameter p giving the maximum occupation number of a state, with $p = 1$ corresponding to a pure Fermi gas, and $p \to \infty$ to a pure Bose gas. In the same spirit is the M-statistics model, with state occupation number M. For any temperature and chemical potential, R eventually transitions in sign from positive to negative as M increases from 1. R thus offers a convincing method of determining when the M-statistics model transitions from Fermionic to Bosonic. The quantum gas of anyons is intrinsically two-dimensional, and has particles with fractional spin α whose variation allows us to change it continuously from a Bose gas to a Fermi gas ($\alpha : 0 \to 1$); the sign of R changes correspondingly from negative to positive.

10.2.2 R for Fluid and Solid Systems, Lennard-Jones Potential

A major element in the study of fluid and solid systems is the Lennard-Jones type potential between particles, shown schematically in Fig. 10.3. This potential approximates the interaction between particles in real fluids and solids. The Lennard-Jones type potential is strongly repulsive at short range and weakly attractive at long range. There is a minimum in the potential where repulsion and attraction balance, and where particles in a condensed liquid or solid phase like to reside. Fluid phases typically posses average separation distances between particles greater than that corresponding to the bottom of the potential well, and so the attractive part of the potential usually dominates. Hence, R should be mostly negative for real pure fluids, which is indeed the case. The study of the Lennard-Jones type interaction supplements that for the simple models above, and takes us a long way towards completing the picture for R for fluid and solid systems.

Let me present results from fluid studies based on experimental fluid data [32, 33], and on computer simulations in fluids and solids on particles interacting via an actual Lennard-Jones potential [34, 35]. In each case R was determined by (10.6), differentiating $f(T, \rho)$ obtained from fits to numerical experimental or computer data. Results of this effort, and the results for the simple models shown above, are summarized in Fig. 10.4. Figure 10.4 shows schematic graphs of R as a function of

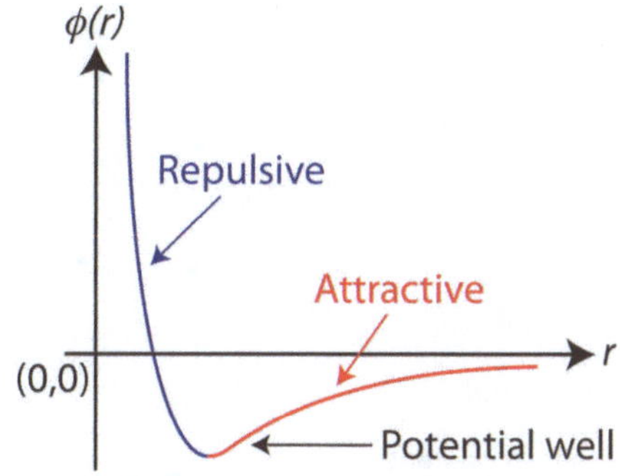

Fig. 10.3 The Lennard-Jones type potential, in which two particles separated by a distance r experience a potential $\phi(r)$, repulsive at short range and attractive at long range

T along curves with the specified v. Particle configurations corresponding to each situation are also shown alongside the schematic graphs.

Figure 10.4a shows the ideal gas, which has $R = 0$. Figure 10.4b shows the behavior of R perhaps more typical of weakly interacting systems. Here, widely spaced particles interact via the attractive tail of the Lennard-Jones potential. Typically R is negative, and $0 < |R| \ll v$. I characterize such situations as having "small" $|R|$, even in cases such as near ideal gases where v might get very large. The idea is that at size scales of one molecular volume, the system gets "grainy," and thermodynamic properties such as R based on averages have increasing difficulty being accurate.

The liquid state is shown in Fig. 10.4c. We have a compactly arranged, disorganized system of particles held together by attractive interactions, and with negative R, and $|R| \sim v$. On compressing the liquid state, there is the possibility of the system organizing into a crystalline solid state, where the predominant interaction is repulsive in character, with R changing sign to positive, and $|R| \sim v$, as shown in Fig. 10.4d. Typical is a discontinuous jump from the liquid into the solid state [35].

An essential case is the critical point regime, with R shown in Fig. 10.4e. There are two curves for R, separated by the critical temperature T_c. The curve at lower temperature represents R along the coexistence curve for both liquid and vapor phases, and the curve at higher temperature represents R along the critical isochore $v = v_c$, where v_c is the critical molar volume. R diverges to negative infinity at the critical point along both curves. On the right side of Fig. 10.4e, I sketch a near critical point particle configuration where a loose cluster has been formed by the attractive long-range tail of the Lennard-Jones type potential. The size of this cluster is given by the correlation length ξ, with $|R| \sim \xi^3$. Another critical point theme is shown in Fig. 10.4f, where we have equal values of R for the coexisting liquid and vapor phases, $R_l = R_v$, as the two phases have identical organized droplet sizes [32–34].

Figure 10.4g shows a somewhat subtle vapor phase theme [32]. Attractive interactions have formed a tight cluster of solid, which is then pressed together by impacts from surrounding particles. Repulsive interactions hold the structure up and R is positive, with $|R| \sim$ cluster size. Such clusters have been reported in computer simulations in the vapor phase of Water near the critical point [36].

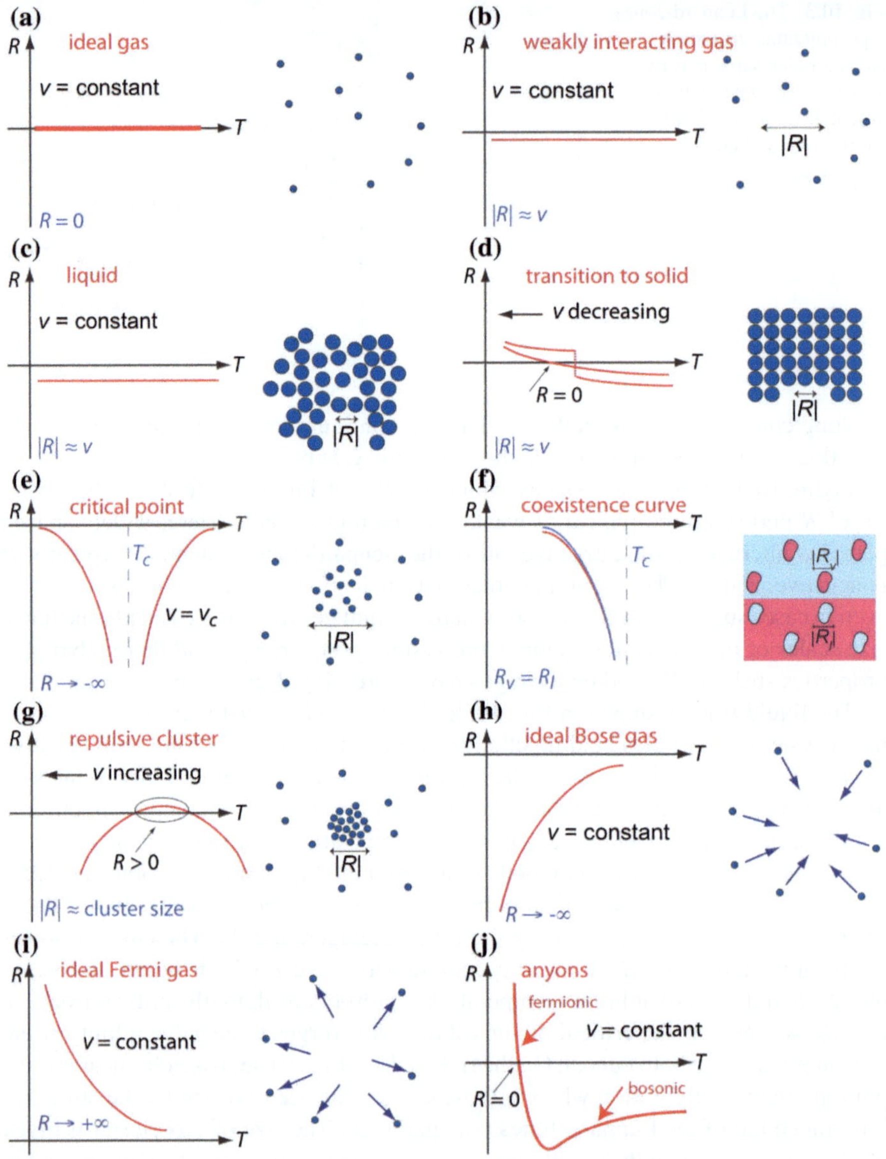

Fig. 10.4 Schematic graphs for R and corresponding particle configurations: **a** the ideal gas, with $R = 0$; **b** the weakly interacting gas, with negative R and $0 < |R| \ll v$, where v is the molecular volume; **c** the liquid, with negative R and $|R| \sim v$; **d** the transition from liquid to solid, with R changing sign to positive in the solid, typically discontinuously; **e** the *critical point*, with $R \to -\infty$ and $|R| \sim \xi^3$; **f** the coexisting gas and liquid phases, with R equal in the vapor and the liquid phases very near the critical point, $R_l = R_v$; **g** an organized compact repulsive cluster held up by the repulsive part of the interparticle interactions, with positive R and $|R| \sim$ cluster size; **h** the ideal Bose gas, with $R \to -\infty$ as $T \to 0$; **i** the ideal 2D or 3D Fermi gas, with $R \to +\infty$ as $T \to 0$; and **j** the anyon gas, with a transition from Bose to Fermi behavior as T decreases at fixed v

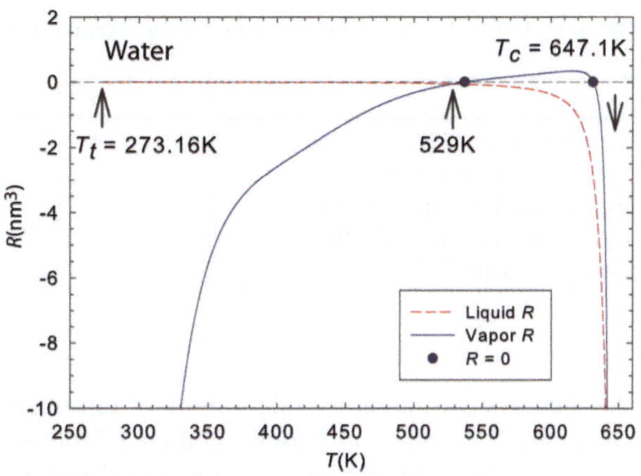

Fig. 10.5 *R* for Water in the coexisting liquid and vapor phases from the triple point to the critical point. Demonstrated are the points made in Fig. 10.4b–g

Figure 10.4h shows the ideal Bose gas, with R always negative, and with R diverging to negative infinity as $T \to 0$. Figure 10.4i shows the ideal Fermi gas, with R always positive, and with R diverging to positive infinity as $T \to 0$. The ideal Fermi gas shows the same qualitative behavior in 3D [18, 19] or 2D [29]. Figure 10.4j shows the gas of anyons with $0 < \alpha < 1$. As we cool at constant v, starting from a high T, R starts with the Bosonic negative sign, but eventually there is a transition to the Fermionic positive sign. Aside from its intrinsic interest, the natural spatial dimension, two, of the anyon gas matches the dimension of black hole event horizons.

Lest the reader think that this is all theoretical, I show R for Water in Fig. 10.5, along the coexistence curve in both the liquid and vapor phases. Figure 10.5 was worked out with data from the NIST Chemistry WebBook [37, 38]. R is in units of cubic nanometers, and is shown from the critical point $T = T_c$ to the triple point $T = T_t$, where T_t is the triple point temperature. Demonstrated are a number of the principles sketched in Fig. 10.4. The predominant sign of R is negative, as the attractive tail of the Lennard-Jones type potential dominates in the fluid.

In conclusion, for fluid and solid systems major elements of the thermodynamic curvature seem to be understood in principle, at least for cases with $n = 2$ independent thermodynamic parameters. Cases with $n > 2$, such as fluid mixtures, are largely unexplored.

10.2.3 R for Discrete Systems

The thermodynamic curvature for discrete systems has been less investigated. Spin systems with ferromagnetic interactions tend to have *aligned* adjacent spins, and to have critical point properties analogous to those for fluid systems. Indeed, R tends to be nicely negative for ferromagnetic spin systems, with $|R| \sim \xi^d$. By analogy with

Table 10.3 The thermodynamic curvature R for several simple spin models for which R has only one sign

System	n	d	R sign	$\|R\|$ divergence
Ising ferromagnet [22, 40]	2	1	−	$T \to 0$
Ising on Bethe lattice [41]	2	...	−	Critical point
Ising on random graph [13, 42]	2	2	−	Critical point
Ising antiferromagnet [22, 40]	2	1	−	$\|R\|$ small
Ideal paramagnet [22, 40]	2	...	0	$\|R\|$ small

the fluid systems, then, we might think of ferromagnetic interactions as somehow "attractive." We might also logically think of antiferromagnetic interactions, which tend to *disalign* adjacent spins, as "repulsive," and with positive R. But there is little evidence that it works out like this. Mirza and Talaei [39] worked out R for a model with *frustrated* spins, and found a regime with large positive R. Perhaps the presence or absence of frustration is the key to interpreting the sign of R for spin systems. More calculations in spin systems would appear indicated before any definitive judgement could be made.

For spin systems, we commonly have a temperature T and a magnetic field H (more than one magnetic field may be present, but this possibility is not explored here). For such models, the partition function gets worked out in terms of $\beta = 1/T$ and $h = -H/T$; namely, $Z = Z(\beta, h)$. The partition function leads to the thermodynamic potential per spin $\phi(\beta, h) = \ln \mathrm{Z}$, and the metric elements $g_{\alpha\beta} = \phi_{,\alpha\beta}$, in coordinates $\left(x^1, x^2\right) = (\beta, h)$ [9]. Here, we set $k_B = 1$. It is fashionable in magnetic models to write R as

$$R = \frac{\begin{vmatrix} \phi_{,11} & \phi_{,12} & \phi_{,22} \\ \phi_{,111} & \phi_{,112} & \phi_{,122} \\ \phi_{,112} & \phi_{,122} & \phi_{,222} \end{vmatrix}}{2\begin{vmatrix} \phi_{,11} & \phi_{,12} \\ \phi_{,12} & \phi_{,22} \end{vmatrix}^2}. \tag{10.9}$$

Many of the results for discrete models were worked out with this formula.

Table 10.3 lists R for some spin models simple enough that R has only one sign (or $R = 0$). The first three models in Table 10.3 have ferromagnetic nearest neighbor interactions. For these, R is negative, with a divergence $R \to -\infty$ either as $T \to 0$ (for $d = 1$), or at a critical point with $T > 0$ (for $d \neq 1$), as interspin coupling brings about a long-range ordering of aligned spins. This situation would appear analogous to the fluid critical point regime. The Ising antiferromagnet has a negative R with magnitude of the order of a lattice spacing, and is similar in this sense to the liquid state of the previous section.

Table 10.4 shows four discrete systems for which R has both signs. The sign of R for the one-dimensional q-state Potts model is related to the number of states

Table 10.4 The thermodynamic curvature R for discrete models for which R has both signs

System	n	d	R sign	$R = 0$	$\|R\|$ divergence
Potts model ($d > 2$) [13, 43]	2	1	$\pm$	Yes	$T \to 0$
Finite Ising ferromagnet [44]	2	1	$\pm$	Yes	$T \to 0$
Ising-Heisenberg [45]	2	1	$\pm$	Yes	$T \to 0$
Kagome Ising lattice [39]	2	2	$\pm$	No	Critical line

per spin q. For $q > 2$, and nonzero magnetic field, there are significant regimes of positive R at low temperature. An abrupt change in the sign of R is present in the one-dimensional Ising ferromagnet of *finite* N spins. R is appropriately negative for large N, but sharply increases to large positive values as N is decreased through a volume $N^* \sim |R(N \to \infty)|$. Work calculating R is in progress for the one-dimensional Ising-Heisenberg model, which shows ferromagnetism, antiferromagnetism, ferrimagnetism, and frustration. The ferrimagnetic phases show substantial regimes of positive R. R for the kagome Ising lattice has recently been worked out, mostly in zero magnetic field. This model has a critical line $T = T_c(H)$ in (T, H) space along which R diverges on both sides, negative on the high T side with dominant ferromagnetic interactions, and positive on the low T side with dominant ferrimagnetic interactions.

The physical interpretation of R for discrete systems is less conclusive than that for the fluid and solid systems. More worked examples are clearly necessary.

10.3 *R* for Black Hole Thermodynamics

This section discusses black hole thermodynamics, mostly in the context of general relativity [46]. String theory and other quantum black holes are beyond the scope of this talk.

10.3.1 Introduction

The classical (nonquantum) properties of black holes date to Schwarzchild's solution of Einstein's field equations [47]. This solution obtains on assuming a static, charge free, spherically symmetric point mass M, located at a central singularity. The solution yields a spherical event horizon, centered on the mass, and with radius $r = 2M$ (in geometrized units). This event horizon bounds an interior from which there may be no escape, even by light. Einstein's field equations may also be solved if we add charge Q (the Reissner-Nordström solution), angular momentum J (the Kerr solution), or if we have all three quantities (M, J, Q) (the Kerr-Newman solution). Hawking, Penrose, and others proved the celebrated uniqueness theorems, that if the collapsing matter is dense enough, then we inevitably approach one of these solutions.

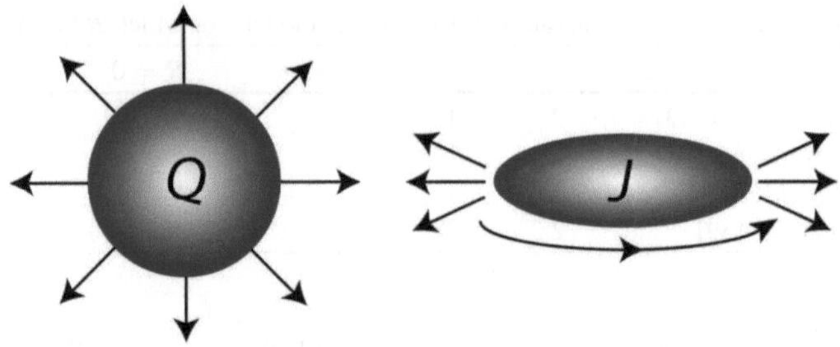

Fig. 10.6 Extremal black holes have so much charge Q that they are on the verge of exploding out under their electrostatic repulsion, or so much angular momentum J that they are on the verge of spinning apart. Both of these scenarios, or any combination of them, are forbidden by cosmic censorship

Frequently discussed is the idea of extremal black holes. Could we add enough charge to a black hole so that it explodes outward under its electrostatic repulsion, as in Fig. 10.6? Or could we add enough angular momentum so that it tears apart under its spin? Cosmic censorship forbids both these scenarios, or any combination of them. We refer to black holes as extremal if they are as close as possible to these mechanical limits. For the Kerr-Newman black hole, the condition of mechanical stability is

$$M^4 - J^2 - M^2 Q^2 > 0. \tag{10.10}$$

At the extremal limit, the black hole temperature $T = 0$, and cosmic censorship is a way of expressing the unattainability of absolute zero temperature. It is important to appreciate, however, that the third law of thermodynamics will not always hold for black holes, as extremal black holes do not always have zero area. Hence, the black hole entropy does not always go to zero at zero temperature. This marks an important difference between black hole thermodynamics and ordinary thermodynamics.

An oft quoted principle of black holes is the "no-hair theorem" [47]. After matter collapses to form a black hole, there is a brief period of settling down during which the history of the black hole's creation is forgotten. The final equilibrium state depends only on (M, J, Q). Such a reduction of complexity is essential for black hole thermodynamics. Taken to its logical extreme, however, and the no-hair conjecture denies the possibility of any form of a distribution of equilibrium black hole microstates. A distribution of microstates is central to statistical mechanics, as well as for thermodynamic fluctuations, and thermodynamic fluctuations are arguably logically necessary to any thermodynamic formalism [48]. My working assumption is then certainly to consider fluctuations about the black hole equilibrium thermodynamic state. If we magnify the regime around the black hole event horizon we might expect to see a fluctuating structure perhaps like in Fig. 10.7. And if we have fluctuations in some quantum structure, would there not be associated fundamental particles?

Fig. 10.7 Fluctuating event horizon. Are particles associated with these fluctuations?

Table 10.5 Comparison between pure fluid thermodynamics and Kerr-Newman black hole thermodynamics

	Pure fluid	Kerr-Newman
Conserved variables	(U, N, V)	(M, J, Q)
Conjugate variables	$(T, \mu, -p)$	(T, Ω, Φ)
Entropy?	Yes	Yes
Thermodynamic laws (0,1,2)?	Yes	Yes
Third law (3)?	Yes	No
Extensive?	Yes	No
Thermodynamically stable?	Yes	No
Statistical mechanics?	Yes	Unclear

10.3.2 Kerr-Newman Black Hole Thermodynamics

Consider now the Kerr-Newman black hole thermodynamics, beginning with a comparison to pure fluids; see Table 10.5. First, I identify the conserved variables; these play a special role in thermodynamic fluctuation theory [49]. For Kerr-Newman black hole thermodynamics, the conserved variables are (M, J, Q), with corresponding conjugate quantities temperature T, angular velocity Ω, and electric potential Φ [50]. Like pure fluid thermodynamics, black hole thermodynamics has well established notions of entropy, and zeroth, first, and second laws (0, 1, 2) of thermodynamics. The third law (3) of thermodynamics, however, is not obeyed in Kerr-Newman black hole thermodynamics since the entropy does not go to zero at zero temperature.

A clear difference between fluid and black hole thermodynamics is that black hole thermodynamics is not extensive [51]. Namely, we cannot scale the mass of the black hole up in such a way as to leave all of the conjugate variables fixed. However, this point poses few difficulties for the black hole thermodynamic fluctuation theory employed here.

Significant is the frequent absence of black hole thermodynamic stability. One manifestation of this are negative heat capacities, which are a fixture of gravitational thermodynamic problems. A black hole lacking thermodynamic stability cannot reach thermodynamic equilibrium with its environment, a significant deficit for the physical interpretation of any quantity, such as R, coming from thermodynamic fluctuation theory. The Kerr-Newman black hole thermodynamics is not stable for any set of values of (M, J, Q) [52]. However, stable black hole cases do exist. These result on either restricting the number of fluctuating variables, adding an AdS background, or altering the assumptions about the black hole's topology. Stable thermodynamic cases get most of the attention in this talk.

There is no consensus on the question of the correct microstructure supporting black hole thermodynamics. String theorists have attempted to calculate such microstructures, particularly for near extremal black holes, starting with Strominger and Vafa [53]; see Bellucci and Tiwari [54, 55] and Wei et al. [56] for recent references.[2] In string theory calculations, the microscopic model is always explicit. By contrast, for general relativity solutions there is no evident microscopic foundation, and I direct my efforts to these in this talk.

10.3.3 Laws of Black Hole Thermodynamics

The Bekenstein-Hawking area law [57, 58] sets the black hole entropy S_{BH} proportional to the area A of the event horizon:

$$\frac{S_{BH}}{k_B} = \frac{1}{4}\left(\frac{A}{L_p^2}\right), \tag{10.11}$$

where

$$L_p = \sqrt{\frac{\hbar G}{c^3}} \tag{10.12}$$

is the Planck length. Here, $\hbar$ is Planck's constant divided by 2π, G is the universal gravitation constant, and c is the speed of light. The area A may be calculated in terms of the conserved variables, given a black hole solution from general relativity. Such a calculation yields the full black hole thermodynamics. For example, here is the formula for A for the Kerr-Newman black hole [59]:

---

[2] If a paper starts with a spacetime metric, and calculates the thermodynamic from the area of the event horizon, it is a general relativistic solution. If the paper starts with a Lagrangian, and a quantum action then it is beyond the scope of my talk.

$$A = 4\pi\left(2M^2 - Q^2 + 2\sqrt{M^4 - J^2 - M^2Q^2}\right). \tag{10.13}$$

The black hole entropy may be added to the ordinary entropy S_o to get the total entropy of the universe:

$$S_{universe} = S_{BH} + S_o. \tag{10.14}$$

We generalize the second law of thermodynamics in the obvious way, that in any process starting from some initial state and going to some final state:

$$\Delta S_{universe} \geq 0. \tag{10.15}$$

Drop now the subscript "BH" ($S_{BH} \to S$), and turn to the first law of black hole thermodynamics. Writing $M = M(S, J, Q)$ leads to

$$dM = TdS + \Omega dJ + \Phi dQ, \tag{10.16}$$

where we define the temperature

$$T = \left(\frac{\partial M}{\partial S}\right)_{J,Q}, \tag{10.17}$$

the angular velocity

$$\Omega = \left(\frac{\partial M}{\partial J}\right)_{S,Q}, \tag{10.18}$$

and the electric potential

$$\Phi = \left(\frac{\partial M}{\partial Q}\right)_{S,J}. \tag{10.19}$$

The first law of black hole thermodynamics (10.16) expresses the change in black hole energy dM to mechanical work terms, ΩdJ and ΦdQ, and a heat term TdS.

Also essential is the 0'th law of black hole thermodynamics, which equates T to the effective surface tension of the event horizon. Calculations show this quantity to be constant over the event horizon, resulting in a unique value for the black hole temperature T. Ω and Φ are similarly constant over the event horizon [59].

Let me make one more observation about the correspondences in Table 10.5 before discussing black hole thermodynamic curvature. While there are natural correspondences between fluid and black hole energy, temperature, entropy, and (I argue) thermodynamic curvature R, there is always uncertainty in making correspondences among other thermodynamic variables. For example, if we have some fluid critical point property, say a divergence in the heat capacity at constant volume, one could not naturally say how this property translates to black hole thermodynamics. This point will be discussed further below in connection with black hole phase transitions.

10.3.4 Black Hole Thermodynamic Curvature R

Black hole thermodynamics leads naturally to corresponding rules for black hole thermodynamic fluctuations, described by an information metric [49, 60–62] of the type in (10.2). In conserved independent coordinates $(x^1, x^2, x^3, \ldots) = (X^1, X^2, X^3, \ldots) = (M, J, Q, \cdots)$, the thermodynamic metric for black hole fluctuations is (in appropriate units)

$$g_{\alpha\beta} = -\frac{\partial^2 S}{\partial X^\alpha \partial X^\beta}, \tag{10.20}$$

where S is the black hole entropy. The form of the thermodynamic metric in (10.20) requires us to know $S = S(X^1, X^2, X^3, \ldots)$. Frequently, however, we know instead $M = M(Y^1, Y^2, Y^3, \ldots)$, where $(Y^1, Y^2, Y^3, \ldots) = (S, J, Q, \ldots)$. In this event, simplification results on writing the thermodynamic metric in the Weinhold energy form, with an additional prefactor $1/T$ [9, 63]:

$$g_{\alpha\beta} = \frac{1}{T}\frac{\partial^2 M}{\partial Y^\alpha \partial Y^\beta}. \tag{10.21}$$

No matter how the thermodynamic metric is written, however, we will get the same value for R for a given thermodynamic state, since R is a thermodynamic invariant.[3]

Thermodynamic fluctuation metrics must be positive definite for thermodynamic stability. With two independent fluctuating variables, this requires *three* conditions:

$$g_{11} > 0, \tag{10.22}$$

$$g_{22} > 0, \tag{10.23}$$

and

$$g_{11}g_{22} - g_{12}^2 > 0. \tag{10.24}$$

[3] The line element (10.2) transforms as a scalar, since probability is a scalar quantity. Hence, the metric elements $g_{\alpha\beta}$ transform as the elements of a second-rank tensor, which the relation between (10.20) and (10.21) satisfies. The resulting thermodynamic curvature R transforms as a scalar. These transformation properties hold under all coordinate transformations, including those resulting from Legendre transformations. This is the case in both ordinary and black hole thermodynamics, despite erroneous claims to the contrary [64].

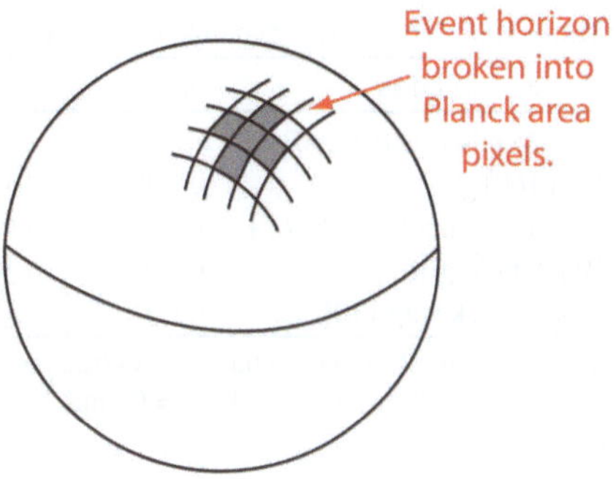

Fig. 10.8 The event horizon broken up into *Planck area pixels*. The *dark pixels* are portrayed as correlated. I propose that $|R|$ measures the average number of *correlated pixels*

Pioneering papers introducing thermodynamic curvature R into the black hole arena are [60, 65, 66]. In particular, Åman and Pidokrajt [60] first evaluated R for several solutions from general relativity. For nondiagonal thermodynamic metrics with $n = 2$, such as those in (10.20) and (10.21):

$$R = -\frac{1}{\sqrt{g}}\left[\frac{\partial}{\partial x^1}\left(\frac{g_{12}}{g_{11}\sqrt{g}}\frac{\partial g_{11}}{\partial x^2} - \frac{1}{\sqrt{g}}\frac{\partial g_{22}}{\partial x^1}\right) + \frac{\partial}{\partial x^2}\left(\frac{2}{\sqrt{g}}\frac{\partial g_{12}}{\partial x^1} - \frac{1}{\sqrt{g}}\frac{\partial g_{11}}{\partial x^2} - \frac{g_{12}}{g_{11}\sqrt{g}}\frac{\partial g_{11}}{\partial x^1}\right)\right], \tag{10.25}$$

where

$$g = g_{11}g_{22} - g_{12}^2. \tag{10.26}$$

But what is the physical interpretation of the black hole thermodynamic curvature R? In my view there is only one rational way to approach this question, and that is to follow the ideas developed in ordinary thermodynamics. It has been argued [29] that the natural units of the thermodynamic curvature are the square of the Planck length L_p^2. Figure 10.8 shows the event horizon broken up into Planck area pixels. Perhaps $|R|$ measures the correlation between fluctuating Planck length pixels? Since I bring no microscopic theory of black holes into play in this talk, I have no direct evidence for such a conjecture. But, by analogy with the case in ordinary thermodynamics, how else could we interpret the black hole thermodynamic curvature?

The picture in Fig. 10.8 assumes that all the black hole statistical activity takes place on the two-dimensional event horizon. This assumption is an element of the black hole membrane paradigm [67]. The motivation of the membrane paradigm is that if you cannot in principle know what is going on inside the black hole, then assume that all of the interesting stuff must be happening on the event horizon. One element of this idea is that if we are going to associate black hole statistics with some familiar model in statistical mechanics, then perhaps we should look most closely at two-dimensional models.

Table 10.6 The thermodynamic curvature R for black hole solutions from general relativity

Name of solution	Dimension	Variables	Stable	R sign	$R = 0$	$\|R\|$ divergence
Reissner-Nordström[60]	$3+1$	(M, Q)	None	0	–	None
Kerr [60]	$3+1$	(M, J)	None	$+$	No	Extremal
Kerr-Newman [29, 60, 69]	$3+1$	(M, J, Q)	None	$+$	No	Extremal
Black hole [70]	$4+1$	(M, J)	None	$+$	No	Extremal
Small black ring [70]	$4+1$	(M, J)	None	$\pm$	Yes	Ext + crit line

The solutions shown here have no regimes of thermodynamic stability. "Extremal" denotes a curve in the space of variables with $T = 0$, and "crit" denotes a critical line with $T \neq 0$, along which $|R|$ diverges

10.3.5 Solutions from General Relativity

The thermodynamic curvature R for black holes has been worked out for a number of systems, and I make no attempt to be complete in my reporting below. Rather, I present some thoughts about how results from various general relativity solutions might be compared with one another, and to solutions from ordinary thermodynamics. In Tables 10.6 and 10.7, I consider only thermodynamic states with $S > 0$, $M > 0$, and $T > 0$. Within this physical range of variables, the solutions divide into two categories, those for which there are no regimes satisfying thermodynamic stability (10.22)–(10.24), and those for which there are such regimes. In either category, R can be readily worked out from (10.25); it is real in all the cases I calculated. However, the physical interpretation I have presented for R for ordinary thermodynamics is based on fluctuation theory, and this assumes thermodynamic stability. I key on the stable cases below.

Table 10.6 shows results for R for several general relativity solutions having no stable thermodynamic states. Tabulated are the dimension (spatial + time), the fluctuating conserved variables, whether or not there are regimes of thermodynamic stability (no cases in Table 10.6), the sign of R (or an indication "0" if R is identically zero), whether or not there are places where the sign of R changes through zero, and whether or not there are divergences $|R| \to \infty$. Of the older solutions: Reissner-Nordström, Kerr, and Kerr-Newman, none are thermodynamically stable for any thermodynamic state. Also, not thermodynamically stable are the two solutions listed with a higher dimension $= 4+1$. Some of the older solutions have been worked out in higher dimensions, but with no reports of thermodynamically stable cases [68].

Table 10.7 shows results for R for several general relativity solutions with "some" or "all" states thermodynamically stable. Thermodynamic stability results on either adding an AdS background, restricting the number of fluctuating variables, or altering the assumptions about the black hole's topology.

Black holes in an AdS background have significant regimes of thermodynamic stability. The simplest member of this category is the BTZ black hole, which is thermodynamically stable for all of its states, and has identically zero R. This behavior

Table 10.7 The thermodynamic curvature R for black hole solutions from general relativity

Name of solution	Dimension	Variables	Stable	R sign	$R = 0$	$\|R\|$ divergence
BTZ [60, 66]	$2+1$	(M, J)	All	0	0	None
RN-AdS [60, 71–73]	$3+1$	(M, Q)	Some	$\pm$	Yes	Ext + crit line
K-AdS [72, 74, 75]	$3+1$	(M, J)	Some	$-$	No	Critical line
Restricted KN [29, 49]	$3+1$	(J, Q)	All	$+$	No	Extremal
Large black ring [70]	$4+1$	(M, J)	All	$-$	No	Ext + crit line

These solutions all have at least some thermodynamically stable regimes. The characterization of R is based only on states in the stable regime. "Extremal" denotes a curve in the space of variables with $T = 0$, and "crit" denote a critical line with $T \neq 0$ along which $|R|$ diverges

is shown schematically in Fig. 10.9a, and it resembles the behavior for the ideal gas in Fig. 10.4a.

In the thermodynamically stable regime, Reissner-Nordström-AdS black holes have an extremal curve $T = 0$, as well as a line of critical points where $|R|$ diverges to infinity. This critical line obtains for $Q < Q_c$, where the critical value Q_c depends on the cosmological constant. For a fixed $Q > Q_c$, as we reduce T from a large value, R diverges to positive infinity at the extremal curve. However, for fixed $Q < Q_c$, as we reduce T from a large value, R diverges to *negative* infinity along the critical line. As T is decreased further, we enter a thermodynamically unstable regime followed by a stable regime where R increases.

The general black hole thermodynamic behavior for RN-AdS has been associated with a phase transition analogous to a van der Waals model by Chamblin et al. [76, 77]. A number of researchers have calculated R for this case [60, 71–73]. The behavior of R for RN-AdS is shown schematically in Fig. 10.9b. For $Q > Q_c$ the behavior of R resembles that in the Fermi gas, shown in Fig. 10.4i. For $Q < Q_c$, R resembles the critical point behavior in Fig. 10.4e. This correspondence is certainly consistent with the association with the van der Waals model. I add that the $Q < Q_c$ curve in Fig. 10.9b has a bump resembling the one in Fig. 10.4g.

Kerr-AdS black holes have no extremal curve in the thermodynamically stable regime. However, a critical line depending on the cosmological constant bounds the thermodynamically stable regime at low T. Along this critical line, R diverges to negative infinity. For the Kerr-AdS black hole thermodynamics, R is always negative. Figure 10.9c sketches R as T is decreased from a large value at constant J. The sketch resembles the critical point behavior in Fig. 10.4e. However, once we cross the critical line, there are no more thermodynamically stable regimes. Banerjee et al. [78, 79] have discussed phase transitions in AdS black holes using the Ehrenfest relations, with special attention to the orders of the phase transitions.

Special thermodynamically stable cases result from the Kerr-Newman solution when we fix one of the three parameters (M, J, Q), and allow the other two to fluctuate. This restriction is not just a mathematical convenience; it has a physical basis. For example, consider adding an electron to the black hole, and calculate the contribution of each of changing (M, J, Q) to the change in the total entropy. We expect one of (M, J, Q) to contribute least to the changing entropy, and if it

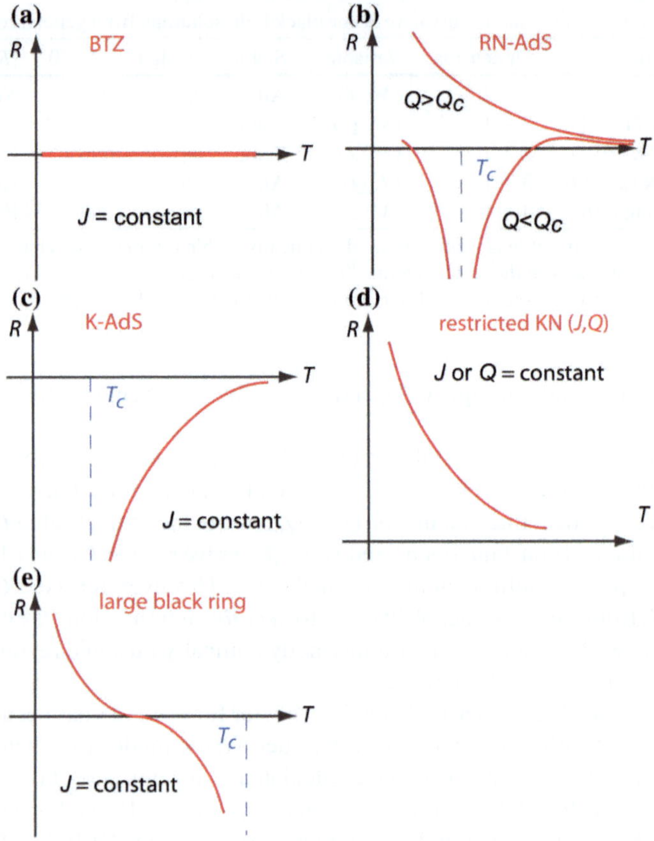

Fig. 10.9 Schematic graphs for R for thermodynamically stable general relativistic black hole solutions: **a** the BTZ solution, with $R = 0$; **b** the RN-AdS solution, with R diverging to positive infinity at the *extremal curve* for $Q > Q_c$, and with R diverging to negative infinity, at temperature $T_c > 0$, along the *critical line* for $Q < Q_c$; **c** the K-AdS solution, with R diverging to negative infinity at the *critical line*, **d** the restricted KN (J, Q) solution, with R diverging to positive infinity at the extremal limit; **e** the large *black* ring solution, with R diverging to positive infinity at the *extremal curve*, and with R diverging to negative infinity along the critical line

contributes much less, we could just ignore the change in that parameter, and let the other two parameters fluctuate. For a black hole with mass on the order of the Planck mass (a quantum black hole), contributions to the changing entropy from the electron mass are hugely less important to the change in total entropy than the changes resulting from its (J, Q). This restricted KN (J, Q) solution has some highly desirable properties, as Table 10.7 shows. This solution is sketched in Fig. 10.9d. In addition, there are some detailed analogies to the 2D ideal Fermi gas in the extremal limit, which may be interesting.

The large black ring solution also has significant regimes of stability, bounded by an extremal curve and a critical line. R diverges to positive infinity at the extremal curve, and to negative infinity along the critical line.

A few patterns present themselves for the thermodynamically stable general relativity solutions considered in this section. In all cases, the divergence of R at the extremal curve is to positive infinity, resembling in this sense the divergence for the ideal Fermi gasses from ordinary thermodynamics. Where there are critical lines (with $|R|$ diverging with $T \neq 0$), the divergence of R is to negative infinity, resembling the critical point divergences in ordinary thermodynamics. But I have considered too few cases here to assert with any confidence that these patterns are general. Further study is obviously necessary.

10.3.6 Discussion of "Inconsistencies"

Much debated in black hole thermodynamics has been the possibility raised by Davies [50] that the curve of diverging heat capacity $C_{J,Q} = T(\partial S/\partial T)_{J,Q}$ in the Kerr-Newman black hole solution corresponds to a phase transition. Diverging heat capacities are a feature of second-order phase transitions in ordinary fluid and spin systems, so Davies' association would appear logical.

Closer examination, however, raises some questions about Davies' correspondence. First, an ordinary thermodynamic system generally has at its foundation some *known* microscopic model. Such a model offers direct insight not only into the character of the thermodynamic variables, but into the microscopic signatures of any thermodynamic anomaly. In the absence of a known microscopic model we have difficulty answering basic questions. If some heat capacity diverges, how could we be sure that we have not just made an inappropriate choice of thermodynamic variables, which reveals infinities with no really fundamental significance? What do we make of curves in thermodynamic state space where one heat capacity diverges, but the other heat capacities stay regular? What if various heat capacities diverge along different curves, as happens in the Kerr-Newman black hole [29, 52], as in Fig. 10.10. Which curve corresponds to a true phase transition? One of them? All of them? Perhaps it is safer to associate curves of diverging R with black hole phase transitions. R has a unique status in identifying microscopic order from thermodynamics. and ordering at the microscopic level is at the foundation of phase transitions.

Black hole solutions with R identically zero, of which Tables 10.6 and 10.7 each have one, have also given rise to debate; see [80] for a review. If R measures in some sense the range of interactions, then one might expect $|R|$ to always be large for black hole thermodynamics, reflecting the concentrated gravitational forces present in these objects. But such reasoning need not obtain. In a classical black hole, the gravitating particles have collapsed to a central singularity, shrinking the interactions between them to zero volume. The statistics underlying the thermodynamics might reside on the event horizon, where unknown constituents might interact with each other by forces perhaps not gravitational. In this scenario, gravity might merely be

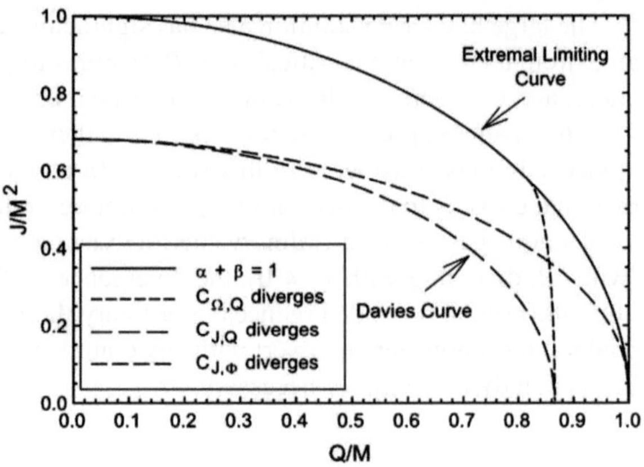

Fig. 10.10 Characteristic curves for the Kerr-Newman black hole. The *curve* along which $C_{J,Q}$ diverges is the *Davies curve*. R diverges at the *extremal limit* and along *curves* corresponding to a change of thermodynamic stability, which have diverging $C_{J,\Phi}$ and $C_{\Omega,Q}$. R does not diverge along the *Davies curve*. Here $\alpha = J^2/M^4$ and $\beta = Q^2/M^2$

a nonstatistical force holding the assembly together, and a result $R = 0$, where the unknown constituents move independently of each other, would make perfect sense.

10.4 Conclusions

What are black holes made out of? This question has not been answered here. However, one way to address this question is by following an agenda of matching the statistical mechanics of known microscopic models to black hole thermodynamic solutions from general relativity, or other theories of gravity. I hope that I have convinced the audience that the thermodynamic curvature R has a contribution to make to this game.

I have given a broad survey of thermodynamic curvature R, one spanning results in fluids and solids, spin systems, and black hole thermodynamics. R results from the unique thermodynamic information metric giving thermodynamic fluctuations. R has a unique status in thermodynamics as being a geometric invariant, the same for any given thermodynamic state no matter what coordinates we calculate in. In ordinary thermodynamics, the sign of R indicates the character of microscopic interactions, and $|R|$ indicates the average size of organized fluctuations. Although I have given no direct evidence that this interpretation holds for black hole thermodynamics, if we believe in the broad generality of thermodynamic principles, this interpretation of R should transcend specific scenarios.

Most incomplete in this talk has been the presentation of spin systems. Frustration in spin systems may be necessary as a way to deal with the frequent failure of the third law of thermodynamics for black holes. Missing entirely from this talk have been results on string theory models, which were simply beyond reach of the speaker. These may ultimately yield the best picture of what is going on in the black hole.

Acknowledgments I thank George Skestos for travel support, and I thank the conference organizer Stefano Bellucci of INFN for giving me the opportunity to present my work.

References

1. F. Weinhold, Thermodynamics and geometry. Phys. Today **29**(3), 23 (1976)
2. G. Ruppeiner, Thermodynamics: a Riemannian geometric model. Phys. Rev. A **20**, 1608 (1979)
3. B. Andresen, P. Salamon, R.S. Berry, Thermodynamics in finite time. Phys. Today **37**(9), 62 (1984)
4. G. Ruppeiner, Thermodynamic curvature: pure fluids to black holes. J. Phys.: Conf. Series **410**, 012138 (2013)
5. L. Landau, E. Lifshitz, *Statistical Physics* (Pergamon, New York, 1977)
6. R.K. Pathria, *Statistical Mechanics* (Butterworth-Heinemann, Oxford, 1996)
7. A. Einstein, On the general molecular theory of heat. Ann. Phys. (Leipzig) **14**, 354 (1904)
8. A. Einstein, The theory of the opalescence of homogeneous fluids and liquid mixtures near the critical state. Ann. Phys. (Leipzig) **33**, 1275 (1910)
9. G. Ruppeiner, Riemannian geometry in thermodynamic fluctuation theory. Rev. Mod. Phys. **67**, 605 (1995)
10. G. Ruppeiner, Riemannian geometry in thermodynamic fluctuation theory. Rev. Mod. Phys. **68**, 313(E) (1996)
11. D. Laugwitz, *Differential and Riemannian Geometry* (Academic, New York, 1965)
12. G. Ruppeiner, Thermodynamic curvature measures interactions. Am. J. Phys. **78**, 1170 (2010)
13. D.A. Johnston, W. Janke, R. Kenna, Information geometry, one, two, three (and four). Acta Phys. Pol. B **34**, 4923 (2003)
14. G. Ruppeiner, New thermodynamic fluctuation theory using path integrals. Phys. Rev. A **27**, 1116 (1983)
15. G. Ruppeiner, Thermodynamic critical fluctuation theory? Phys. Rev. Lett. **50**, 287 (1983)
16. L. Diósi, B. Lukács, Covariant evolution equation for the thermodynamic fluctuations. Phys. Rev. A **31**, 3415 (1985)
17. S. Weinberg, *Gravitation and Cosmology* (Wiley, New York, 1972)
18. H. Janyszek, R. Mrugała, Riemannian geometry and stability of ideal quantum gases. J. Phys. A: Math. Gen. **23**, 467 (1990)
19. H. Oshima, T. Obata, H. Hara, Riemann scalar curvature of ideal quantum gases obeying Gentile's statistics. J. Phys. A **32**, 6373 (1999)
20. B. Mirza, H. Mohammadzadeh, Thermodynamic geometry of deformed bosons and fermions. J. Phys. A: Math. Gen. **44**, 475003 (2011)
21. D. Brody, N. Rivier, Geometrical aspects of statistical mechanics. Phys. Rev. E **51**, 1006 (1995)
22. H. Janyszek, R. Mrugała, Riemannian geometry and the thermodynamics of model magnetic systems. Phys. Rev. A **39**, 6515 (1989)
23. D.C. Brody, D.W. Hook, Information geometry in vapour–liquid equilibrium. J. Phys. A: Math. Theor. **42**, 023001 (2009)
24. W. Janke, D.A. Johnston, R. Kenna, Information geometry of the spherical model. Phys. Rev. E **67**, 046106 (2003)
25. G. Ruppeiner, J. Chance, Thermodynamic curvature of a one-dimensional fluid. J. Chem. Phys. **92**, 3700 (1990)
26. J. Nulton, P. Salamon, Geometry of the ideal gas. Phys. Rev. A **31**, 2520 (1985)
27. G. Ruppeiner, C. Davis, Thermodynamic curvature of the multicomponent ideal gas. Phys. Rev. A **41**, 2200 (1990)
28. K. Kaviani, A. Dalafi-Rezaie, Pauli paramagnetic gas in the framework of Riemannian geometry. Phys. Rev. E **60**, 3520 (1999)

29. G. Ruppeiner, Thermodynamic curvature and phase transitions in Kerr-Newman black holes. Phys. Rev. D **78**, 024016 (2008)
30. M.R. Ubriaco, Stability and anyonic behavior of systems with M-statistics. Physica A **392**, 4868 (2013). arXiv:1302.4719 [hep-th]
31. B. Mirza, H. Mohammadzadeh, Nonperturbative thermodynamic geometry of anyon gas. Phys. Rev. E **80**, 011132 (2009)
32. G. Ruppeiner, Thermodynamic curvature from the critical point to the triple point. Phys. Rev. E **86**, 021130 (2012)
33. G. Ruppeiner, A. Sahay, T. Sarkar, G. Sengupta, Thermodynamic geometry, phase transitions, and the Widom line. Phys. Rev. E **86**, 052103 (2012). arXiv:1106.2270 [hep-th]
34. H.-O. May, P. Mausbach, Riemannian geometry study of vapor–liquid phase equilibria and supercritical behavior of the Lennard–Jones fluid. Phys. Rev. E **85**, 031201 (2012)
35. H.-O. May, P. Mausbach, G. Ruppeiner, Thermodynamic curvature for attractive and repulsive intermolecular forces. Phys. Rev. E. **88**, 032123 (2013)
36. E. Johansson, K. Bolton, P. Ahlström, Simulations of vapor water clusters at vapor–liquid equilibrium. J. Chem. Phys. **123**, 024504 (2005)
37. NIST Chemistry WebBook, available at the Web Site http://webbook.nist.gov/chemistry/
38. W. Wagner, A. Pruss, The IAPWS formulation 1995 for the thermodynamic properties of ordinary water substance for general and scientific use. J. Phys. Chem. Ref. Data **31**, 387 (2002)
39. B. Mirza, Z. Talaei, Thermodynamic geometry of a kagome Ising model in a magnetic field. Phys. Lett. A **377**, 513 (2013)
40. G. Ruppeiner, Application of Riemannian geometry to the thermodynamics of a simple fluctuating magnetic system. Phys. Rev. A **24**, 488 (1981)
41. B.P. Dolan, Geometry and thermodynamic fluctuations of the Ising model on a Bethe lattice. Proc. R. Soc. London A **454**, 2655 (1998)
42. W. Janke, D.A. Johnston, R.P.K.C. Malmini, Information geometry of the Ising model on planar random graphs. Phys. Rev. E **66**, 056119 (2002)
43. B.P. Dolan, D.A. Johnston, R. Kenna, The information geometry of the one-dimensional Potts model. J. Phys. A: Math. Gen. **35**, 9025 (2002)
44. D.C. Brody, A. Ritz, Information geometry of finite Ising models. J. Geom. Phys. **47**, 207 (2003)
45. To be submitted.
46. J.D. Bekenstein, Black-hole thermodynamics. Phys. Today **33**(1), 24 (1980)
47. C.W. Misner, K.S. Thorne, J.A. Wheeler, *Gravitation* (Freeman, San Francisco, 1973)
48. G.N. Lewis, Generalized thermodynamics including the theory of fluctuations. J. Am. Chem. Soc. **53**, 2578 (1931)
49. G. Ruppeiner, Stability and fluctuations in black hole thermodynamics. Phys. Rev. D **75**, 024037 (2007)
50. P. Davies, The thermodynamic theory of black holes. Proc. R. Soc. London A **353**, 499 (1977)
51. P.T. Landsberg, D. Tranah, Thermodynamics of non-extensive entropies. Collect. Phenom. **3**, 73 (1980)
52. D. Tranah, P.T. Landsberg, Thermodynamics of non-extensive entropies. Collect. Phenom. **3**, 81 (1980)
53. A. Strominger, C. Vafa, Microscopic origin of the Bekenstein–Hawking entropy. Phys. Lett. B **379**, 99 (1996)
54. S. Bellucci, B.N. Tiwari, State-space correlations and stabilities. Phys. Rev. D **82**, 084008 (2010)
55. S. Bellucci, B.N. Tiwari, Thermodynamic geometry and topological Einstein-Yang-Mills black holes. Entropy **14**, 1045 (2012)
56. S.-W. Wei, Y.-X. Liu, Critical phenomena and thermodynamic geometry of charged Gauss–Bonnet AdS black holes. Phys. Rev. D **87**, 044014 (2013)
57. J. Bekenstein, Generalized second law of thermodynamics in black-hole physics. Phys. Rev. D **9**, 3292 (1974)

58. S. Hawking, Black holes and thermodynamics. Phys. Rev. D **13**, 191 (1976)
59. L. Smarr, Mass formula for Kerr black holes. Phys. Rev. Lett. **30**, 71 (1973)
60. J. Åman, I. Bengtsson, N. Pidokrajt, Geometry of black hole thermodynamics. Gen. Relativ. Gravit. **35**, 1733 (2003)
61. D. Pavón, J.M. Rubí, Thermodynamical fluctuations of a massive Schwarzschild black hole. Phys. Lett. **99A**, 214 (1983)
62. D. Pavón, J.M. Rubí, Kerr black hole thermodynamical fluctuations. Gen. Relativ. Gravit. **17**, 387 (1985)
63. P. Salamon, J.D. Nulton, E. Ihrig, On the relation between entropy and energy versions of thermodynamic length. J. Chem. Phys. **80**, 436 (1984)
64. H. Quevedo, Geometrothermodynamics of black holes. Gen. Relativ. Gravit. **40**, 971 (2008)
65. S. Ferrara, G. Gibbons, R. Kallosh, Black holes and critical points in moduli space. Nuc. Phys. B **500**, 75 (1997)
66. R.G. Cai, J.H. Cho, Thermodynamic curvature of the BTZ black hole. Phys. Rev. D **60**, 067502 (1999)
67. K.S. Thorne, R.H. Price, D.A. Macdonald, *Black Holes: The Membrane Paradigm* (Yale University Press, New Haven, 1986)
68. J.E. Åman, N. Pidokrajt, Geometry of higher-dimensional black hole thermodynamics. Phys. Rev. D **73**, 024017 (2006)
69. B. Mirza, M. Zamaninasab, Ruppeiner geometry of RN black holes: flat or curved? JHEP **1006**, 059 (2007)
70. G. Arcioni, E. Lozano-Tellechea, Stability and critical phenomena of black holes and black rings. Phys. Rev. D **72**, 104021 (2005)
71. J. Shen, R.-G. Cai, B. Wang, R.-K. Su, Thermodynamic geometry and critical behavior of black holes. Int. J. Mod. Phys. A **22**, 11 (2007)
72. A. Sahay, T. Sarkar, G. Sengupta, On the thermodynamic geometry and critical phenomena of AdS black holes. JHEP **1007**, 82 (2010)
73. C. Niu, Y. Tian, X.-N. Wu, Critical phenomena and thermodynamic geometry of Reissner-Nordström-anti-de Sitter black holes. Phys. Rev. D **85**, 024017 (2012)
74. R. Banerjee, S.K. Modak, S. Samanta, Second order phase transition and thermodynamic geometry in Kerr-AdS black holes. Phys. Rev. D **84**, 064024 (2011)
75. Y.-D. Tsai, X.-N. Wu, Y. Yang, Phase structure of the Kerr-AdS black hole. Phys. Rev. D **85**, 044005 (2012)
76. A. Chamblin, R. Emparan, C.V. Johnson, R.C. Myers, Charged AdS black holes and catastrophic holography. Phys. Rev. D **60**, 064018 (1999)
77. A. Chamblin, R. Emparan, C.V. Johnson, R.C. Myers, Holography, thermodynamics, and fluctuations of charged AdS black holes. Phys. Rev. D **60**, 104026 (1999)
78. R. Banerjee, S. Ghosh, D. Roychowdhury: New type of phase transition in Reissner-Nordström-AdS black hole and its thermodynamic geometry. Phys. Lett. B **696**, 156 (2011)
79. R. Banerjee, S. Modak, D. Roychowdhury, A unified picture of phase transition: from liquid–vapour systems to AdS black holes. JHEP **1210**, 125 (2012)
80. A. Medved, A commentary on Ruppeiner metrics for black holes. Mod. Phys. Lett. A **23**, 2149 (2008)

Chapter 11
Coset Approach to the Partial Breaking of Global Supersymmetry

S. Bellucci, S. Krivonos and A. Sutulin

11.1 Introduction

The characteristic feature of the theories with a partial breaking of global supersymmetries is the appearance of the Goldstone fermionic fields, associated with the broken supertranslations, as the components of Goldstone supermultiplets of unbroken supersymmetry. The natural description of such theories is achieved within the coset approach [1–6]. The usefulness of the coset approach in the applications to the theories with partial breaking of the supersymmetry have been demonstrated by many authors [4–24]. The presence of the unbroken supersymmetry makes quite reasonable the idea to choose the corresponding superfields as the basic ones, and many interesting superspace actions describing different patterns of supersymmetry breaking have been constructed in such a way [7, 10–12]. However, the standard methods of coset approach fail to construct the superfield action, because the superspace Lagrangian is weakly invariant with respect to supersymmetry—it is shifted by the full space–time or spinor derivatives under broken/unbroken supersymmetry transformations. Another rather technical difficulty is the explicit construction of the component action from the superspace one, which is written in terms of the superfields subjected to highly nonlinear constraints. Finally, in some cases the covariantization of the irreducibility constraints with respect to the broken supersymmetry is not evident, if at all possible. For example, it has been demonstrated in [7] that such constraints for the vector supermultiplet can be covariantized only together with the equations of motion.

It turned out that one can gain more information about component off-shell actions if an attention is shifted to the broken supersymmetry. It was demonstrated in [21–23]

S. Bellucci (✉) · A. Sutulin
INFN-Laboratori Nazionali di Frascati, Via E. Fermi 40, 00044 Frascati, Italy
e-mail: Stefano.Bellucci@lnf.infn.it

S. Krivonos · A. Sutulin
Bogoliubov Laboratory of Theoretical Physics, JINR, 141980 Dubna, Russia

S. Bellucci (ed.), *Breaking of Supersymmetry and Ultraviolet Divergences in Extended Supergravity*, Springer Proceedings in Physics 153,
DOI: 10.1007/978-3-319-03774-5_11, © Springer International Publishing Switzerland 2014

that with a suitable choice of the parametrization of the coset, the θ-coordinates of unbroken supersymmetry and the physical bosonic components do not transform under broken supersymmetry. Moreover, the physical fermions transform as the Goldstino of the Volkov-Akulov model [5] with respect to broken supersymmetry. Therefore, the physical fermions can enter the component on-shell action only (1) through the determinant of the vielbein (to compensate the variation of the volume $d^d x$), (2) through the covariant space–time derivatives, or (3) through the Wess-Zumino term, if it exists. The first two ingredients can be easily constructed within the coset method, while the Wess-Zumino term can be also constructed from Cartan forms following the recipe of [25]. As a result, we will have the Ansatz for the action with several constant parameters, which have to be fixed by the invariance with respect to unbroken supersymmetry. The pleasant feature of such an approach is that the fermions are "hidden" inside the covariant derivatives and determinant of the vielbein, making the whole action short, with the explicit geometric meaning of each term. In the present paper we review this procedure in applications to the actions of the superparticle in $D = 3$ realizing $N = 4 \cdot 2^k \to N = 2 \cdot 2^k$ pattern of supersymmetry breaking, the action of superparticle in $D = 5$ with the $N = 16$ supersymmetry broken down to $N = 8$ one, the on-shell component actions of $N = 1$, $D = 5$ supermembrane and its dual cousins and the component action of $N = 1$ supermembrane in $D = 4$. All these explicit actions confirm our conjecture about the structure of the component action. Finally, we briefly discuss some related questions and further possible applications of our method.

11.2 Basics of the Method

In this section we present main features of the coset approach, applying to supersymmetric models in which one half of the global supersymmetries are spontaneously broken. Before going to supersymmetric systems, we will consider how this method works in the purely bosonic case.

Let us split the generators of the target space of the D-dimensional Poincaré group, which is supposed to be spontaneously broken on the world volume down to the d-dimensional Poincaré subgroup, into the generators of unbroken $\{P, M, N\}$ and spontaneously broken $\{Z, K\}$ symmetries. The generators P and Z form D-dimensional translations, M generators span the $so(1, d-1)$—Lorentz algebra on the world volume, the generators N rotate broken translations Z among themselves and thus they span $so(D-d)$ algebra, while generators K belong to the coset $so(1, D-1)/so(1, d-1) \times so(D-d)$. All transformations of the D-dimensional Poincaré group can be realized by the left action of different group elements on the coset space[1]

$$g = e^{xP} e^{q(x)Z} e^{\Lambda(x)K}. \tag{11.2.1}$$

[1] For the sake of brevity we suppress here all space–time indices.

The spontaneous breaking of Z and K symmetries is reflected in the character of corresponding coset coordinates which are Goldstone fields $q(x)$ and $\Lambda(x)$ in the present case. The transformation properties of coordinates x and fields $\{q(x), \Lambda(x)\}$ may be easily found in this approach, while all needed information about the geometry of the coset space (11.2.1) is contained in the Cartan forms

$$g^{-1}dg = \Omega_P P + \Omega_M M + \Omega_Z Z + \Omega_K K + \Omega_N N. \tag{11.2.2}$$

All Cartan forms except for Ω_M and Ω_N are transformed homogeneously under all symmetries. Due to the general theorem [17] not all of the above Goldstone fields have to be treated as independent. In the present case the fields $\Lambda(x)$ can be covariantly expressed through x-derivatives of $q(x)$ by imposing the constraint

$$\Omega_Z = 0. \tag{11.2.3}$$

Equations encoded in the conditions (11.2.3), do not contain dynamic restrictions and are purely kinematic. Thus, we are dealing with the fields $q(x)$ only. It is very important that the form Ω_P defines the vielbein E (d-bein in the present case), connecting the covariant world volume coordinate differentials Ω_P and the world volume coordinate differential dx as

$$\Omega_P = E \cdot dx. \tag{11.2.4}$$

Combining all these ingredients, one may immediately write the action

$$S = -\int d^d x + \int d^d x \ \det E, \tag{11.2.5}$$

which is invariant under all symmetries. In (11.2.5) we have added the trivial first term to fulfill the condition $S_{q=0} = 0$. The action (11.2.5) is just the static gauge form of the action of $p = (d-1)$-branes.

The supersymmetric generalization of the coset approach involves into the game new spinor generators Q and S which extend the D-dimensional Poincaré group to the supersymmetric one

$$\{Q, Q\} \sim P, \quad \{S, S\} \sim P, \quad \{Q, S\} \sim Z. \tag{11.2.6}$$

The most interesting cases are those when the Q supersymmetry is kept unbroken, while the S supersymmetry is supposed to be spontaneously broken.[2] When $\#Q = \#S$ we are facing the so-called $1/2$ Partial Breaking of Global Supersymmetry cases

---

[2] If all supersymmetries are considered as spontaneously broken, the corresponding action can be constructed similarly to the bosonic case, resulting in the some synthesis of Volkov and Akulov [5] and Nambu-Goto actions. An enlightening example of such a construction can be found in [24].

(PBGS), which most of all interesting supersymmetric domain walls belong to. Only such cases of supersymmetry breaking will be considered in this paper.

Now, all symmetries can be realized by group elements acting on the coset element

$$g = e^{xP} e^{\theta Q} e^{\boldsymbol{q}(x,\theta)Z} e^{\psi(x,\theta)S} e^{\boldsymbol{\Lambda}(x,\theta)K}. \qquad (11.2.7)$$

The main novel feature of the supersymmetric coset (11.2.7) is the appearance of the Goldstone superfields $\{\boldsymbol{q}(x,\theta), \psi(x,\theta), \boldsymbol{\Lambda}(x,\theta)\}$ which depend on the coordinates of the world volume superspace $\{x,\theta\}$. The rest of the coset approach machinery works in the same manner: one may construct the Cartan forms (11.2.2) for the coset (11.2.7) (which will contain the new forms Ω_Q and Ω_S), one may find the supersymmetric d-bein and corresponding bosonic ∇_P and spinor ∇_Q covariant derivatives, etc. One may even write the proper generalizations of the covariant constraints (11.2.3) as

$$\Omega_Z = 0, \quad \Omega_S| = 0, \qquad (11.2.8)$$

where $|$ means the $d\theta$-projection of the form (see e.g. [13] and references therein). The $d\theta$-parts of these constraints are closely related with the "geometro-dynamical" constraint of the superembedding approach (see e.g. [26]).

Unfortunately, this similarity between purely bosonic and supersymmetric cases is not complete due to the existence of the following important new features of theories with partial breaking of global supersymmetry:

- In contrast with the bosonic case, not all of the physical fields appear among the parameters of the coset. A famous example comes from the supersymmetric space-filling D3-brane (aka $N = 1$ Born-Infeld theory) where the coset element (11.2.7) contains only P, Q and S generators [8, 10], while the field strength F is “hidden” inside the superfield $\psi : F \sim \nabla_Q \psi|$. Nevertheless, *it is true* that the *all physical bosonic components* can be found in the quantity $\nabla_Q \psi|$.
- The supersymmetric generalization (11.2.8) of the bosonic kinematic constraints (11.2.3) in most cases contains not only kinematic conditions, but also dynamic superfield equations of motion. A prominent example again may be found in [8]. Moreover, in many cases it is unknown how to split these constraints into kinematical and dynamical ones.
- But the most unpleasant feature of the supersymmetric cases is that the standard methods of nonlinear realizations fail to construct the superfield action! The main reason for this is simple: all that we have at hands are the covariant Cartan forms, which we can construct the superfield invariants from, while the superspace Lagrangian is not invariant. Instead it is shifted by the full spinor derivatives under unbroken and/or broken supersymmetries.

Nevertheless, we are going to apply a coset approach to the supersymmetric cases and to demonstrate how on-shell component actions can be constructed within it. The main idea is to start with the Ansatz for the action manifestly invariant with respect to *spontaneously broken supersymmetry*. Funny enough, it is rather easy to do, due to the following properties:

- In our parametrization of the coset element (11.2.7) the superspace coordinates θ do not transform under broken supersymmetry. Thus, all components of superfields transform *independently*,
- The covariant derivatives ∇_P and ∇_Q are invariant under broken supersymmetry. Therefore, the bosonic physical components which are contained in $\nabla_Q \psi(x, \theta)|$ can be treated as "matter fields" (together with the field $\boldsymbol{q}(x, \theta)|$ itself) with respect to broken supersymmetry,
- All physical fermionic components are just $\theta = 0$ projections of the superfield $\psi(x, \theta)$ and these components transform as the fermions of the Volkov-Akulov model [5] with respect to broken supersymmetry.

The immediate consequence of these facts is the conclusion that the physical fermionic components can enter the component on-shell action either through the determinant of the d-bein constructed with the help of the Cartan form Ω_P in the limit $\theta = 0$, namely, $\mathcal{E} = E|$, through the space–time derivatives of the "matter fields" $\nabla_P \boldsymbol{q}|$, or through the Wess-Zumino terms if they exist. Thus, the most general Ansatz for the on-shell component action, which is invariant with respect to spontaneously broken supersymmetry, has the form

$$S = \int d^d x - \int d^d x \det \mathcal{E} \mathcal{F}(\nabla_Q \psi|, \nabla_P \boldsymbol{q}|) + S_{WZ}. \tag{11.2.9}$$

Note, that the arguments of the function $\mathcal{F}$ are the bosonic physical components $\nabla_Q \psi|$ and the covariant space–time derivatives of $\boldsymbol{q}$ (which, by the way, are also contained in $\nabla_Q \psi|$). In certain cases, for fixing an explicit form of the function $\mathcal{F}$ it is sufficient that the following two conditions be satisfied

1. The action (11.2.9) should have a proper bosonic limit, which is known in almost all interesting cases. One should note, that this limit for the action (11.2.9) is trivial
$$S_{bos} = \int d^d x \Big(1 - \mathcal{F}(\nabla_Q \psi|, \partial_P q)\Big).$$
2. The action (11.2.9) in the linear limit should possess a linear version of unbroken supersymmetry, i.e. it should be just the sum of the kinetic terms for all bosonic and fermionic components with the relative coefficients fixed by unbroken supersymmetry.

One should note that the Wess-Zimino action, which is invariant under broken supersymmetry, can be also constructed from the Cartan forms following the recipe of [25]. Thus, the role of unbroken supersymmetry is to fix the coefficients in the action (11.2.9) to achieve its invariance with respect to unbroken sypersymmetry.

In the next two sections of the present paper we will show how the coset approach works in the cases of the superparticle in $D = 3$ and $D = 5$ with the chiral and quartet Goldstone supermultiplets, respectively. Then in Sect. 11.5 we will extend our analysis to the cases of $N = 1$ supermembrane in $D = 4$ as well as of the dual system—$N = 1$ supersymmetric space filling D2-brane. In Sect. 11.6 we will show that in order to construct $N = 2$ supersymmetric action for the supermembrane action in $D = 4$, one needs to add the corresponding Wess-Zumino term. In Appendices we collect the technical details, notation and explicit proof of invariance of the supermembrane action with respect to both, broken and unbroken supersymmetries. We conclude with some comments and perspectives.

11.3 Superparticle in D=3

The main goal of this section is to provide the detailed structure of the component on-shell actions for the one-dimensional system realizing a one half breaking of the global supersymmetry. As an example, we consider a system with $N = 16 \to N = 8$ pattern of supersymmetry breaking based on the superalgebra with two "semi-central charges" $(Z, \overline{Z})$. We show that the resulting component action describes a superparticle in $D = 3$.

11.3.1 Superparticle in D=3: Kinematics

It is a well known fact that the action for the given pattern of the supersymmetry breaking is completely defined by the choice of the corresponding Goldstone supermultiplet [7–14, 16]. The bosonic scalars of the supermultiplet are associated with the "semi-central charges" in the supersymmetry algebra (11.2.6). To describe a system with one complex boson (or two real bosons) one has to choose $N = 16, d = 1$ Poincaré superalgebra with two "semi-central charges" $(Z, \overline{Z})$

$$\left\{Q^{ia}, \overline{Q}_{jb}\right\} = 2\delta^a_b\delta^i_j P, \quad \left\{S^{ia}, \overline{S}_{jb}\right\} = 2\delta^a_b\delta^i_j P,$$
$$\left\{Q^{ia}, S^{jb}\right\} = 2i\varepsilon^{ab}\varepsilon^{ij} Z, \quad \left\{\overline{Q}_{ia}, \overline{S}_{jb}\right\} = -2i\varepsilon_{ab}\varepsilon_{ij}\overline{Z}. \tag{11.3.1}$$

Here $i, a = 1, 2$ refer to the indices of the fundamental representations of two commuting $SU(2)$ groups. In (11.3.1) P is the generator of one-dimensional translation, while Q^{ia}, $\overline{Q}_{ia}$ and S^{ia}, $\overline{S}_{ia}$ are the generators of unbroken and spontaneously broken $N = 8$ supersymmetries, respectively. As we already explain in the Introduction, in the coset approach the statement that S supersymmetry and $(Z, \overline{Z})$ translations are spontaneously broken is reflected in the structure of the element of the coset space

$$g = e^{itP}\, e^{\theta_{ia}Q^{ia}+\bar{\theta}^{ia}\overline{Q}_{ia}}\, e^{i(\boldsymbol{q}Z+\bar{\boldsymbol{q}}\overline{Z})}\, e^{\psi_{ia}S^{ia}+\bar{\psi}^{ia}\overline{S}_{ia}}. \tag{11.3.2}$$

Once we state that the coordinates ψ and $\boldsymbol{q}$ are the superfields depending on the $N = 8, d = 1$ superspace coordinates $(t, \theta, \bar{\theta})$, then we are dealing with the spontaneously breaking of the corresponding symmetries. Thus, in our case we will treat $\psi(t,\theta,\bar\theta)$, $\boldsymbol{q}(t,\theta,\bar\theta)$ as $N=8, d=1$ Goldstone superfields accompanying $N=16\rightarrow N=8$ breaking of supersymmetry in one dimension.

The transformation properties of coordinates and superfields under both unbroken and broken supersymmetries are induced by the left multiplications of the group element g_0 on the coset (11.3.2)

$$g_0\, g = g'.$$

Thus, for the unbroken supersymmetry with $g_0 = e^{\varepsilon_{ia}Q^{ia}+\bar\varepsilon^{ia}\overline{Q}_{ia}}$ one gets

$$\delta_Q t = i\left(\varepsilon_{ia}\bar\theta^{ia}+\bar\varepsilon^{ia}\theta_{ia}\right),\quad \delta_Q\theta_{ia}=\varepsilon_{ia},\quad \delta_Q\bar\theta^{ia}=\bar\varepsilon^{ia}, \tag{11.3.3}$$

while for the broken supersymmetry with $g_0 = e^{\eta_{ia}S^{ia}+\bar\eta^{ia}\overline{S}_{ia}}$ the transformations read

$$\begin{aligned}&\delta_S t = i\left(\eta_{ia}\bar\psi^{ia}+\bar\eta^{ia}\psi_{ia}\right),\quad \delta_S\psi_{ia}=\eta_{ia},\quad \delta_S\bar\psi^{ia}=\bar\eta^{ia},\\ &\delta_S\boldsymbol{q} = -2\eta_{ia}\theta^{ia},\quad \delta_S\bar{\boldsymbol{q}} = 2\bar\eta^{ia}\bar\theta_{ia}.\end{aligned} \tag{11.3.4}$$

The local geometric properties of the system are specified by the left-invariant Cartan forms

$$g^{-1}dg = i\omega_P P+(\omega_Q)_{ia}Q^{ia}+(\bar\omega_Q)^{ia}\overline{Q}_{ia}+i\omega_Z Z+i\bar\omega_Z\overline{Z}+(\omega_S)_{ia}S^{ia}+(\bar\omega_S)^{ia}\overline{S}_{ia} \tag{11.3.5}$$

which can be explicitly written in the considered case as

$$\begin{aligned}&\omega_P = dt - i\left(\bar\theta^{ia}d\theta_{ia}+\theta_{ia}d\bar\theta^{ia}+\bar\psi^{ia}d\psi_{ia}+\psi_{ia}d\bar\psi^{ia}\right),\quad (\omega_Q)_{ia}=d\theta_{ia},\\ &(\bar\omega_Q)^{ia}=d\bar\theta^{ia},\quad (\omega_S)_{ia}=d\psi_{ia},\quad (\bar\omega_S)^{ia}=d\bar\psi^{ia},\quad \omega_Z = d\boldsymbol{q}+2\psi^{ia}d\theta_{ia},\\ &\bar\omega_Z = d\bar{\boldsymbol{q}}-2\bar\psi_{ia}d\bar\theta^{ia}.\end{aligned} \tag{11.3.6}$$

Using the covariant differentials $(\omega_P, d\theta_{ia}, d\bar\theta^{ia})$ (11.3.6), one may construct the covariant derivatives

$$\begin{aligned}&\partial_t = E\,\nabla_t,\quad E = 1-i\left(\psi_{ia}\dot{\bar\psi}^{ia}+\bar\psi^{ia}\dot\psi_{ia}\right),\quad E^{-1}=1+i\left(\psi_{ia}\nabla_t\bar\psi^{ia}+\bar\psi^{ia}\nabla_t\psi_{ia}\right),\\ &\nabla^{ia} = D^{ia}-i\left(\psi_{kb}D^{ia}\bar\psi^{kb}+\bar\psi^{kb}D^{ia}\psi_{kb}\right)\nabla_t = D^{ia}-i\left(\psi_{kb}\nabla^{ia}\bar\psi^{kb}+\bar\psi^{kb}\nabla^{ia}\psi_{kb}\right)\partial_t,\\ &\overline{\nabla}_{ia} = \overline{D}_{ia}-i\left(\psi_{kb}\overline{D}_{ia}\bar\psi^{kb}+\bar\psi^{kb}\overline{D}_{ia}\psi_{kb}\right)\nabla_t = \overline{D}_{ia}-i\left(\psi_{kb}\overline{\nabla}_{ia}\bar\psi^{kb}+\bar\psi^{kb}\overline{\nabla}_{ia}\psi_{kb}\right)\partial_t,\end{aligned} \tag{11.3.7}$$

where

$$D^{ia}=\frac{\partial}{\partial\theta_{ia}}-i\bar\theta^{ia}\partial_t,\quad \overline{D}_{ia}=\frac{\partial}{\partial\bar\theta^{ia}}-i\theta_{ia}\partial_t,\quad \left\{D^{ia},\overline{D}_{jb}\right\}=-2i\delta^a_b\delta^i_j\partial_t. \tag{11.3.8}$$

The covariant derivatives (11.3.7) satisfy the following (anti)commutation relations

$$\begin{aligned}
\left\{\nabla^{ia}, \nabla^{jb}\right\} &= -2i\left(\nabla^{ia}\psi_{kc}\nabla^{jb}\bar{\psi}^{kc} + \nabla^{ia}\bar{\psi}^{kc}\nabla^{jb}\psi_{kc}\right)\nabla_t,\\
\left\{\overline{\nabla}_{ia}, \overline{\nabla}_{jb}\right\} &= -2i\left(\overline{\nabla}_{ia}\psi_{kc}\overline{\nabla}_{jb}\bar{\psi}^{kc} + \overline{\nabla}_{ia}\bar{\psi}^{kc}\overline{\nabla}_{jb}\psi_{kc}\right)\nabla_t,\\
\left[\nabla_t, \nabla^{ia}\right] &= -2i\left(\nabla_t\psi_{kc}\nabla^{ia}\bar{\psi}^{kc} + \nabla_t\bar{\psi}^{kc}\nabla^{ia}\psi_{kc}\right)\nabla_t,\\
\left[\nabla_t, \overline{\nabla}_{ia}\right] &= -2i\left(\nabla_t\psi_{kc}\overline{\nabla}_{ia}\bar{\psi}^{kc} + \nabla_t\bar{\psi}^{kc}\overline{\nabla}_{ia}\psi_{kc}\right)\nabla_t,\\
\left\{\nabla^{ia}, \overline{\nabla}_{jb}\right\} &= -2i\delta^a_b\delta^i_j\nabla_t - 2i\left(\nabla^{ia}\psi_{kc}\overline{\nabla}_{jb}\bar{\psi}^{kc} + \nabla^{ia}\bar{\psi}^{kc}\overline{\nabla}_{jb}\psi_{kc}\right)\nabla_t.
\end{aligned} \tag{11.3.9}$$

To reduce the number of independent Goldstone superfields let us impose the conditions on the $d\theta$-projections of the Cartan forms $(\omega_Z, \bar{\omega}_Z)$ (11.3.6)

$$\begin{cases} \omega_Z|_\theta = 0, \\ \overline{\omega}_Z|_\theta = 0, \end{cases} \quad \Rightarrow \quad \begin{cases} \overline{\nabla}_{ia}\boldsymbol{q} = 0, & \nabla^{ia}\boldsymbol{q} - 2\psi^{ia} = 0, \\ \nabla^{ia}\bar{\boldsymbol{q}} = 0, & \overline{\nabla}_{ia}\bar{\boldsymbol{q}} + 2\bar{\psi}_{ia} = 0. \end{cases} \tag{11.3.10}$$

One part of these kinematical constraints can be recognized as the covariant chirality conditions on the superfields $\boldsymbol{q}$ and $\bar{\boldsymbol{q}}$, while the remaining two equations express the fermionic Goldstone superfields ψ^{ia} and $\bar{\psi}_{ia}$ as the spinor derivatives of the bosonic superfields $\boldsymbol{q}$ and $\bar{\boldsymbol{q}}$, thereby realizing the Inverse Higgs phenomenon [17].

11.3.2 Superparticle in D=3: Dynamics

It is well known that the standard chirality conditions are not enough to select an irreducible $N = 8, d = 1$ supermultiplet: one has impose additional, second order in the spinor derivatives constraints on the superfield $\{\boldsymbol{q}, \bar{\boldsymbol{q}}\}$ [27]. Unfortunately, as it often happened in the coset approach, the direct covariantization of the irreducibility constraints is not covariant [8], while the simultaneous covariantization of the constraints and the equations of motion works perfectly. That is why we propose the following equations which should describe our superparticle

$$\nabla^{ia}\psi_{jb} = 0, \qquad \overline{\nabla}_{ia}\bar{\psi}^{jb} = 0. \tag{11.3.11}$$

These equations are covariant with respect to both unbroken and broken supersymmetries. One should wonder whether the equations (11.3.11) are self-consistent? Indeed, due to (11.3.10) from (11.3.11) we have

$$\nabla^{ia}\psi_{jb} = \frac{1}{2}\nabla^{ia}\nabla_{jb}\boldsymbol{q} = 0 \quad \Rightarrow \quad \{\nabla^{ia}, \nabla_{jb}\}\boldsymbol{q} = 0. \tag{11.3.12}$$

So, one may expect some additional conditions on the superfield $\boldsymbol{q}$ due to the relations (11.3.9). However, on the constraints surface in (11.3.11) we have

$$\left\{\nabla^{ia},\nabla^{jb}\right\}=0,\qquad \left\{\overline{\nabla}_{ia},\overline{\nabla}_{jb}\right\}=0, \tag{11.3.13}$$

and thus the equations (11.3.11) are perfectly self-consistent.

It is worth mentioning that the rest of the commutators in (11.3.9) are also simplified, when (11.3.11) are satisfied. Indeed, on the constraints (11.3.11) surface they read

$$\begin{gathered}\left\{\nabla^{ia},\overline{\nabla}_{jb}\right\}=-2i\delta^i_j\delta^a_b(1+\boldsymbol{\lambda}\bar{\boldsymbol{\lambda}})\nabla_t,\quad \left[\nabla_t,\nabla^{ia}\right]=2i\bar{\boldsymbol{\lambda}}\nabla_t\boldsymbol{\psi}^{ia}\nabla_t,\\ \left[\nabla_t,\overline{\nabla}_{ia}\right]=2i\boldsymbol{\lambda}\nabla_t\bar{\boldsymbol{\psi}}_{ia}\nabla_t,\end{gathered} \tag{11.3.14}$$

where we introduced the superfields $\{\boldsymbol{\lambda},\bar{\boldsymbol{\lambda}}\}$

$$\begin{cases}\overline{\nabla}_{ia}\boldsymbol{\psi}_{jb}+\varepsilon_{ij}\varepsilon_{ab}\boldsymbol{\lambda}=0,\\ \nabla^{ia}\bar{\boldsymbol{\psi}}^{jb}+\varepsilon^{ij}\varepsilon^{ab}\bar{\boldsymbol{\lambda}}=0,\end{cases}\quad\Rightarrow\quad\begin{cases}\nabla_t\boldsymbol{q}+\frac{i\boldsymbol{\lambda}}{1+\boldsymbol{\lambda}\bar{\boldsymbol{\lambda}}}=0,\\ \nabla_t\bar{\boldsymbol{q}}-\frac{i\bar{\boldsymbol{\lambda}}}{1+\boldsymbol{\lambda}\bar{\boldsymbol{\lambda}}}=0.\end{cases} \tag{11.3.15}$$

The superfield equations (11.3.11) lead in the bosonic limit to the following equation of motion for the complex scalar field $q=\boldsymbol{q}|_{\theta=0}$:

$$\frac{d}{dt}\left[\frac{\dot q}{\sqrt{1-4\dot q\dot{\bar q}}}\right]=0. \tag{11.3.16}$$

The last equation can be easy deduced from the bosonic action

$$S_{bos}=\int dt\left(1-\sqrt{1-4\dot q\dot{\bar q}}\right). \tag{11.3.17}$$

Thus, the bosonic action for a particle in $D=3$ space–time is known.

11.3.3 Superparticle in D = 3: Component Action

Despite the explicit construction of the proper equations of motion within the superfield version of the coset approach, it is poorly adapted for the construction of the action. That is why in the paper [21] the component version of the coset approach to construct the actions has been proposed. In the application to the present case, the basic steps of this method can be formulated as follows:

- Firstly, on-shell our $N=8$ supermultiplet $\{\boldsymbol{q},\bar{\boldsymbol{q}}\}$ contains the following physical components:

$$q = \boldsymbol{q}|_{\theta=0}, \quad \bar{q} = \bar{\boldsymbol{q}}|_{\theta=0}, \quad \psi_{ia} = \boldsymbol{\psi}_{ia}|_{\theta=0}, \quad \bar{\psi}^{ia} = \bar{\boldsymbol{\psi}}^{ia}|_{\theta=0}.$$

They are just the first components of the superfields parameterizing the coset (11.3.2).

- Secondly, with respect to broken supersymmetry $\delta\theta = \delta\bar{\theta} = 0$ (11.3.4). This means, that the transformation properties of the physical components $\{q, \bar{q}, \psi_{ia}, \bar{\psi}^{ia}\}$ under broken supersymmetry can be extracted from the coset

$$g|_{\theta=0} = e^{itP}\, e^{i(qZ+\bar{q}\overline{Z})}\, e^{\psi_{ia}S^{ia}+\bar{\psi}^{ia}\overline{S}_{ia}}. \tag{11.3.18}$$

In other words, the fields $\{q, \bar{q}, \psi_{ia}, \bar{\psi}^{ia}\}$ parameterize the coset (11.3.18) which is responsible for full breaking of the S supersymmetry. Moreover, with respect to this supersymmetry the fields $\{q, \bar{q}\}$ are just "matter fields", because $\delta_S q = \delta_S \bar{q} = 0$, while the fermions $\{\psi_{ia}, \bar{\psi}^{ia}\}$ are just Goldstone fermions. This means that the component action has to be of the Volkov-Akulov type [5], i.e. the fermions $\{\psi_{ia}, \bar{\psi}^{ia}\}$ may enter the action through the einbein $\mathcal{E}$ or through the covariant derivatives $\mathcal{D}_t q, \mathcal{D}_t \bar{q}$ only, with

$$\partial_t = \mathcal{E}\mathcal{D}_t, \quad \mathcal{E} = E|_{\theta=0} = 1 - i\left(\psi_{ia}\dot{\bar{\psi}}^{ia} + \bar{\psi}^{ia}\dot{\psi}_{ia}\right),$$
$$\mathcal{E}^{-1} = 1 + i\left(\psi_{ia}\mathcal{D}_t\bar{\psi}^{ia} + \bar{\psi}^{ia}\mathcal{D}_t\psi_{ia}\right). \tag{11.3.19}$$

Thus, the unique candidate to be the component on-shell action, invariant with respect to spontaneously broken S supersymmetry reads

$$S = \alpha\int dt + \int dt\mathcal{E}\mathcal{F}\left[\mathcal{D}_t q \mathcal{D}_t \bar{q}\right] \tag{11.3.20}$$

with an arbitrary, for the time being, function $\mathcal{F}$ and a constant parameter α.

- Finally, considering the bosonic limit of the action (11.3.20) and comparing it with the known bosonic action (11.3.17) one may find the function $\mathcal{F}$:

$$\int dt\left(\alpha + \mathcal{F}\left[\dot{q}\dot{\bar{q}}\right]\right) = \int dt\left(1 - \sqrt{1 - 4\dot{q}\dot{\bar{q}}}\right) \Rightarrow$$
$$\mathcal{F} = \left(1 - \alpha - \sqrt{1 - 4\dot{q}\dot{\bar{q}}}\right). \tag{11.3.21}$$

Therefore, the most general component action possessing the proper bosonic limit (11.3.17) and invariant under spontaneously broken supersymmetry has the form

$$S = \alpha\int dt + (1-\alpha)\int dt\,\mathcal{E} - \int dt\,\mathcal{E}\sqrt{1 - 4\mathcal{D}_t q \mathcal{D}_t \bar{q}}. \tag{11.3.22}$$

In principle, the invariance of the action (11.3.22) under broken supersymmetry is evident. Nevertheless, it should be explicitly checked.

From (11.3.4) we obtain the total variations of our components and the time coordinate t:

$$\delta_S t = i\left(\eta_{ia}\bar{\psi}^{ia} + \bar{\eta}^{ia}\psi_{ia}\right), \quad \delta_S\psi_{ia} = \eta_{ia}, \quad \delta_S\bar{\psi}^{ia} = \bar{\eta}^{ia}, \quad \delta_S q = 0, \quad \delta_S\bar{q} = 0. \tag{11.3.23}$$

Therefore, the transformations of the components in the fixed point read

$$\delta^*_S q = \delta_S q - \delta_S t\,\dot{q}, \quad \delta^*_S\psi_{ia} = \delta_S\psi_{ia} - \delta_S t\,\dot{\psi}_{ia}. \tag{11.3.24}$$

Then, it immediately follows from (11.3.24) and definitions (11.3.19) that

$$\delta^*_S\left(\mathcal{E}\mathcal{F}\left[\mathcal{D}_t q\mathcal{D}_t\bar{q}\right]\right) = -i\partial_t\left[\left(\eta_{ia}\bar{\psi}^{ia} + \bar{\eta}^{ia}\psi_{ia}\right)\mathcal{E}\mathcal{F}\left[\mathcal{D}_t q\mathcal{D}_t\bar{q}\right]\right]. \tag{11.3.25}$$

Thus, the two last terms in the action (11.3.22) are invariant, while the invariance of the first term is evident.

The final step is to check the invariance of the action (11.3.22) under unbroken supersymmetry which is realized on the components as follows:

$$\begin{aligned}\delta^*_Q q &= -2\varepsilon^{ia}\psi_{ia} + i\left(\varepsilon^{ia}\psi_{ia}\bar{\lambda} + \bar{\varepsilon}^{ia}\bar{\psi}_{ia}\lambda\right)\partial_t q\\ \delta^*_Q\psi_{ia} &= \bar{\varepsilon}_{ia}\lambda + i\left(\varepsilon^{jb}\psi_{jb}\bar{\lambda} + \bar{\varepsilon}^{jb}\bar{\psi}_{jb}\lambda\right)\partial_t\psi_{ia}.\end{aligned} \tag{11.3.26}$$

Here, λ is the first component of the superfield $\boldsymbol{\lambda}$ defined in (11.3.15)

$$\lambda = \frac{2i\mathcal{D}_t q}{1 + \sqrt{1 - 4\mathcal{D}_t q\mathcal{D}_t\bar{q}}}. \tag{11.3.27}$$

From (11.3.26) and the definitions (11.3.19) one may easily find the transformation properties of the main ingredients

$$\begin{aligned}\delta^*_Q\mathcal{E} &= i\partial_t\left[\left(\varepsilon^{jb}\psi_{jb}\bar{\lambda} + \bar{\varepsilon}^{jb}\bar{\psi}_{jb}\lambda\right)\mathcal{E}\right] - 2i\left(\varepsilon^{jb}\dot{\psi}_{jb}\bar{\lambda} + \bar{\varepsilon}^{jb}\dot{\bar{\psi}}_{jb}\lambda\right),\\ \delta^*_Q\mathcal{D}_t q &= i\left(\varepsilon^{jb}\psi_{jb}\bar{\lambda} + \bar{\varepsilon}^{jb}\bar{\psi}_{jb}\lambda\right)\partial_t(\mathcal{D}_t q) - 2\varepsilon^{jb}\mathcal{D}_t\psi_{jb}\\ &\quad + 2i\left(\varepsilon^{jb}\mathcal{D}_t\psi_{jb}\bar{\lambda} + \bar{\varepsilon}^{jb}\mathcal{D}_t\bar{\psi}_{jb}\lambda\right)\mathcal{D}_t q.\end{aligned} \tag{11.3.28}$$

Now, one may calculate the variation of the integrand in the action (11.3.20)

$$
\begin{aligned}
\delta^*_Q(\mathcal{E}\,\mathcal{F}) = 2\partial_t \Bigg[\mathcal{E}\,\frac{\varepsilon^{jb}\psi_{jb}\mathcal{D}_t\bar{q} - \bar{\varepsilon}^{jb}\bar{\psi}_{jb}\mathcal{D}_t q}{1+\sqrt{1-4\mathcal{D}_t q\mathcal{D}_t\bar{q}}}\,\mathcal{F}\Bigg] \\
+ \frac{\varepsilon^{jb}\dot{\psi}_{jb}\mathcal{D}_t\bar{q} - \bar{\varepsilon}^{jb}\dot{\bar{\psi}}_{jb}\mathcal{D}_t q}{1+\sqrt{1-4\mathcal{D}_t q\mathcal{D}_t\bar{q}}} \\
\Big[-4\mathcal{F} - 2\mathcal{F}'\Big(1+\sqrt{1-4\mathcal{D}_t q\mathcal{D}_t\bar{q}} - 4\mathcal{D}_t q\mathcal{D}_t\bar{q}\Big)\Big].
\end{aligned}
\tag{11.3.29}
$$

Substituting the function $\mathcal{F}$ (11.3.21) and its derivative over its argument $\mathcal{D}_t q\mathcal{D}_t\bar{q}$, we find that the second term in the variation (11.3.29) cancels out, provided $\alpha = 2$. Keeping in mind that the first term in the action (11.3.22) is trivially invariant under unbroken supersymmetry, we conclude that the unique component action invariant under both unbroken and broken $N = 8$ supersymmetries reads

$$
S = 2\int dt - \int dt\,\mathcal{E}\Big(1+\sqrt{1-4\mathcal{D}_t q\mathcal{D}_t\bar{q}}\Big). \tag{11.3.30}
$$

We end this section with two comments.
Firstly, one should note that the construction of the component action, we considered in the previous section, has two interesting peculiarities:

- It is based on the coset realization of the $N = 16$ superalgebra (11.3.1)
- In the component action (11.3.30) the summation over indices $\{i, a\}$ of two $SU(2)$ groups affected only physical fermions $\{\psi_{ia}, \bar{\psi}^{ia}\}$.

It is quite clear, that in such a situation one may consider two subalgebras of $N = 16$ superalgebra:

- $N = 8$ supersymmetry, by choosing the corresponding supercharges as

$$
\tilde{Q}^i \equiv Q^{i1}, \quad \tilde{\overline{Q}}_i \equiv \overline{Q}_{i1}, \quad \tilde{S}^i \equiv S^{i2}, \quad \tilde{\overline{S}}_i \equiv \overline{S}_{i2}, \tag{11.3.31}
$$

- $N = 4$ supersymmetry with the supercharges

$$
\hat{Q} \equiv Q^{11}, \quad \hat{\overline{Q}} \equiv \overline{Q}_{11}, \quad \hat{S} \equiv S^{22}, \quad \hat{\overline{S}} \equiv \overline{S}_{22}. \tag{11.3.32}
$$

It is evident that the corresponding component actions will be given by the same expression (11.3.30), in which the "new" einbeins and covariant derivatives read

$$
N = 8\,\text{case}: \Big\{\partial_t = \tilde{\mathcal{E}}\tilde{\mathcal{D}}_t,\, \tilde{\mathcal{E}} = 1 - i\Big(\psi_{i2}\dot{\bar{\psi}}^{i2} + \bar{\psi}^{i2}\dot{\psi}_{i2}\Big), \quad \tilde{\mathcal{E}}^{-1} = 1 + i\Big(\psi_{i2}\tilde{\mathcal{D}}_t\bar{\psi}^{i2} + \bar{\psi}^{i2}\tilde{\mathcal{D}}_t\psi_{i2}\Big), \tag{11.3.33}
$$

$$
N = 4\,\text{case}: \Big\{\partial_t = \hat{\mathcal{E}}\hat{\mathcal{D}}_t,\, \hat{\mathcal{E}} = 1 - i\Big(\psi_{22}\dot{\bar{\psi}}^{22} + \bar{\psi}^{22}\dot{\psi}_{22}\Big), \quad \hat{\mathcal{E}}^{-1} = 1 + i\Big(\psi_{22}\hat{\mathcal{D}}_t\bar{\psi}^{22} + \bar{\psi}^{22}\hat{\mathcal{D}}_t\psi_{22}\Big). \tag{11.3.34}
$$

Thus, we see that the action (11.3.30) has a universal character, describing the series of theories with the following patterns of global supersymmetry breaking $N = 16 \to N = 8$, $N = 8 \to N = 4$ and $N = 4 \to N = 2$.

Secondly, it is almost evident, that the universality of the action (11.3.30) can be used to extend our construction to the cases of $N = 4 \cdot 2^k$ supersymmetries by adding the needed numbers of $SU(2)$ indices to the superscharges as

$$Q \to Q^{\alpha_1 \ldots \alpha_k},\ \overline{Q} \to \overline{Q}_{\alpha_1 \ldots \alpha_k},\quad S \to S^{\alpha_1 \ldots \alpha_k},\ \overline{S} \to \overline{S}_{\alpha_1 \ldots \alpha_k}, \tag{11.3.35}$$

obeying the $N = 4 \cdot 2^k$ Poincaré superalgebra

$$\begin{aligned}
&\left\{Q^{\alpha_1 \ldots \alpha_k}, \overline{Q}_{\beta_1 \ldots \beta_k}\right\} = 2\delta^{\alpha_1}_{\beta_1} \ldots \delta^{\alpha_k}_{\beta_k} P, \quad \left\{S^{\alpha_1 \ldots \alpha_k}, \overline{S}_{\beta_1 \ldots \beta_k}\right\} = 2\delta^{\alpha_1}_{\beta_1} \ldots \delta^{\alpha_k}_{\beta_k} P,\\
&\left\{Q^{\alpha_1 \ldots \alpha_k}, S^{\beta_1 \ldots \beta_k}\right\} = 2i\varepsilon^{\alpha_1\beta_1} \ldots \varepsilon^{\alpha_k\beta_k} Z,\\
&\left\{\overline{Q}_{\alpha_1 \ldots \alpha_k}, \overline{S}_{\beta_1 \ldots \beta_k}\right\} = -2i\varepsilon_{\alpha_1\beta_1} \ldots \varepsilon_{\alpha_k\beta_k} \overline{Z}.
\end{aligned} \tag{11.3.36}$$

Once again, the component action describing superparticles in $D = 3$ space with $N = 4 \cdot 2^k$ Poincaré supersymmetry partially broken down to the $N = 2 \cdot 2^k$ one will be given by the same expression (11.3.30) with the following substitutions

$$\psi \to \psi_{\alpha_1 \ldots \alpha_k},\ \bar{\psi} \to \bar{\psi}^{\alpha_1 \ldots \alpha_k},\ \mathcal{E} = 1 - i\left(\psi_{\alpha_1 \ldots \alpha_k}\dot{\bar{\psi}}^{\alpha_1 \ldots \alpha_k} + \bar{\psi}^{\alpha_1 \ldots \alpha_k}\dot{\psi}_{\alpha_1 \ldots \alpha_k}\right). \tag{11.3.37}$$

11.4 Superparticle in D=5

In this section we will apply our approach to $N = 16$ superparticle in $D = 5$. The corresponding superfield equations of motion for this system, which possesses 8 manifest and 8 spontaneously broken supersymmetries, have been constructed within the coset approach in [14], while the action is still unknown.

To describe the superparticle in $D = 5$ with 16 supersymmetries one has to start with the following superalgebra

$$\{Q^i_\alpha, Q^j_\beta\} = \varepsilon^{ij}\Omega_{\alpha\beta}P, \quad \{Q^i_\alpha, S^{b\beta}\} = \delta^\beta_\alpha Z^{ib}, \quad \{S^{a\alpha}, S^{b\beta}\} = -\varepsilon^{ab}\Omega^{\alpha\beta}P,$$
$$(i, a = 1, 2;\ \alpha, \beta = 1, 2, 3, 4) \tag{11.4.1}$$

where the invariant $Spin(5)$ symplectic metric $\Omega_{\alpha\beta}$, allowing to raise and lower the spinor indices, obeys the conditions[3]

[3] We use the following convention: $\varepsilon^{\alpha\beta\lambda\sigma}\varepsilon_{\alpha\beta\lambda\sigma} = 24,\ \varepsilon^{\alpha\beta\lambda\sigma}\varepsilon_{\alpha\beta\mu\rho} = 2(\delta^\lambda_\mu\delta^\sigma_\rho - \delta^\lambda_\rho\delta^\sigma_\mu)$.

$$\Omega_{\alpha\beta} = -\Omega_{\beta\alpha}, \quad \Omega^{\alpha\beta} = -\frac{1}{2}\,\varepsilon^{\alpha\beta\lambda\sigma}\Omega_{\lambda\sigma}, \quad \Omega_{\alpha\beta} = -\frac{1}{2}\,\varepsilon_{\alpha\beta\lambda\sigma}\Omega^{\lambda\sigma}, \quad \Omega_{\alpha\beta}\Omega^{\beta\gamma} = \delta_\alpha^\gamma. \tag{11.4.2}$$

From the one-dimensional perspective this algebra is $N = 16$ super Poincaré algebra with four central charges Z^{ia}. If we are going to treat S supersymmetry to be spontaneously broken, then we have to consider the following element of the coset[4]:

$$g = e^{tP}\, e^{\theta_i^\alpha Q_\alpha^i}\, e^{\boldsymbol{q}_{ia} Z^{ia}}\, e^{\psi_{a\alpha} S^{a\alpha}}. \tag{11.4.3}$$

Here (t, θ_i^α) are the coordinates of $N = 8, d = 1$ superspace while $\boldsymbol{q}_{ia} = \boldsymbol{q}_{ia}(t, \theta_i^\alpha)$, $\psi_{a\alpha} = \psi_{a\alpha}(t, \theta_i^\alpha)$, are the Goldstone superfields.

Similarly to the case considered in the previous section, one may find the transformation properties of the coordinates and superfields, by acting from the left on the coset element (11.4.3) by different elements of the group with constant parameters. So, for the unbroken supersymmetry $[g_0 = \exp(\varepsilon_i^\alpha Q_\alpha^i)]$ one gets

$$\delta_Q t = -\frac{1}{2}\,\varepsilon_i^\alpha \theta^{i\beta}\Omega_{\alpha\beta}, \quad \delta_Q \theta_i^\alpha = \varepsilon_i^\alpha, \tag{11.4.4}$$

while for the broken supersymmetry $[g_0 = \exp(\eta_{a\alpha} S^{a\alpha})]$ the corresponding transformations read

$$\delta_S t = -\frac{1}{2}\,\eta_\alpha^a \psi_{a\beta}\Omega^{\alpha\beta}, \quad \delta_S \psi_{a\alpha} = \eta_{a\alpha}, \quad \delta_S \boldsymbol{q}_{ia} = -\eta_{a\alpha}\theta_i^\alpha. \tag{11.4.5}$$

The last needed ingredient is the Cartan forms, defined in a standard way as

$$g^{-1}dg = \omega_P P + (\omega_Q)_i^\alpha\, Q_\alpha^i + (\omega_Z)_{ia}\, Z^{ia} + (\omega_S)_{a\alpha}\, S^{a\alpha}, \tag{11.4.6}$$

with

$$\omega_P = dt - \frac{1}{2}\,d\theta_i^\alpha \theta^{i\beta}\Omega_{\alpha\beta} + \frac{1}{2}\,d\psi_{a\alpha}\psi_\beta^a\Omega^{\alpha\beta}, \quad (\omega_Z)_{ia} = d\boldsymbol{q}_{ia} - d\theta_i^\alpha \psi_{a\alpha},$$
$$(\omega_Q)_i^\alpha = d\theta_i^\alpha, \quad (\omega_S)_{a\alpha} = d\psi_{a\alpha}. \tag{11.4.7}$$

Using the covariant differentials $\{\omega_P, (\omega_Q)_i^\alpha\}$ one may construct the covariant derivatives ∇_t and ∇_α^i

$$\partial_t = E\,\nabla_t, \quad E = 1 + \frac{1}{2}\,\Omega^{\beta\gamma}\psi_\beta^a\partial_t\psi_{a\gamma}, \quad E^{-1} = 1 - \frac{1}{2}\,\Omega^{\beta\gamma}\psi_\beta^a\nabla_t\psi_{a\gamma}, \tag{11.4.8}$$

$$\nabla_\alpha^i = D_\alpha^i + \frac{1}{2}\,\Omega^{\beta\gamma}\psi_\beta^a D_\alpha^i\psi_{a\gamma}\nabla_t = D_\alpha^i + \frac{1}{2}\,\Omega^{\beta\gamma}\psi_\beta^a\nabla_\alpha^i\psi_{a\gamma}\partial_t, \tag{11.4.9}$$

[4] Here, we strictly follow the notations adopted in [14] which are slightly different with those we used in the previous sections.

where

$$D^i_\alpha = \frac{\partial}{\partial \theta^\alpha_i} + \frac{1}{2}\,\theta^{i\beta}\Omega_{\alpha\beta}\partial_t, \quad \left\{D^i_\alpha, D^j_\beta\right\} = \varepsilon^{ij}\,\Omega_{\alpha\beta}\,\partial_t. \tag{11.4.10}$$

These covariant derivatives satisfy the following (anti)commutation relations:

$$\begin{aligned}\left\{\nabla^i_\alpha, \nabla^j_\beta\right\} &= \varepsilon^{ij}\,\Omega_{\alpha\beta}\,\nabla_t + \Omega^{\lambda\sigma}\,\nabla^i_\alpha\psi^b_\lambda\,\nabla^j_\beta\psi_{b\sigma}\,\nabla_t,\\ \left[\nabla_t, \nabla^i_\alpha\right] &= \Omega^{\beta\gamma}\,\nabla_t\psi^b_\beta\,\nabla^i_\alpha\psi_{b\gamma}\nabla_t.\end{aligned} \tag{11.4.11}$$

Now, in a full analogy with the previously considered case, we impose the following invariant condition on the $d\theta$-projections of Cartan form $(\omega_Z)_{ia}$ (11.4.7):

$$(\omega_Z)_{ia}|_\theta = 0 \quad \Rightarrow \quad \left\{\begin{array}{ll}\nabla^{(j}_\alpha\,\boldsymbol{q}^{i)}_a = 0, & \text{(a)}\\ \nabla^i_\alpha\,\boldsymbol{q}_{ia} - 2\psi_{a\alpha} = 0. & \text{(b)}\end{array}\right. \tag{11.4.12}$$

The condition (11.4.12b) identifies the fermionic superfield $\psi_{a\alpha}$ with the spinor derivatives of the superfield $\boldsymbol{q}_{ia}$, just reducing the independent superfields to bosonic $\boldsymbol{q}_{ia}$ ones (this is again the Inverse Higgs Phenomenon [17]). The conditions (11.4.12a) are more restrictive—they nullify all auxiliary components in the superfield $\boldsymbol{q}_{ia}$. Indeed, it immediately follows from (11.4.12) that

$$\frac{3}{2}\nabla^j_\beta\psi_{a\alpha} = \left\{\nabla^j_\beta, \nabla^i_\alpha\right\}\boldsymbol{q}_{ia} - \frac{1}{2}\left\{\nabla^j_\alpha, \nabla^i_\beta\right\}\boldsymbol{q}_{ia}. \tag{11.4.13}$$

Using anti-commutators (11.4.11), one may solve this equation as follows:

$$\nabla^j_\beta\,\psi_{a\alpha} + \frac{1}{2}\,\boldsymbol{\lambda}^j_a\Omega_{\alpha\beta} = 0, \tag{11.4.14}$$

where the superfield $\boldsymbol{\lambda}^{ia}$ is defined as ($\boldsymbol{\lambda}^2 = \boldsymbol{\lambda}^{ia}\boldsymbol{\lambda}_{ia}$)

$$\nabla_t\,\boldsymbol{q}^{ia} - \frac{1}{2}\,\frac{\boldsymbol{\lambda}^{ia}}{1+\frac{\boldsymbol{\lambda}^2}{8}} = 0. \tag{11.4.15}$$

Thus, we have the on-shell situation. In [14] the corresponding bosonic equation of motion has been found to be

$$\frac{d}{dt}\left(\frac{\dot{q}_{ia}}{\sqrt{1-2\dot{q}^{jb}\dot{q}_{jb}}}\right) = 0, \tag{11.4.16}$$

where $q_{ia} = \boldsymbol{q}_{ia}|_{\theta=0}$ are the first components of the superfield $\boldsymbol{q}_{ia}$. The equation of motion (11.4.14) corresponds to the static-gauge form of Nambu-Goto action for the massive particle in $D = 5$ space–time

$$S_{bos} \sim \int dt \left(1 - \sqrt{1 - 2\dot{q}^{ia}\dot{q}_{ia}}\right). \tag{11.4.17}$$

To construct the on-shell component action we will follow the same procedure which was described above in full details. So, we will omit unessential details concentrating only on the new features.

If we are interested in the invariance with respect to broken S supersymmetry, then we may consider the reduced coset element

$$g|_{\theta=0} = e^{tP}\, e^{q_{ia}Z^{ia}}\, e^{\psi_{a\alpha}S^{a\alpha}}. \tag{11.4.18}$$

Here, q_{ia} and $\psi_{a\alpha}$ are the first components of the superfields $\boldsymbol{q}_{ia}$ and $\boldsymbol{\psi}_{a\alpha}$. Similarly to the discussion in Sect. 11.3, the Goldstone fermions $\psi_{a\alpha}$ may enter the component action only through the einbein $\mathcal{E}$ and the covariant derivatives $\mathcal{D}_t q_{ia}$, defined now as

$$\partial_t = \mathcal{E}\,\mathcal{D}_t, \quad \mathcal{E} = 1 + \frac{1}{2}\,\Omega^{\beta\gamma}\psi^a_\beta \partial_t \psi_{a\gamma}, \quad \mathcal{E}^{-1} = 1 - \frac{1}{2}\,\Omega^{\beta\gamma}\psi^a_\beta \mathcal{D}_t \psi_{a\gamma}, \tag{11.4.19}$$

Keeping in the mind the known bosonic limit of the action (11.4.17), we come to the unique candidate of the component on-shell action

$$S = \alpha \int dt + (1-\alpha) \int dt\, \mathcal{E} - \int dt\, \mathcal{E}\sqrt{1 - 2\mathcal{D}_t q^{ia} \mathcal{D}_t q_{ia}}. \tag{11.4.20}$$

This action is perfectly invariant with respect to broken S supersymmetry, realized on the physical components and their derivatives as

$$\delta^*_S q_{ia} = \frac{1}{2}\,\eta^b_\alpha \psi_{b\beta}\Omega^{\alpha\beta}\partial_t q_{ia}, \quad \delta^*_S(\mathcal{D}_t\, q_{ia}) = \frac{1}{2}\,\eta^b_\alpha \psi_{b\beta}\Omega^{\alpha\beta}\partial_t(\mathcal{D}_t\, q_{ia}),$$
$$\delta^*_S \psi_{a\alpha} = \eta_{a\alpha} + \frac{1}{2}\,\eta^b_\beta \psi_{b\lambda}\Omega^{\beta\lambda}\partial_t \psi_{a\alpha}. \tag{11.4.21}$$

From (11.4.21) one may find the transformation properties of the einbein $\mathcal{E}$

$$\delta^*_S \mathcal{E} = \frac{1}{2}\,\eta^a_\alpha \partial_t \left(\mathcal{E}\Omega^{\alpha\beta}\psi_{a\beta}\right). \tag{11.4.22}$$

Now, combining (11.4.21) and (11.4.22), we will get

$$\delta^*_S \left(\mathcal{E}\mathcal{F}\left[\mathcal{D}_t\, q^{jb}\mathcal{D}_t\, q_{jb}\right]\right) = \frac{1}{2}\,\eta^a_\alpha \partial_t \left(\Omega^{\alpha\beta}\psi_{a\beta}\, \mathcal{E}\, \mathcal{F}\left[\mathcal{D}_t\, q^{jb}\mathcal{D}_t\, q_{jb}\right]\right), \tag{11.4.23}$$

and, therefore, the second and the third terms in the action (11.4.20) are separately invariant with respect to S supersymmetry. The first term in (11.4.20) is trivially invariant with respect to both, broken and unbroken supersymmetries.

The last step is to impose invariance with respect to unbroken Q supersymmetry. Under the transformations of unbroken supersymmetry taken in the fixed point the variation of any superfield reads

$$\delta_Q^* \boldsymbol{F} = -\varepsilon_i^\alpha Q_\alpha^i \boldsymbol{F}.$$

From this one may find the variations of the components q_{ia} and $\psi_{a\alpha}$ and their covariant derivatives:

$$\begin{aligned}
\delta_Q^* q_{ia} &= -\varepsilon_i^\alpha \psi_{a\alpha} + \frac{1}{4}\varepsilon_j^\alpha \lambda^{jb}\psi_{b\alpha}\partial_t q_{ia},\\
\delta_Q^*(\mathcal{D}_t q_{ia}) &= -\varepsilon_i^\alpha \mathcal{D}_t \psi_{a\alpha} + \frac{1}{4}\varepsilon_j^\alpha \frac{\lambda_{ia}}{1+\frac{1}{8}\lambda^2}\lambda^{jb}\mathcal{D}_t\psi_{b\alpha} + \frac{1}{4}\varepsilon_j^\alpha \lambda^{jb}\psi_{b\alpha}\partial_t(\mathcal{D}_t q_{ia}),\\
\delta_Q^* \psi_{a\alpha} &= \frac{1}{2}\varepsilon_j^\beta \Omega_{\alpha\beta}\lambda_a^j + \frac{1}{4}\varepsilon_j^\beta \lambda^{jb}\psi_{b\beta}\partial_t \psi_{a\alpha}.
\end{aligned} \tag{11.4.24}$$

The variation of the einbein $\mathcal{E}$ can be also computed and it reads

$$\delta_Q^* \mathcal{E} = \frac{1}{4}\varepsilon_j^\beta \partial_t\left(\mathcal{E}\lambda^{jb}\psi_{b\beta}\right) - \frac{1}{2}\varepsilon_j^\beta \lambda^{jb}\partial_t \psi_{b\beta}. \tag{11.4.25}$$

It is a matter of lengthy, but straightforward calculations to check that the action (11.4.20) is invariant under unbroken supersymmetry (11.4.24), (11.4.25) if $\alpha = 2$.

Thus, the component action, invariant under both unbroken and broken $N = 8$ supersymmetries, reads

$$S = \int dt \left[2 - \mathcal{E}\left(1 + \sqrt{1 - 2\mathcal{D}_t q^{ia}\mathcal{D}_t q_{ia}}\right)\right]. \tag{11.4.26}$$

11.5 Supermembrane in D = 4

As an instructive application of our approach we consider in this section as an example two models, namely, the supermembrane in $D = 4$ and the supersymmetric space-filling D2-brane. We will mainly follow the paper [12].

11.5.1 Supermembrane in D=4: Kinematical Constraints, Equations of Motion and the Component Action

The nonlinear realization of the breaking $N = 1, D = 4 \to N = 1, d = 3$ has been constructed in [12]. There, the $N = 1, D = 4$ super Poincaré group has been realized in its coset over the $d = 3$ Lorentz group $SO(1,2)$

$$g = e^{x^{ab}P_{ab}} e^{\theta^a Q_a} e^{\boldsymbol{q} Z} e^{\boldsymbol{\psi}^a S_a} e^{\boldsymbol{\Lambda}^{ab} K_{ab}}. \tag{11.5.1}$$

Here, x^{ab}, θ^a are $N = 1, d = 3$ superspace coordinates, while the remaining coset parameters are Goldstone superfields, $\boldsymbol{\psi}^a = \boldsymbol{\psi}^a(x,\theta)$, $\boldsymbol{q} = \boldsymbol{q}(x,\theta)$, $\boldsymbol{\Lambda}^{ab} = \boldsymbol{\Lambda}^{ab}(x,\theta)$. To reduce the number of independent superfields one has to impose the constraints[5]

$$\Omega_Z = 0 \quad \Rightarrow \quad \begin{cases} \nabla_{ab}\boldsymbol{q} + \frac{4}{1+2\boldsymbol{\lambda}^2}\boldsymbol{\lambda}_{ab} = 0, & \text{(a)} \\ \nabla_a \boldsymbol{q} - \boldsymbol{\psi}_a = 0. & \text{(b)} \end{cases} \tag{11.5.2}$$

Equation (11.5.2) allow us to express $\boldsymbol{\lambda}_{ab}(x,\theta)$ and $\boldsymbol{\psi}^a(x,\theta)$ through covariant derivatives of $\boldsymbol{q}(x,\theta)$. Thus, the bosonic superfield $\boldsymbol{q}(x,\theta)$ is the only essential Goldstone superfield we need for this case of the partial breaking of the global supersymmetry. The constraints (11.5.2) are covariant under all symmetries and they do not imply any dynamics and leave $\boldsymbol{q}(x,\theta)$ off-shell.

The last step we can make within the coset approach is to write the covariant superfield equations of motion. It was shown in [12] that this can be achieved by imposing the following constraint on the Cartan form:

$$\Omega_S| = 0 \quad \Rightarrow \quad \begin{cases} \nabla^a \boldsymbol{\psi}_a = 0, & \text{(a)} \\ \nabla_{(a}\boldsymbol{\psi}_{b)} = -2\boldsymbol{\lambda}_{ab}. & \text{(b)} \end{cases} \tag{11.5.3}$$

where $|$ denotes the ordinary $d\theta$-projection of the form Ω_S.

Equation (11.5.3) imply the proper dynamical equation of motion

$$\nabla^a \nabla_a \boldsymbol{q} = 0. \tag{11.5.4}$$

This equation is also covariant with respect to all symmetries, and its bosonic limit for $q(x) = \boldsymbol{q}(x,\theta)|_{\theta=0}$ reads

$$\partial_{ab}\left(\frac{\partial^{ab} q}{\sqrt{1 - \frac{1}{2}\partial q \cdot \partial q}}\right) = 0, \tag{11.5.5}$$

[5] We collect the exact expressions for the covariant derivatives ∇_{ab}, ∇_a and their properties, constructed in [12], in Appendix A.

which corresponds to the "static gauge" form of the $D = 4$ membrane Nambu-Goto action

$$S = \int d^3x \left(1 - \sqrt{1 - \frac{1}{2}\partial^{ab}q\partial_{ab}q} \right). \tag{11.5.6}$$

Thus, the (11.5.3) indeed describe the supermembrane in $D = 4$.

Until now we just repeated the standard coset approach steps from the paper [12] in the application to the $N = 1$, $D = 4$ supermembrane. As was already mentioned in Sect. 11.2, the nonlinear realization approach fails to construct the superfield action. That is why, to construct the superfield action one has to involve some additional arguments/scheme as it has been done, for example, in [12].

Funny enough, if we instead will be interested in the component action, then it can be constructed almost immediately within the nonlinear realization approach. One may check that all important features of the on-shell [i.e. with (11.5.3) taken into account] component action we summarized in Sect. 11.2, are present in the case at hands. Indeed,

- All physical components, i.e. $\boldsymbol{q}|_{\theta=0}$ and $\psi^a|_{\theta=0}$, are among the "coordinates" of our coset (11.5.1) as the $\theta = 0$ parts of the corresponding superfields,
- Under spontaneously broken supersymmetry the superspace coordinates θ^a do not transform at all (A.5). Therefore, the corresponding transformation properties of the fermionic components $\psi^a|_{\theta=0}$ are *the same as in the Volkov-Akulov model* [5], where all supersymmetries are supposed to be spontaneously broken,
- Finally, the $\theta = 0$ component of our essential Goldstone superfield $\boldsymbol{q}(x,\theta)$ does not transform under spontaneously broken supersymmetry and, therefore, it behaves like a "matter" field within the Volkov-Akulov scheme.

As the immediate consequences of these features we conclude that

- The fermionic components $\psi^a|_{\theta=0}$ may enter the component action either through $\det \mathcal{E}$ (A.14) (to compensate the transformation of volume d^3x under (A.5) or through the covariant derivatives $\mathcal{D}_{ab}$ (A.12), only,
- The "matter" field $q = \boldsymbol{q}|_{\theta=0}$ may enter the action only through covariant derivatives $\mathcal{D}_{ab}q$.

Thus, the unique candidate to be the component on-shell action, invariant with respect to spontaneously broken supersymmetry S reads

$$S = \alpha \int d^3x + \beta \int d^3x \, \det \mathcal{E}\mathcal{F}(\mathcal{D}^{ab}q\, \mathcal{D}_{ab}q), \tag{11.5.7}$$

with an arbitrary, for the time being, function $\mathcal{F}$. All other interactions between the bosonic component q and the fermions of spontaneously broken supersymmetry ψ^a are forbidden!

Note, that the first, trivial term in (11.5.7) is independently invariant under broken (and unbroken!) supersymmetries, because, in virtue of (A.5)

$$\delta_S \int d^3x \sim \int d^3x\, \partial_{ab}\left(\xi^a \psi^b\right) \quad \text{and, therefore} \quad \delta_S \int d^3x = 0. \tag{11.5.8}$$

As we already said in Sect. 11.2, this term in the action (11.5.7) ensures the validity of the limit $S_{q=0,\psi=0} = 0$.

The action (11.5.7) is the most general component action invariant with respect to broken supersymmetry. But in the present case we explicitly know its bosonic limit—it should be just the Nambu-Goto action (11.5.6). Some additional information about its structure comes from the linearized form of the action, which, according with its invariance with respect to unbroken supersymmetry, has to be

$$S_{lin} \sim \psi^a \partial_{ab} \psi^b - \frac{1}{4}\partial^{ab} q \partial_{ab} q. \tag{11.5.9}$$

Combining all these ingredients, which completely fix the parameters α and β in (11.5.7), we can write the component action of $N = 1, D = 4$ supermembrane as

$$S = \int d^3x \left[2 - \det \mathcal{E} \left(1 + \sqrt{1 - \frac{1}{2}\mathcal{D}^{ab} q \mathcal{D}_{ab} q}\right)\right]. \tag{11.5.10}$$

The explicit expression for $\det \mathcal{E}$ has the form

$$\begin{aligned}\det \mathcal{E} &= 1 + \frac{1}{2}\psi^a \mathcal{D}_{ab}\psi^b - \frac{1}{16}\psi^d \psi_d\, \mathcal{D}^{ab}\psi^c \mathcal{D}_{ab}\psi_c = \\ &= 1 + \frac{1}{2}\psi^a \partial_{ab}\psi^b + \frac{1}{8}\psi^d \psi_d \left(\partial^{ab}\psi_b \partial_{ac}\psi^c + \frac{1}{2}\partial^{ab}\psi^c \partial_{ab}\psi_c\right).\end{aligned} \tag{11.5.11}$$

Let us stress, that such a simple form of the component action is achieved only in the rather specific basis, where the bosonic q and fermionic fields ψ^a are the Goldstone fields of the nonlinear realization. Surely, this choice is not unique and in different bases the explicit form of action could drastically change. The most illustrative example is given by the action in [28], where the on-shell component action for the supermembrane has been constructed for the first time.

The detailed proof that the action (11.5.10) is invariant with respect to both, broken and unbroken supersymmetries, can be found in Appendix B.

11.5.2 Supersymmetric Space-Filling D2-Brane

Due to the duality between scalar field and gauge field strength in $d = 3$, the action for D2-brane can be easily constructed within the coset approach. The idea of the construction is similar to the purely bosonic case. The crucial step is to treat the first, bosonic component of $\boldsymbol{\lambda}_{ab}$ as an independent component [i.e. to ignore the (a) part of (11.5.2)]. Now, the generalized variant of the action (11.5.10) reads

$$S = \int d^3x \left[2 - \det\mathcal{E} - \det\mathcal{E}\left(1 + 2\frac{\lambda^{ab}(\mathcal{D}_{ab}q + 2\lambda_{ab})}{1-2\lambda^2}\right)\right]. \tag{11.5.12}$$

All these summands have a description in terms of $\theta = 0$ parts of the Cartan forms (A.9). The first term is just a volume form constructed from ordinary differentials dx^{ab}. The second terms is a volume form constructed from semi-covariant differentials $d\hat{x}^{ab}$

$$d\hat{x}^{ab} = dx^{ab} + \frac{1}{4}\psi^a d\psi^b + \frac{1}{4}\psi^b d\psi^a.$$

Finally, the last term in (11.5.12) is a volume form constructed from the $\theta = 0$ component of the form Ω_P^{ab} (A.9)

$$d\tilde{x}^{ab} = d\hat{x}^{ab} + \frac{2}{1-2\lambda^2}\lambda^{ab}\left(\mathcal{D}_{cd}q + 2\lambda_{cd}\right)d\hat{x}^{cd}.$$

Since the action (11.5.12) depends only on λ^{ab} and not on its derivatives, the λ-equation of motion

$$\mathcal{D}_{ab}q = -\frac{4\lambda_{ab}}{1+2\lambda^2} \tag{11.5.13}$$

can be used to eliminate λ^{ab} in favor of $\mathcal{D}_{ab}q$. Clearly, the (11.5.13) is just the (*a*) part of the constraints (11.5.2), we ignored while introducing the action (11.5.12). Plugging λ_{ab} expressed through $\mathcal{D}_{ab}q$ back into (11.5.12) gives us the action (11.5.10).

Alternatively, the equation of motion for q

$$\partial_{ab}\left[\frac{\det\mathcal{E}\ \lambda^{cd}\ \left(\mathcal{E}^{-1}\right)_{cd}{}^{ab}}{1-2\lambda^2}\right] = 0 \tag{11.5.14}$$

has the form of the $d = 3$ Bianchi identity for the field strength F^{ab}

$$F^{ab} \equiv \frac{\det\mathcal{E}\ \lambda^{cd}\ \left(\mathcal{E}^{-1}\right)_{cd}{}^{ab}}{1-2\lambda^2} \quad\Rightarrow\quad \partial_{ab}F^{ab} = 0. \tag{11.5.15}$$

Substituting this into the action (11.5.12) and integrating by parts, one may bring it to the supersymmetric D2-brane action

$$S = \int d^3x \left[2 - \det\mathcal{E}\left(1 + \sqrt{1+8\widetilde{F}^2}\right)\right], \tag{11.5.16}$$

where

$$\widetilde{F}_{ab} \equiv \frac{\mathcal{E}_{ab}{}^{cd}\ F_{cd}}{\det\mathcal{E}} = \frac{\lambda_{ab}}{1-2\lambda^2}. \tag{11.5.17}$$

Therefore,

$$S = 2\int d^3x \left[1 - \det \mathcal{E}\, \frac{1}{1-2\lambda^2}\right]. \tag{11.5.18}$$

Clearly, in the bosonic limit $\widetilde{F}_{ab} = F_{ab}$ and thus, the bosonic part of the (11.5.16) is the standard Born-Infeld action for D2-brane, as it should be.

11.6 Supermembrane in D = 5

In this section we construct the on-shell component action for $N = 1, D = 5$ supermembrane and its dual versions, corresponding to a vector and a double vector supermultiplets. We demonstrate that the proper choice of the components and using the covariant (with respect to broken supersymmetry) derivatives drastically simplify the action: it can be represented as the sum of four terms each having an explicit geometric meaning.

11.6.1 Supermembrane

In the present case we are dealing with spontaneous breaking of $N = 1, D = 5$ Poincaré supersymmetry down to $N = 2, d = 3$ one. From the $d = 3$ standpoint the $N = 1, D = 5$ supersymmetry algebra is a central-charges extended $N = 4$ Poincaré superalgebra with the following basic anticommutation relations:

$$\left\{Q_a, \overline{Q}_b\right\} = 2P_{ab}, \quad \left\{S_a, \overline{S}_b\right\} = 2P_{ab}, \quad \left\{Q_a, S_b\right\} = 2\epsilon_{ab}Z, \quad \left\{\overline{Q}_a, \overline{S}_b\right\} = 2\epsilon_{ab}\overline{Z}. \tag{11.6.1}$$

The $d = 3$ translations generator P_{ab} and the central charge generators $Z, \overline{Z}$ form $D = 5$ translation generators. We will also split the generators of $D = 5$ Lorentz algebra $so(1,4)$ into $d = 3$ Lorentz algebra generators M_{ab}, the generators K_{ab} and $\overline{K}_{ab}$ belonging to the coset $SO(1,4)/SO(1,2) \times U(1)$ and $U(1)$ generator J. The full set of (anti)commutation relations can be found in the Appendix C.

Keeping $d = 3$ Lorentz and, commuting with it, $U(1)$ subgroups of $D = 5$ Lorentz group $SO(1,4)$ linearly realized, we will choose the coset element as

$$g = e^{ix^{ab}P_{ab}} e^{\theta^a Q_a + \bar{\theta}^a \overline{Q}_a} e^{i(\boldsymbol{q}Z + \bar{\boldsymbol{q}}\overline{Z})} e^{\psi^a S_a + \bar{\psi}^a \overline{S}_a} e^{i(\Lambda^{ab}K_{ab} + \bar{\Lambda}^{ab}\overline{K}_{ab})}. \tag{11.6.2}$$

Here, $\left\{x^{ab}, \theta^a, \bar{\theta}^a\right\}$ are $N = 2, d = 3$ superspace coordinates, while the remaining coset parameters are $N = 2$ Goldstone superfields. The whole $N = 1, D = 5$ Poincaré supergroup can be realized in this coset by the left acting on (11.6.2) of the different elements of the supergroup. We summarize in Appendix C the resulting transformation properties of the coordinates and superfields with respect to unbroken (C.6), broken (C.7) supersymmetries and automorphism (C.8), as well as a pure technical calculation of Cartan forms, semi-covariant derivatives and their superalgebra (C.11), (C.14), (C.18).

Similarly to the previously considered cases, to reduce the number of independent superfields one has to impose the constraints

$$\Omega_Z = 0 \quad \Rightarrow \quad \begin{cases} \nabla_{ab}\boldsymbol{q} = -2i\, \frac{(1+\boldsymbol{l}\cdot\bar{\boldsymbol{l}})\boldsymbol{l}_{ab} - \boldsymbol{l}^2 \bar{\boldsymbol{l}}_{ab}}{(1+\boldsymbol{l}\cdot\bar{\boldsymbol{l}})^2 - \boldsymbol{l}^2\bar{\boldsymbol{l}}^2}, \\ \nabla_a \boldsymbol{q} = -2i\, \psi_a, \quad \overline{\nabla}_a \boldsymbol{q} = 0, \end{cases}$$

$$\overline{\Omega}_Z = 0 \quad \Rightarrow \quad \begin{cases} \nabla_{ab}\bar{\boldsymbol{q}} = 2i\, \frac{(1+\boldsymbol{l}\cdot\bar{\boldsymbol{l}})\bar{\boldsymbol{l}}_{ab} - \bar{\boldsymbol{l}}^2 \boldsymbol{l}_{ab}}{(1+\boldsymbol{l}\cdot\bar{\boldsymbol{l}})^2 - \boldsymbol{l}^2\bar{\boldsymbol{l}}^2}, \\ \overline{\nabla}_a \bar{\boldsymbol{q}} = -2i\, \bar{\psi}_a, \quad \nabla_a \bar{\boldsymbol{q}} = 0. \end{cases} \tag{11.6.3}$$

Here, in order to simplify the expressions, we have passed to the some variant of the stereographic parametrization of the coset $SO(1,4)/SO(1,2) \times U(1)$

$$\boldsymbol{l}_{ab} = \left(\frac{\tanh \sqrt{\boldsymbol{Y}}}{\sqrt{\boldsymbol{Y}}} \right)_{ab}^{cd} \boldsymbol{\Lambda}_{cd}, \quad \bar{\boldsymbol{l}}_{ab} = \left(\frac{\tanh \sqrt{\boldsymbol{Y}}}{\sqrt{\boldsymbol{Y}}} \right)_{ab}^{cd} \overline{\boldsymbol{\Lambda}}_{cd}. \tag{11.6.4}$$

The Equation (11.6.3) allow us to express superfields $\boldsymbol{\Lambda}_{ab}$, $\overline{\boldsymbol{\Lambda}}_{ab}$ and ψ^a, $\bar{\psi}^a$ through covariant derivatives of $\boldsymbol{q}(x,\theta,\bar{\theta})$ and $\bar{\boldsymbol{q}}(x,\theta,\bar{\theta})$. Thus, the bosonic superfields $\boldsymbol{q}(x,\theta,\bar{\theta})$,
$\bar{\boldsymbol{q}}(x,\theta,\bar{\theta})$ are the only essential Goldstone superfields needed for this case of the partial breaking of the global supersymmetry. The constraints (11.6.3) are covariant under all symmetries, they do not imply any dynamics and leave $\boldsymbol{q}(x,\theta,\bar{\theta})$ and $\bar{\boldsymbol{q}}(x,\theta,\bar{\theta})$ off-shell.

Within the coset approach we may also write the covariant superfield equations of motion. This can be achieved by imposing the proper constraint on the Cartan forms for broken supersymmetry. In the present case these constraints read

$$\Omega_S| = 0 \Rightarrow (a)\ \nabla_a \psi_b = 0, \quad (b)\ \overline{\nabla}_b \psi^a = -i\, \boldsymbol{\Lambda}_b^c \left(\frac{\tan 2\sqrt{\overline{\boldsymbol{T}}}}{\sqrt{\overline{\boldsymbol{T}}}} \right)_c^a \equiv -i\, \boldsymbol{\lambda}_b^a$$

$$\overline{\Omega}_S\big| = 0 \Rightarrow (a)\ \overline{\nabla}_a \bar{\psi}_b = 0, \quad (b)\ \nabla_b \bar{\psi}^a = i\, \boldsymbol{\Lambda}_b^c \left(\frac{\tan 2\sqrt{\boldsymbol{T}}}{\sqrt{\boldsymbol{T}}} \right)_c^a \equiv i\, \bar{\boldsymbol{\lambda}}_b^{a},$$

(11.6.5)

where | means the $d\theta$-projection of the forms.

Let us make a few comments concerning the constraints given above:

- The easiest way to check that the (11.6.3), (11.6.5) put the theory on-shell is to consider these equations in the linearized form

$$\partial_{ab}\boldsymbol{q} = -2i\,\boldsymbol{\Lambda}_{ab}\ (a),\quad D_a\boldsymbol{q} = -2i\,\boldsymbol{\psi}_a\ (b),\quad \overline{D}_a\boldsymbol{q} = 0\ (c), \tag{11.6.6}$$

$$D_a\boldsymbol{\psi}_b = 0\ (a),\quad \overline{D}_b\boldsymbol{\psi}^a = -2i\,\boldsymbol{\Lambda}^a_b\ (b). \tag{11.6.7}$$

Acting on (11.6.6b) by $\overline{D}_b$ and using the (11.6.6c) and the algebra of spinor derivatives (C.15) we immediately conclude that (11.6.7b) follows from (11.6.6). In addition, the (11.6.7a) means that the auxiliary component of the superfield $\boldsymbol{q}$ is zero and, therefore, our system is on-shell

$$D_a\boldsymbol{\psi}_b = 0 \Rightarrow D^2\boldsymbol{q} = 0 \Rightarrow \partial_{ab}D^b\boldsymbol{q} = 0 \Rightarrow \Box\boldsymbol{q} = 0. \tag{11.6.8}$$

- It turns out that the variables $\{\boldsymbol{\lambda}^b_a, \bar{\boldsymbol{\lambda}}^b_a\}$ defined in (11.6.5), are more suitable than the $\{\boldsymbol{l}_{ab}, \bar{\boldsymbol{l}}_{ab}\}$ (11.6.4) one. Using the algebra of covariant derivatives (C.18) it is easy to find the following relations from (11.6.3) and (11.6.5):

$$\nabla_{ab}\boldsymbol{q} = -i\,\frac{\boldsymbol{\lambda}_{ab} - \frac{1}{2}\boldsymbol{\lambda}^2\bar{\boldsymbol{\lambda}}_{ab}}{1 - \frac{1}{4}\boldsymbol{\lambda}^2\bar{\boldsymbol{\lambda}}^2},\quad \nabla_{ab}\bar{\boldsymbol{q}} = i\,\frac{\bar{\boldsymbol{\lambda}}_{ab} - \frac{1}{2}\bar{\boldsymbol{\lambda}}^2\boldsymbol{\lambda}_{ab}}{1 - \frac{1}{4}\boldsymbol{\lambda}^2\bar{\boldsymbol{\lambda}}^2}. \tag{11.6.9}$$

These equations play the same role as those in (11.6.3), relating the superfields $\{\boldsymbol{\lambda}_{ab}, \bar{\boldsymbol{\lambda}}_{ab}\}$ (and, therefore, the superfields $\{\boldsymbol{\Lambda}_{ab}, \overline{\boldsymbol{\Lambda}}_{ab}\}$) with the space–time derivatives of the superfields $\{\boldsymbol{q}, \bar{\boldsymbol{q}}\}$.

Now we present two different ways to construct the bosonic action. The first of them is based on the consideration of the bosonic coset related to (11.6.2) and on the invariance of constraints (11.6.3), (11.6.5) with respect to all $N = 1$, $D = 5$ Poincaré supergroup. Thus we have

$$g_{bos} = e^{i\,x^{ab}P_{ab}}e^{i\,(qZ+\bar{q}\overline{Z})}e^{i\,(\Lambda^{ab}K_{ab}+\overline{\Lambda}^{ab}\overline{K}_{ab})}. \tag{11.6.10}$$

Clearly, the corresponding bosonic Cartan forms can be easily extracted from their superfields version (C.10). The bosonic version of the constraints (11.6.3) results in the relations

$$\partial_{ab}q = -2i\,\frac{(1 + l\cdot\bar{l})l_{ab} - l^2\bar{l}_{ab}}{(1 + l\cdot\bar{l})^2 - l^2\bar{l}^2},\quad \partial_{ab}\bar{q} = 2i\,\frac{(1 + l\cdot\bar{l})\bar{l}_{ab} - \bar{l}^2 l_{ab}}{(1 + l\cdot\bar{l})^2 - l^2\bar{l}^2}, \tag{11.6.11}$$

while the bosonic vielbein $\mathcal{B}_{ab}{}^{cd} = \mathcal{E}_{ab}{}^{cd}|_{\psi=0}$

$$\left(\Omega_P^{bos}\right) = dx^{ab}\mathcal{B}_{ab}{}^{cd}P_{cd} \tag{11.6.12}$$

acquires the form

$$\mathcal{B}^{cd}_{ab} = \delta^{(c}_a\delta^{d)}_b - \frac{2}{(1 + l\cdot\bar{l})^2 - l^2\,\bar{l}^2}\left[(1 + l\cdot\bar{l})\left(\bar{l}^{cd}l_{ab} + l^{cd}\bar{l}_{ab}\right) - \bar{l}^2\,l^{cd}l_{ab} - l^2\,\bar{l}^{cd}\bar{l}_{ab}\right].$$

Therefore, the simplest invariant bosonic action reads

$$S_{bos} = \int d^3x \det \mathcal{B} = \int d^3x \frac{(1 - l \cdot \bar{l})^2 - l^2 \bar{l}^2}{(1 + l \cdot \bar{l})^2 - l^2 \bar{l}^2}, \tag{11.6.13}$$

or in terms of $\{q, \bar{q}\}$

$$S_{bos} = \int d^3x \sqrt{\left(1 - \partial_{ab}q\, \partial^{ab}\bar{q}\right)^2 - \left(\partial_{ab}q\, \partial^{ab}q\right)\left(\partial_{cd}\bar{q}\, \partial^{cd}\bar{q}\right)}. \tag{11.6.14}$$

The latter is the static gauge Nambu-Goto action for membrane in $D = 5$. One can also add the following action, trivially invariant under the transformations $ISO(1, 4)$

$$S_0 = \int d^3x. \tag{11.6.15}$$

Another way to derive the bosonic action is to use automorphism transformation laws. These laws (C.8) in the bosonic limit have the form

$$\delta x^{ab} = 2i\left(\bar{a}^{ab} q - a^{ab}\bar{q}\right), \quad \delta q = -2i\,(ax), \quad \delta\bar{q} = 2i\,(\bar{a}x). \tag{11.6.16}$$

The active form of these transformations reads

$$\delta^* q = -2i\,(ax) - 2i\,\partial_{ab}q\left(\bar{a}^{ab}q - a^{ab}\bar{q}\right), \quad \delta^*\bar{q} = 2i\,(\bar{a}x) - 2i\,\partial_{ab}\bar{q}\left(\bar{a}^{ab}q - a^{ab}\bar{q}\right). \tag{11.6.17}$$

Due to translations, $U(1)$-rotations and $d = 3$ Lorentz invariance, the action may depend only on the following scalar combination of partial derivatives of bosons $\{q, \bar{q}\}$

$$\xi = \partial_{ab}q\, \partial^{ab}\bar{q}, \quad \eta = \partial_{ab}q\, \partial^{ab}q, \quad \bar{\eta} = \partial_{ab}\bar{q}\, \partial^{ab}\bar{q}, \tag{11.6.18}$$

which in accordance with (11.6.17) transforms as

$$\begin{aligned}
\delta^*\xi &= 2i\,(\bar{a}\partial q) - 2i\,(a\partial\bar{q}) - 2i\,(\bar{a}^{ab}q - a^{ab}\bar{q})\partial_{ab}\xi - 2i\,(\bar{a}\partial q)\xi \\
&\quad + 2i(a\partial\bar{q})\xi - 2i\,(\bar{a}\partial\bar{q})\eta + 2i\,(a\partial q)\bar{\eta}, \\
\delta^*(\eta\bar{\eta}) &= 4i(\bar{a}\partial\bar{q})\eta - 4i\,(a\partial q)\bar{\eta} - 2i\,(\bar{a}^{kl}q - a^{kl}\bar{q})\partial_{kl}(\eta\bar{\eta}) \\
&\quad - 4i\,(\bar{a}\partial q)\eta\bar{\eta} + 4i\,(a\partial\bar{q})\eta\bar{\eta} \\
&\quad + 4i\,(a\partial q)\xi\bar{\eta} - 4i\,(\bar{a}\partial\bar{q})\xi\eta.
\end{aligned} \tag{11.6.19}$$

Therefore, the variation of the arbitrary function $\mathcal{F}(\xi, \eta\bar{\eta})$ reads

$$\begin{aligned}\delta^* \mathcal{F} = 2i\ [(a\partial q)\bar\eta - (\bar a\partial\bar q)\eta]\left(\mathcal{F}_\xi + 2(\xi-1)\mathcal{F}_{(\eta\bar\eta)}\right)\\ +\ 2i\ [(\bar a\partial q) - (a\partial\bar q)]\left(\mathcal{F} + (1-\xi)\mathcal{F}_\xi - 2\eta\bar\eta\mathcal{F}_{(\eta\bar\eta)}\right)\\ -\ 2i\ \partial_{ab}\left[\left(q\bar a^{ab} - \bar q a^{ab}\right)\mathcal{F}\right].\end{aligned} \tag{11.6.20}$$

Thus, to achieve the invariance of the action one has impose the following restrictions on the function $\mathcal{F}$:

$$\mathcal{F}_\xi + 2(\xi-1)\mathcal{F}_{(\eta\bar\eta)} = 0, \quad \mathcal{F} + \mathcal{F}_\xi(1-\xi) - 2(\eta\bar\eta)\mathcal{F}_{(\eta\bar\eta)} = 0, \tag{11.6.21}$$

with the evident solution

$$\mathcal{F} = \sqrt{(1-\xi)^2 - \eta\bar\eta}. \tag{11.6.22}$$

Therefore, the invariant action has the form

$$S = \int d^3x \sqrt{(1-\partial_{ab}q\partial^{ab}\bar q)^2 - (\partial_{ab}q\partial^{ab}q)(\partial_{kl}\bar q\partial^{kl}\bar q)},$$

and thus, it coincides with that previously constructed in (11.6.14), as it should be.

Let us now construct the full component action for supermembrane which will be invariant under both broken and unbroken supersymmetries. We begin our analysis with the broken supersymmetry S.

The superspace coordinates $\{\theta, \bar\theta\}$ of the coset (11.6.2) do not transform under S supersymmetry. Therefore, each component of superfields transforms independently under the broken supersymmetry. Thus, from (C.7) one finds

$$\delta x^{ab} = i\left(\varepsilon^{(a}\bar\psi^{b)} + \bar\varepsilon^{(a}\psi^{b)}\right), \quad \delta q = 0, \quad \delta\bar q = 0, \quad \delta\psi^a = \varepsilon^a, \quad \delta\bar\psi^a = \bar\varepsilon^a. \tag{11.6.23}$$

Then, one may easily check that the $\theta = 0$ projections of the covariant differential $\triangle x^{ab}$ (C.10)

$$\hat\triangle x^{ab} \equiv \triangle x^{ab}|_{\theta=0} = dx^{ab} - i\left(\psi^{(a}d\bar\psi^{b)} + \bar\psi^{(a}d\psi^{b)}\right) \equiv \mathcal{E}^{ab}_{cd}\, dx^{cd}, \tag{11.6.24}$$

as well as the covariant derivatives constructed from them

$$\mathcal{D}_{ab} = \left(\mathcal{E}^{-1}\right)_{ab}^{cd}\,\partial_{cd} \tag{11.6.25}$$

are also invariant under broken supersymmetry. From all this it immediately follows that the action possessing the proper bosonic limit (11.6.14) and invariant under broken supersymmetry reads

$$S_1 = \int d^3x\ \det\mathcal{E}\sqrt{(1-\mathcal{D}_{ab}\,q\mathcal{D}^{ab}\bar q)^2 - (\mathcal{D}_{ab}\,q\mathcal{D}^{ab}q)(\mathcal{D}_{cd}\bar q\ \mathcal{D}^{cd}\bar q)}. \tag{11.6.26}$$

The action (11.6.26) reproduces the kinetic terms for the bosonic and fermionic components

$$S_1 = \int d^3x \left[-i \left(\psi^a \partial_{ab} \bar{\psi}^b + \bar{\psi}^a \partial_{ab} \psi^b \right) - \partial_{ab} q \partial^{ab} \bar{q} + \cdots \right], \qquad (11.6.27)$$

but the coefficient between them is strictly fixed. This could be not enough to maintain unbroken supersymmetry. So, one has to add to (11.6.26) the purely fermionic action

$$S_2 = \int d^3x \, \det \mathcal{E}, \qquad (11.6.28)$$

which is trivially invariant under broken supersymmetry. Finally, in order to have a proper limit

$$S_{q \to 0, \psi \to 0} = 0,$$

one has to involve into the game the trivial action S_0 that reads as

$$S_0 = \int d^3x. \qquad (11.6.29)$$

Thus, the Ansatz for the supersymmetric action acquires the form

$$\begin{aligned} S &= (1+\alpha) S_0 - S_1 - \alpha S_2 \\ &= (1+\alpha) \int d^3x - \int d^3x \, \det \mathcal{E} \\ &\left(\alpha + \sqrt{(1 - \mathcal{D}_{ab} q \mathcal{D}^{ab} \bar{q})^2 - (\mathcal{D}_{ab} q \mathcal{D}^{ab} q)(\mathcal{D}_{cd} \bar{q} \mathcal{D}^{cd} \bar{q})} \right), \end{aligned} \qquad (11.6.30)$$

where the constant α has to be defined.

In the previously considered cases in the above sections, the Ansatz, similar to (11.6.30), was completely enough to maintain the second, unbroken supersymmetry. A careful analysis shows that in the present case there is an additional Wess-Zumino term which has to be taken into account

$$S_{WZ} = i \int d^3x \, \det \mathcal{E} \, \left(\psi^m \mathcal{D}_{ab} \bar{\psi}_m - \bar{\psi}^m \mathcal{D}_{ab} \psi_m \right) \mathcal{D}^{ac} \, q \, \mathcal{D}_c{}^b \, \bar{q}. \qquad (11.6.31)$$

The variation of S_{WZ} under S supersymmetry reads (note, that only the variations of $\psi, \bar{\psi}$ without derivatives play a role)

$$\delta S_{WZ} = i \int d^3x \, \det \mathcal{E} \, \left(\varepsilon^m \mathcal{D}_{ab} \bar{\psi}_m - \bar{\varepsilon}^m \mathcal{D}_{ab} \psi_m \right) \mathcal{D}^{ac} q \, \mathcal{D}_c{}^b \bar{q}. \qquad (11.6.32)$$

The simplest way to check that $\delta S_{WZ} = 0$ is to pass to the $d = 3$ vector notations.[6] Then we have

$$\begin{aligned}
\delta S_{WZ} &\sim \int d^3x \; \det\mathcal{E}\; \epsilon^{IJK} \left(\varepsilon^m \mathcal{D}_I \bar{\psi}_m - \bar{\varepsilon}^m \mathcal{D}_I \psi_m\right) \mathcal{D}_J\, q\, \mathcal{D}_K\, \bar{q} \\
&\sim \int d^3x \; \det\mathcal{E}\; \det\mathcal{E}^{-1} \epsilon^{IJK} \left(\varepsilon^m \partial_I \bar{\psi}_m - \bar{\varepsilon}^m \partial_I \psi_m\right) \partial_J\, q\, \partial_K\, \bar{q} \\
&\sim \int d^3x \; \partial_I \left[\epsilon^{IJK} \left(\varepsilon^m \bar{\psi}_m - \bar{\varepsilon}^m \psi_m\right) \partial_J\, q\, \partial_K\, \bar{q}\right] = 0. \qquad (11.6.33)
\end{aligned}$$

Thus, the action S_{WZ} (11.6.31) is invariant under S supersymmetry and our Ansatz for the membrane action is extended to be

$$S = (1+\alpha)\, S_0 - S_1 - \alpha S_2 + \beta S_{WZ}. \qquad (11.6.34)$$

Thus, after imposing broken supersymmetry, the component action (11.6.34) is fixed up to two constants α and β. No other terms or structures are admissible!

Now we are going to demonstrate how the unbroken supersymmetry fixes these constants. In order to maintain the unbroken supersymmetry, one has to find the transformation properties of all objects presented in (11.6.34). Using the transformations of the superspace coordinates (C.6) one gets for the ϵ-part of the transformations

$$\begin{aligned}
\delta\psi_a &= -\epsilon^b \left(D_b \boldsymbol{\psi}_a\right)\big|_{\theta=0} = \epsilon^b \psi^m \bar{\lambda}^n_b \partial_{mn} \psi_a, \\
\delta\mathcal{D}_{ab}\psi_c &= -\epsilon^d \left(D_d \nabla_{ab} \boldsymbol{\psi}_c\right)\big|_{\theta=0} = 2\epsilon^d \mathcal{D}_{ab}\psi^m \bar{\lambda}^n_d \mathcal{D}_{mn}\psi_b + \epsilon^d \psi^m \bar{\lambda}^n_d \partial_{mn} \mathcal{D}_{ab}\psi_c, \\
\delta\mathcal{D}_{ab} q &= -\epsilon^d \left(D_d \nabla_{ab} \boldsymbol{q}\right)|_{\theta=0} = 2\epsilon^d \mathcal{D}_{ab}\psi^m \bar{\lambda}^n_d \mathcal{D}_{mn} q + 2i\, \epsilon^d \mathcal{D}_{ab}\psi_d \\
&\quad + \epsilon^d \psi^m \bar{\lambda}^n_d \partial_{mn} \mathcal{D}_{ab} q, \qquad (11.6.35)
\end{aligned}$$

and, as the consequence,

$$\delta \det\mathcal{E} = \partial_{mn}\left[\epsilon^d \psi^m \bar{\lambda}_d{}^n \det\mathcal{E}\right] - 2\epsilon^d \bar{\lambda}_d{}^n \mathcal{D}_{mn}\psi^n \det\mathcal{E}. \qquad (11.6.36)$$

In order to fix the parameter α one may consider just the kinetic terms in the action (11.6.34)

$$S_{kin} = \int d^3x \left[-i\, (\alpha+1) \left(\psi^a \partial_{ab} \bar{\psi}^b + \bar{\psi}^a \partial_{ab} \psi^b\right) + \partial_{ab} q\, \partial^{ab} \bar{q}\right], \qquad (11.6.37)$$

which has to be invariant under the linearized form of the transformations (11.6.35) [see also (11.6.6), (11.6.7)]

$$\delta\bar{\psi}_a = -i\, \epsilon^b \bar{\lambda}_{ba} \simeq -\epsilon^b \partial_{ba} \bar{q}, \quad \delta\partial_{ab} q = 2i\, \epsilon^d \partial_{ab}\psi_d. \qquad (11.6.38)$$

[6] Our conventions to pass to/from vector indices are summarized in Appendix C, (C.20).

Varying the integrand in (11.6.37) and integrating by parts, we get

$$\delta S_{kin} = \int d^3x \left[2i\,(\alpha+1)\epsilon^c \psi^a \partial_{ab}\partial_c{}^b \bar{q} - 2i\,\epsilon^d \psi_d \Box \bar{q}\right]$$
$$= \int d^3x \left[i\,(\alpha+1)\epsilon^d \psi_d \Box \bar{q} - 2i\,\epsilon^d \psi_d \Box \bar{q}\right]. \qquad (11.6.39)$$

Therefore, we have to fix the constant α as

$$\alpha = 1. \qquad (11.6.40)$$

The fixing of the last parameter β is more involved. Using the transformation properties (11.6.35) one may find

$$\delta\mathcal{F} = 2\left(\epsilon^c \bar{\lambda}^n_c \mathcal{D}_{ab}\psi^m \mathcal{D}_{nm} q + i\,\epsilon^c \mathcal{D}_{ab}\psi_c\right)\frac{\partial\mathcal{F}}{\partial\mathcal{D}_{ab}q}$$
$$+\; 2\epsilon^c \bar{\lambda}^n_c \mathcal{D}_{ab}\psi^m \mathcal{D}_{mn}\bar{q}\,\frac{\partial\mathcal{F}}{\partial\mathcal{D}_{ab}\bar{q}} + \epsilon^c \bar{\lambda}^n_c \psi^m \partial_{mn}\mathcal{F}, \qquad (11.6.41)$$

where

$$\mathcal{F} \equiv \sqrt{(1-\mathcal{D}_{ab}q\,\mathcal{D}^{ab}\bar{q})^2 - (\mathcal{D}_{ab}q\,\mathcal{D}^{ab}q)(\mathcal{D}_{cd}\bar{q}\,\mathcal{D}^{cd}\bar{q})}. \qquad (11.6.42)$$

In order to avoid the appearance of the square roots, it proved to be more convenient to use the equalities

$$\frac{\partial\mathcal{F}}{\partial\mathcal{D}_{ab}q} = -i\,\frac{\bar{\lambda}^{ab}+\frac{1}{2}\bar{\lambda}^2\lambda^{ab}}{1-\frac{1}{4}\lambda^2\bar{\lambda}^2}, \quad \frac{\partial\mathcal{F}}{\partial\mathcal{D}_{ab}\bar{q}} = i\,\frac{\lambda^{ab}+\frac{1}{2}\lambda^2\bar{\lambda}^{ab}}{1-\frac{1}{4}\lambda^2\bar{\lambda}^2}. \qquad (11.6.43)$$

Performing a straightforward calculation one gets

$$\delta\left[-\det\mathcal{E}\,(1+\mathcal{F})\right] = 2i\,\epsilon^c \det\mathcal{E}\left(\mathcal{D}_{ab}\psi_c\,\mathcal{D}^{ab}\bar{q} - 2\mathcal{D}_{am}\psi^m\,\mathcal{D}^a_c\bar{q}\right)$$
$$-\; 2\epsilon^c \det\mathcal{E}\,\bar{\lambda}_{cm}\mathcal{D}_{ab}\psi^m\,\mathcal{D}^{ad}q\,\mathcal{D}^b_d\bar{q}. \qquad (11.6.44)$$

Similarly, one may find the variation of the integrand of the action S_{WZ} (up to the surface terms disappearing after integration over d^3x)

$$\delta\mathcal{L}_{WZ} = -2\beta\epsilon^c \det\mathcal{E}\left[\left(\psi^k\mathcal{D}_{ab}\bar{\psi}_k - \bar{\psi}^k\mathcal{D}_{ab}\psi_k\right)\mathcal{D}^{ad}\psi_c\,\mathcal{D}^b_d\bar{q} - \bar{\lambda}_{cm}\mathcal{D}_{ab}\psi^m\,\mathcal{D}^{ad}q\,\mathcal{D}^b_d\bar{q}\right]. \qquad (11.6.45)$$

Now, it is a matter of quite lengthly, but again straightforward calculations, to check that the sum of variations (11.6.44) and (11.6.45) is a surface term if

$$\beta = 1. \tag{11.6.46}$$

Thus, we conclude that the action of the supermembrane in $D = 5$, which is invariant with respect to both unbroken and broken supersymmetries, has the form

$$S = 2\int d^3x - \int d^3x \, \det\mathcal{E}\left(1 + \sqrt{(1 - \mathcal{D}_{ab}q\,\mathcal{D}^{ab}\bar{q})^2 - (\mathcal{D}_{ab}q\,\mathcal{D}^{ab}q)(\mathcal{D}_{cd}\bar{q}\,\mathcal{D}^{cd}\bar{q})}\right)$$
$$+ \, i\int d^3x \, \det\mathcal{E}\,(\psi^m\mathcal{D}_{ab}\bar{\psi}_m - \bar{\psi}^m\mathcal{D}_{ab}\psi_m)\,\mathcal{D}^{ac}q\,\mathcal{D}_c{}^b\bar{q}. \tag{11.6.47}$$

11.6.2 Dualization of the Scalars: Vector and Double Vector Supermultiplets

Due to the duality between scalar field, entering the action with the space–time derivatives only, and gauge field strength in $d = 3$, the actions for the vector (one scalar dualized) and the double vector (both scalars dualized) supermultiplets can be easily obtained within the coset approach. Before performing such dualizations, let us rewrite the action (11.6.47) in the vector notations. If we introduce the quantity

$$\mathcal{G}_{ab} = \frac{1}{\sqrt{2}}\left(\psi^m\mathcal{D}_{ab}\bar{\psi}_m - \bar{\psi}^m\mathcal{D}_{ab}\psi_m\right), \tag{11.6.48}$$

then only vector indices show up in the action. Passing to the vector notation, we get

$$S = 2\int d^3x - \int d^3x \, \det\mathcal{E}\left(1 + \sqrt{(1 - \mathcal{D}_I q\mathcal{D}_I\bar{q})^2 - (\mathcal{D}_I q\mathcal{D}_I q)(\mathcal{D}_J\bar{q}\mathcal{D}_J\bar{q})}\right)$$
$$+ \, i\int d^3x \, \det\mathcal{E}\epsilon^{IJK}\mathcal{G}_I\mathcal{D}_J q\,\mathcal{D}_K\bar{q}, \tag{11.6.49}$$

where

$$\mathcal{D}_I = \left(\mathcal{E}^{-1}\right)_I{}^J\partial_J, \quad \mathcal{E}_I{}^J = \delta_I^J - \frac{1}{\sqrt{2}}\left(\sigma^J\right)_{ab}\left(\psi^a\partial_I\bar{\psi}^b + \bar{\psi}^a\partial_I\psi^b\right). \tag{11.6.50}$$

11.6.2.1 Vector Supermultiplet

The vector $N = 2, d = 3$ supermultiplet includes one scalar and one gauge fields among the physical bosonic components. Thus, we have to dualize one of the scalar components in the action (11.6.49). To perform dualization, one has to pass to a pair of real bosons $\{u, v\}$

$$q = \frac{1}{2}(u + iv), \quad \bar{q} = \frac{1}{2}(u - iv). \tag{11.6.51}$$

In terms of the newly defined scalars, the action (11.6.49) reads

$$S = 2\int d^3x - \int d^3x \, \det\mathcal{E}\left[1 + \sqrt{\left(1 - \frac{1}{2}\mathcal{D}_I u \mathcal{D}_I u\right)\left(1 - \frac{1}{2}\mathcal{D}_J v \mathcal{D}_J v\right) - \frac{1}{4}\left(\mathcal{D}_I u \mathcal{D}_I v\right)^2}\,\right]$$
$$+ \frac{1}{2}\int d^3x \, \det\mathcal{E}\, \epsilon^{IJK} \mathcal{G}_I \mathcal{D}_J u \, \mathcal{D}_K v. \tag{11.6.52}$$

The equation of motion for the bosonic field v has the form

$$\partial_I \left(\det\mathcal{E}\left(\mathcal{E}^{-1}\right)_J^{\;I} V_J\right) = 0, \quad V_I = \widetilde{V}_I + \frac{1}{2}\epsilon_{IJK} G_J \mathcal{D}_K u, \tag{11.6.53}$$

where

$$\widetilde{V}_I = \frac{\left(1 - \frac{1}{2}\mathcal{D}u \cdot \mathcal{D}u\right)\mathcal{D}_I v + \frac{1}{2}\,\mathcal{D}u \cdot \mathcal{D}v \; \mathcal{D}_I u}{2\sqrt{\left(1 - \frac{1}{2}\mathcal{D}u \cdot \mathcal{D}u\right)\left(1 - \frac{1}{2}\mathcal{D}v \cdot \mathcal{D}v\right) - \frac{1}{4}\left(\mathcal{D}u \cdot \mathcal{D}v\right)^2}}. \tag{11.6.54}$$

Then, one may find that

$$\mathcal{D}_I v = \frac{2\widetilde{V}_I - \widetilde{V} \cdot \mathcal{D}u \; \mathcal{D}_I u}{\sqrt{1 - \frac{1}{2}\mathcal{D}u \cdot \mathcal{D}u + 2\widetilde{V} \cdot \widetilde{V} - \left(\widetilde{V} \cdot \mathcal{D}u\right)^2}}. \tag{11.6.55}$$

Performing the Rauth transformation over the bosonic field v, we finally get

$$\tilde{S} = 2\int d^3x - \int d^3x \, \det\mathcal{E}\left(1 + \sqrt{1 - \frac{1}{2}\mathcal{D}u \cdot \mathcal{D}u + 2\widetilde{V} \cdot \widetilde{V} - \left(\widetilde{V} \cdot \mathcal{D}u\right)^2}\right). \tag{11.6.56}$$

This is the action for the $N = 2, d = 3$ vector supermultiplet which possesses an additional, spontaneously broken $N = 2$ supersymmetry.

One should stress that the real field strength is defined in (11.6.53), but the action has a much simpler structure written in terms of $\widetilde{V}_I$.

11.6.2.2 Double Vector Supermultiplet

In order to obtain a double vector supermultiplet, one may dualize both scalars in the action (11.6.49). As the first step, one has to find the equations of motion for the scalar fields

$$\partial_I \left(\det \mathcal{E} \left(\mathcal{E}^{-1}\right)^I_J V^J\right) = 0, \quad \partial_I \left(\det \mathcal{E} \left(\mathcal{E}^{-1}\right)^I_J \overline{V}^J\right) = 0, \tag{11.6.57}$$

where

$$V_I = \widetilde{V}_I - i\,\epsilon_{IJK} G_J \mathcal{D}_k \bar{q}, \quad \widetilde{V}_I = \frac{(1 - \mathcal{D}q \cdot \mathcal{D}\bar{q})\,\mathcal{D}_I \bar{q} + (\mathcal{D}\bar{q} \cdot \mathcal{D}\bar{q})\,\mathcal{D}_I q}{\sqrt{(1 - \mathcal{D}q \cdot \mathcal{D}\bar{q})^2 - (\mathcal{D}q \cdot \mathcal{D}q)(\mathcal{D}\bar{q} \cdot \mathcal{D}\bar{q})}}. \tag{11.6.58}$$

After a standard machinery with the Rauth transformations we finally get the action

$$\widehat{S} = 2\int d^3x - \int d^3x\, \det \mathcal{E} \left[1 + \sqrt{\left(1 + \widetilde{V} \cdot \overline{\widetilde{V}}\right)^2 - \widetilde{V}^2\, \overline{\widetilde{V}}^2} - i\,\epsilon_{IJK}\, G_I\, \widetilde{V}_J \overline{\widetilde{V}}_K\right]. \tag{11.6.59}$$

The bosonic sector of this action coincides with that constructed in [29]. Again, the simplest form of the action is achieved with the help of $\widetilde{V}_I$ variables which are related with field strengths as in (11.6.57), (11.6.58).

11.7 Conclusion

In this paper, using a remarkable connection between the partial breaking of global supersymmetry, the coset approach, which realized the specific pattern of supersymmetry breaking, and the Nambu-Goto actions for the extended object, we have reviewed the construction of the on-shell component actions for the superparticle in $D = 3$ realizing $N = 4 \cdot 2^k \to N = 2 \cdot 2^k$ pattern of supersymmetry breaking, for the superparticle in $D = 5$ with the $N = 16$ supersymmetry broken down to $N = 8$ one, for the $N = 1$, $D = 5$ supermembrane and its dual cousins, and for the $N = 1$ supermembrane in $D = 4$. Of course, such actions can be obtained by dimensional reduction from the higher dimension actions or even from the known superspace actions. Nevertheless, if we pay more attention to the spontaneously broken supersymmetry and, thus, use the corresponding covariant derivatives, together with the proper choice of the components, the resulting action can be drastically simplified. So, the implications of our results are threefold:

- We demonstrated that the coset approach can be used far beyond the construction of the superfield equations of motion, if we are interested in the component actions,
- We showed that there is a rather specific choice of the superfields and their components which drastically simplifies the component action,
- We argued that the broken supersymmetry fixed the on-shell component action up to some constants, while the role of the unbroken supersymmetry is just to fix these constants.

The application of our approach is not limited to the cases of P-branes only. Different types of D-branes could be also considered in a similar manner. However, once we are dealing with the field strengths, which never show up as the coordinates of the coset space, the proper choice of the components becomes very important. In particular, the Born-Infeld-Nambu-Goto action (11.6.52), we constructed by the dualization of one scalar field, has a nice, compact form in terms of the "covariant" field strength $\widetilde{V}_I$ which is related with the "genuine" field strength, obeying the Bianchi identity, in a rather complicated way (11.6.53). The same is also true for the Born-Infeld type action (11.6.59). In order to clarify the nature of these variables, one has to consider the corresponding patterns of the supersymmetry breaking [with one, or without central charges in the $N = 4, d = 3$ Poincaré superalgebra (A.16)] independently. In this respect, the detailed analysis of $N = 2 \rightarrow N = 1$ supersymmetry breaking in $d = 4$ seems to be much more interesting, being a preliminary step to the construction of $N = 4$ Born-Infeld action [16, 18, 20] and/or to the action describing partial breaking of $N = 1, D = 10$ supersymmetry with the hypermultiplet as the Goldstone superfield.

In this paper we also showed that the on-shell component actions for superparticle have the universal form

$$S = \alpha \int dt + (1-\alpha) \int dt\, \mathcal{E} - \int dt\, \mathcal{E}\sqrt{1 - \beta \mathcal{D}_t q \mathcal{D}_t q}.$$

With our approach, we explicitly constructed such actions for the superparticles in $D = 3$ realizing $N = 4 \cdot 2^k \rightarrow N = 2 \cdot 2^k$ pattern of supersymmetry breaking, and in $D = 5$ with the $N = 16$ supersymmetry broken down to $N = 8$ one. It was shown that the corresponding component on-shell actions are invariant under both unbroken and broken supersymmetry. In the considered models only the equality of both unbroken and broken supersymmetries was essential, and their number did not play any role, we expect that all superparticle models with one half partial breaking of global supersymmetry can be constructed similarly, confirming, thereby, its universality.

One possible application of this method is the construction of models with partial breaking of global supersymmetry in cases when $d > 2$, where the superspace actions are known (see e.g. [7–10]). We assume that these actions derived with our method will have a more simple and understandable form.

It would be quite instructive to understand which new features will appear when we will replace the trivial, flat target space by, for example, the AdS one [30]. It seems that the strategy will be the same, and we are planning to report the corresponding results elsewhere.

Acknowledgments We thank our co-authors Nikolay Kozyrev and Armen Yeranyan with whom some of the results mentioned here were obtained. We wish to acknowledge useful discussions with them, as well as with S. Kuzenko and D. Sorokin. This work was partially supported by RFBR grants 12-02-00517-a, 13-02-91330-NNIO-a and 13-02-90602 Apm-a, as well as by the ERC Advanced Grant no. 226455 *"Supersymmetry, Quantum Gravity and Gauge Fields"* (*SUPERFIELDS*).

Appendix A: Superalgebra, Coset Space, Transformations and Cartan Forms

In this appendix we collected some formulas from the paper [12] where the nonlinear realization of $N = 1, D = 4$ Poincaré group in its coset over $d = 3$ Lorentz group $SO(1, 2)$ was constructed.

In $d = 3$ notation the $N = 1, D = 4$ Poincaré superalgebra contains the following set of generators:

$$\mathrm{N} = 2,\ \mathrm{d} = 3\ \mathrm{SUSY} \quad \propto \quad \{Q_a, P_{ab}, S_a, Z, M_{ab}, K_{ab}\}, \tag{A.1}$$

$a, b = 1, 2$ being the $d = 3$ $SL(2, R)$ spinor indices.[7] Here, P_{ab} and Z are $D = 4$ translation generators, Q_a and S_a are the generators of super-translations, the generators M_{ab} form $d = 3$ Lorentz algebra $so(1, 2)$, while the generators K_{ab} belong to the coset $SO(1, 3)/SO(1, 2)$. The basic anticommutation relations read

$$\{Q_a, Q_b\} = P_{ab}, \quad \{Q_a, S_b\} = \epsilon_{ab} Z, \quad \{S_a, S_b\} = P_{ab}. \tag{A.2}$$

The coset element was defined in [12] as

$$g = e^{x^{ab} P_{ab}} e^{\theta^a Q_a} e^{\boldsymbol{q} Z} e^{\psi^a S_a} e^{\boldsymbol{\Lambda}^{ab} K_{ab}}. \tag{A.3}$$

Here, x^{ab}, θ^a are $N = 1, d = 3$ superspace coordinates, while the remaining coset parameters are Goldstone superfields, $\boldsymbol{q} = \boldsymbol{q}(x, \theta)$, $\psi^a \equiv \psi^a(x, \theta)$, $\boldsymbol{\Lambda}^{ab} = \boldsymbol{\Lambda}^{ab}(x, \theta)$.

The transformation properties of the coordinates and superfields with respect to all symmetries can be found by acting from the left on the coset element (A.3) by the different elements of $N = 1, D = 4$ supergroup. They have the following explicit form:

- Translations and Unbroken supersymmetry $[g_0 = \exp\,(a^{ab} P_{ab} + \eta^a Q_a)]$

$$\delta x^{ab} = a^{ab} - \frac{1}{4}\eta^a \theta^b - \frac{1}{4}\eta^b \theta^a, \quad \delta\theta^a = \eta^a. \tag{A.4}$$

- Broken supersymmetry $[g_0 = \exp\,(\xi^a S_a)]$

$$\delta x^{ab} = -\frac{1}{4}\xi^a \psi^b - \frac{1}{4}\xi^b \psi^a, \quad \delta\boldsymbol{q} = \xi^a \theta_a, \quad \delta\psi^a = \xi^a. \tag{A.5}$$

- K transformations $[g_0 = \exp\,(r^{ab} K_{ab})]$

[7] The indices are raised and lowered as follows: $V^a = \epsilon^{ab} V_b$, $V_b = \epsilon_{bc} V^c$, $\epsilon_{ab}\epsilon^{bc} = \delta_a^c$.

$$\begin{aligned}
\delta x^{ab} &= -2\boldsymbol{q} r^{ab} - \frac{1}{2}\theta_c r^{ca}\psi^b - \frac{1}{2}\theta_c r^{cb}\psi^a + \frac{1}{2}\theta^a r^{bc}\psi_c + \frac{1}{2}\theta^b r^{ac}\psi_c,\\
\delta\theta^a &= -2r^{ab}\psi_b, \quad \delta\boldsymbol{q} = -4r_{ab}x^{ab}, \quad \delta\psi^a = 2r^{ab}\theta_b,\\
\delta\boldsymbol{\lambda}^{ab} &= r^{ab} - 4\boldsymbol{\lambda}^{ac}r_{cd}\boldsymbol{\lambda}^{db}.
\end{aligned} \tag{A.6}$$

- Broken Z-translations $[g_0 = \exp(cZ)]$

$$\delta\boldsymbol{q} = c. \tag{A.7}$$

- The $d = 3$ Lorentz group $SO(1,2) \sim SL(2,R)$ acts as rotations of the spinor indices.

In (A.6) the coordinates of the stereographic parametrization of the coset $SO(1,3)/SO(1,2)$ have been defined as

$$\boldsymbol{\lambda}^{ab} = \frac{\tanh\left(\sqrt{2\boldsymbol{\Lambda}^2}\right)}{\sqrt{2\boldsymbol{\Lambda}^2}}\boldsymbol{\Lambda}^{ab}, \quad \tanh^2\left(\sqrt{2\boldsymbol{\Lambda}^2}\right) \equiv 2\boldsymbol{\lambda}^2, \quad \boldsymbol{\Lambda}^2 \equiv \boldsymbol{\Lambda}_{ab}\boldsymbol{\Lambda}^{ab}, \quad \boldsymbol{\lambda}^2 \equiv \boldsymbol{\lambda}_{ab}\boldsymbol{\lambda}^{ab}. \tag{A.8}$$

The most important objects in the coset are the Cartan forms

$$g^{-1}dg = \Omega_Q + \Omega_P + \Omega_Z + \Omega_S + \Omega_K + \Omega_M.$$

In what follows we will need only the forms $\Omega_Q, \Omega_P, \Omega_Z$ and Ω_S which were constructed in [12]

$$\begin{aligned}
\Omega_Z &= \frac{1+2\boldsymbol{\lambda}^2}{1-2\boldsymbol{\lambda}^2}\left[d\hat{\boldsymbol{q}} + \frac{4}{1+2\boldsymbol{\lambda}^2}\boldsymbol{\lambda}_{ab}d\hat{x}^{ab}\right]Z,\\
\Omega_P &\equiv \Omega_P^{ab}P_{ab} = \left[d\hat{x}^{ab} + \frac{2}{1-2\boldsymbol{\lambda}^2}\boldsymbol{\lambda}^{ab}\left(d\hat{\boldsymbol{q}} + 2\boldsymbol{\lambda}_{cd}d\hat{x}^{cd}\right)\right]P_{ab},\\
\Omega_Q &\equiv \Omega_Q^a Q_a = \frac{1}{\sqrt{1-2\boldsymbol{\lambda}^2}}\left[d\theta^a + 2\boldsymbol{\lambda}^{ab}d\psi_b\right]Q_a,\\
\Omega_S &\equiv \Omega_S^a S_a = \frac{1}{\sqrt{1-2\boldsymbol{\lambda}^2}}\left[d\psi^a - 2\boldsymbol{\lambda}^{ab}d\theta_b\right]S_a.
\end{aligned} \tag{A.9}$$

$$\begin{aligned}
d\hat{x}^{ab} &\equiv dx^{ab} + \frac{1}{4}\theta^a d\theta^b + \frac{1}{4}\theta^b d\theta^a + \frac{1}{4}\psi^a d\psi^b + \frac{1}{4}\psi^b d\psi^a,\\
d\hat{\boldsymbol{q}} &\equiv d\boldsymbol{q} + \psi_a d\theta^a.
\end{aligned} \tag{A.10}$$

Note, that all Cartan forms, except for Ω_M, transform homogeneously under all symmetries.

Having at hands the Cartan forms, one may construct the "semi-covariant" (covariant with respect to $d = 3$ Lorentz, unbroken and broken supersymmetries only) as

$$d\hat{x}^{ab}\nabla_{ab} + d\theta^a\nabla_a = dx^{ab}\frac{\partial}{\partial x^{ab}} + d\theta^a\frac{\partial}{\partial\theta^a}. \tag{A.11}$$

Explicitly, they read [12]

$$\nabla_{ab} = (E^{-1})_{ab}^{cd}\,\partial_{cd}, \quad \nabla_a = D_a + \frac{1}{2}\psi^b D_a\psi^c\,\nabla_{bc} = D_a + \frac{1}{2}\psi^b\nabla_a\psi^c\,\partial_{bc}, \tag{A.12}$$

where

$$D_a = \frac{\partial}{\partial\theta^a} + \frac{1}{2}\theta^b\partial_{ab}, \quad \{D_a, D_b\} = \partial_{ab}, \tag{A.13}$$

$$E_{ab}^{cd} = \frac{1}{2}(\delta_a^c\delta_b^d + \delta_a^d\delta_b^c) + \frac{1}{4}(\psi^c\partial_{ab}\psi^d + \psi^d\partial_{ab}\psi^c), \tag{A.14}$$

$$(E^{-1})_{ab}^{cd} = \frac{1}{2}(\delta_a^c\delta_b^d + \delta_a^d\delta_b^c) - \frac{1}{4}(\psi^c\nabla_{ab}\psi^d + \psi^d\nabla_{ab}\psi^c). \tag{A.15}$$

These derivatives obey the following algebra:

$$[\nabla_{ab}, \nabla_{cd}] = -\nabla_{ab}\psi^m\nabla_{cd}\psi^n\nabla_{mn}, \qquad [\nabla_{ab}, \nabla_c] = \nabla_{ab}\psi^m\nabla_c\psi^n\nabla_{mn},$$
$$\{\nabla_a, \nabla_b\} = \nabla_{ab} + \nabla_a\psi^m\nabla_b\psi^n\nabla_{mn}. \tag{A.16}$$

Appendix B

In this Appendix we will prove the invariance of the supermembrane action (11.5.10) under broken and unbroken supersymmetries. The proof for the broken supersymmetry is the easiest one and we will start with this invariance.

Broken Supersymmetry

Under spontaneously broken S^a supersymmetry our coordinates and the physical components transform as in (A.5)

$$\delta x^{ab} = -\frac{1}{4}\xi^a\psi^b - \frac{1}{4}\xi^b\psi^a, \quad \delta q = 0, \quad \delta\psi^a = \xi^a. \tag{B.1}$$

One may immediately check that the $\theta = 0$ part of the covariant differential $d\hat{x}^{ab}$, defined in (A.10)

$$d\hat{x}^{ab} = dx^{ab} + \frac{1}{4}\psi^a d\psi^b + \frac{1}{4}\psi^b d\psi^a \tag{B.2}$$

is invariant under the transformations (B.1). Therefore, the covariant derivatives $\mathcal{D}_{ab} = \nabla_{ab}|_{\theta=0}$ (A.12) are also invariant under broken supersymmetry transformations. Now, for the active form of the transformations $[\delta^*\phi = \phi'(x) - \phi(x)]$ we have

$$\delta_S^* \mathcal{D}^{ab} q = \frac{1}{2}\xi^c \psi^d \partial_{cd} \mathcal{D}^{ab} q \quad \Rightarrow \quad \delta_S^* \mathcal{F}(\mathcal{D}q \cdot \mathcal{D}q) = \frac{1}{2}\xi^a \psi^b \partial_{ab} \mathcal{F},$$

$$\delta_S^* \psi^a = \xi^a + \frac{1}{2}\xi^c \psi^d \partial_{cd} \psi^a, \qquad \delta_S^* \mathcal{D}^{ab} \psi^c = \frac{1}{2}\xi^d \psi^e \partial_{de} \mathcal{D}^{ab} \psi^c, \tag{B.3}$$

and, therefore,

$$\delta_S^* \det \mathcal{E} = \frac{1}{2}\xi^a \mathcal{D}_{ab} \psi^b - \frac{1}{8}\xi^d \psi_d \mathcal{D}^{ab} \psi^c \mathcal{D}_{ab} \psi_c + \frac{1}{2}\xi^c \psi^d \partial_{cd} \det \mathcal{E}. \tag{B.4}$$

Thus, the integrand in the action (11.5.7) transforms as follows:

$$\begin{aligned}\delta_S^*\Big(\det \mathcal{E}\mathcal{F}\Big) &= \left(\frac{1}{2}\xi^a \mathcal{D}_{ab}\psi^b - \frac{1}{8}\xi^d \psi_d \mathcal{D}^{ab}\psi^c \mathcal{D}_{ab}\psi_c\right)\mathcal{F} + \frac{1}{2}\xi^c \psi^d \partial_{cd}\Big(\det \mathcal{E}\mathcal{F}\Big) \\ &= \left(\frac{1}{2}\xi^a \mathcal{D}_{ab}\psi^b - \frac{1}{8}\xi^d \psi_d \mathcal{D}^{ab}\psi^c \mathcal{D}_{ab}\psi_c - \frac{1}{2}\xi^c \partial_{cd}\psi^d \det \mathcal{E}\right)\mathcal{F}.\end{aligned} \tag{B.5}$$

It is a matter of direct calculations to check that the expression in the parentheses in (B.5) is zero. Thus, the action (11.5.7), as well as the action (11.5.10), are indeed invariant under spontaneously broken supersymmetry.

Unbroken Supersymmetry

It is funny, but in contrast with the superfield approach in which unbroken supersymmetry is manifest, to prove the invariance of the component action (11.5.10) under unbroken supersymmetry is a rather complicated task.

Under unbroken Q^a supersymmetry the covariant derivatives ∇_{ab}, ∇_a (A.12) are invariant by construction. Therefore, the objects $\nabla_{ab}\psi_c$, $\nabla_{ab}q$ are the superfields with the standard transformation properties

$$\delta_Q^* \psi^a = -\eta^b (D_b \psi^a)|_{\theta=0} = 2\eta^b \left(\lambda_b{}^a - \frac{1}{2}\psi^m \lambda_b{}^n \partial_{mn} \psi^a\right), \tag{B.6}$$

$$\begin{aligned}\delta_Q^* \mathcal{D}_{ab}\psi_c &= -\eta^d (D_d \nabla_{ab}\psi_c)|_{\theta=0} \\ &= -\eta^d \left(2\mathcal{D}_{ab}\psi^m \lambda_d{}^n \mathcal{D}_{mn}\psi_c - 2\mathcal{D}_{ab}\lambda_{dc} + \psi^m \lambda_d{}^n \partial_{mn} \mathcal{D}_{ab}\psi_c\right), \\ \delta_Q^* \mathcal{D}_{ab} q &= -\eta^c (D_c \nabla_{ab}\, \boldsymbol{q})|_{\theta=0}\end{aligned} \tag{B.7}$$

$$= -\eta^c \left(\frac{1-2\lambda^2}{1+2\lambda^2} \mathcal{D}_{ab}\psi_c + \psi^m \lambda_c{}^n \partial_{mn} \mathcal{D}_{ab}\, q \right). \tag{B.8}$$

Therefore,

$$\begin{aligned} \delta_Q^* \det \mathcal{E} &= \eta^c \lambda_c{}^a \mathcal{D}_{ab}\psi^b - \eta^c \mathcal{D}_{ab}\lambda^b{}_c \psi^a + \eta^c \lambda_c^n \psi^a \mathcal{D}_{ab}\psi^m \mathcal{D}_{mn}\psi^b \\ &\quad - \frac{1}{4}\eta^b \lambda_b^a \psi_a \mathcal{D}^{mn}\psi^k \mathcal{D}_{mn}\psi_k - \frac{1}{8}\psi^2 \eta^d \lambda_d{}^b \mathcal{D}_{bc}\psi^c \mathcal{D}^{mn}\psi^k \mathcal{D}_{mn}\psi_k \\ &\quad + \frac{1}{4}\psi^2 \eta^d \mathcal{D}_{ab}\lambda_{dc}\mathcal{D}^{ab}\psi^c - \eta^c \lambda_c{}^n \psi^m \partial_{mn} \det \mathcal{E}, \end{aligned} \tag{B.9}$$

and

$$\delta_Q^* \mathcal{F} = -2\,\frac{1-2\lambda^2}{1+2\lambda^2}\,\eta^c \mathcal{D}_{ab}\psi_c \mathcal{D}^{ab}\, q\, \mathcal{F}' - \eta^c \lambda_c{}^n \psi^m \partial_{mn}\mathcal{F}. \tag{B.10}$$

The $\mathcal{F}'$ in (B.10) denotes the derivative $\mathcal{F}$ over its argument (i.e. over $\mathcal{D}q \cdot \mathcal{D}q$ in our case).

Combining these expressions we will get the following variation of the integrand of our action (11.5.10):

$$\delta_Q^* \mathcal{L} = \delta_Q^* \Big(\det \mathcal{E}\, \mathcal{F} \Big) = \delta_Q^* \det \mathcal{E}\, \mathcal{F} + \det \mathcal{E}\, \delta_Q^* \mathcal{F}. \tag{B.11}$$

In (B.11) the last terms from $\delta_Q^* \det \mathcal{E}$ (B.9) and $\delta_Q^* \mathcal{F}$ (B.10) combine together to produce

$$-\eta^a \lambda_a{}^b \psi^c \partial_{bc} \Big(\det \mathcal{E}\mathcal{F} \Big).$$

Therefore, after integration by parts in this term we will get

$$\begin{aligned} \delta_Q^* \mathcal{L} = &\Big(\eta^c \lambda_c{}^a \mathcal{D}_{ab}\psi^b - \eta^c \mathcal{D}_{ab}\lambda^b{}_c \psi^a + \eta^c \lambda_c{}^n \psi^a \mathcal{D}_{ab}\psi^m \mathcal{D}_{mn}\psi^b - \frac{1}{4}\eta^b \lambda_b{}^a \psi_a \mathcal{D}^{mn}\psi^k \mathcal{D}_{mn}\psi_k \\ &- \frac{1}{8}\psi^2 \eta^d \lambda_d{}^b \mathcal{D}_{bc}\psi^c \mathcal{D}^{mn}\psi^k \mathcal{D}_{mn}\psi_k + \frac{1}{4}\psi^2 \eta^d \mathcal{D}_{ab}\lambda_{dc}\mathcal{D}^{ab}\psi^c \Big) \mathcal{F} \\ &- 2\,\frac{1-2\lambda^2}{1+2\lambda^2}\,\eta^c \mathcal{D}_{ab}\psi_c \mathcal{D}^{ab} q \mathcal{F}' \det \mathcal{E} + \eta^c \partial_{mn}\lambda_c{}^n \psi^m \mathcal{F} \det \mathcal{E} + \eta^c \lambda_c{}^n \partial_{mn}\psi^m \mathcal{F} \det \mathcal{E}. \end{aligned} \tag{B.12}$$

Now, one may check that terms with the derivatives of λ_{ab} in (B.12) just canceled.

The next step is to substitute into (B.12) the explicit expressions for λ_{ab} (11.5.2) and for $\mathcal{F}$ (11.5.10)

$$\lambda_{ab} = \frac{-\frac{1}{2}\mathcal{D}_{ab}\, q}{1+\sqrt{1-\frac{1}{2}\mathcal{D}q \cdot \mathcal{D}q}}, \qquad \mathcal{F} = 1 + \sqrt{1 - \frac{1}{2}\mathcal{D}q \cdot \mathcal{D}q}. \tag{B.13}$$

If we note that

$$\lambda_{ab} = \frac{-\frac{1}{2}\mathcal{D}_{ab}q}{\mathcal{F}} \quad \text{and} \quad \frac{1-2\lambda^2}{1+2\lambda^2} = -\frac{1}{4\,F'}, \tag{B.14}$$

it will be not so strange that after substitution of (B.13) into (B.12), the variation $\delta^*_Q\mathcal{L}$ will not contain any square roots. So, it will read

$$\begin{aligned}\delta^*_Q\mathcal{L} = &-\frac{1}{2}\eta^c\mathcal{D}_c{}^a q\mathcal{D}_{ab}\psi^b - \frac{1}{2}\eta^c\mathcal{D}_c{}^n q\psi^a\mathcal{D}_{ab}\psi^m\mathcal{D}_{mn}\psi^b + \frac{1}{8}\eta^b\mathcal{D}_b{}^a q\psi_a\mathcal{D}^{cd}\psi^e\mathcal{D}_{cd}\psi_e\\ &+\frac{1}{16}\psi^2\eta^a\mathcal{D}_a{}^b q\mathcal{D}_{bc}\psi^c\mathcal{D}^{de}\psi^f\mathcal{D}_{de}\psi_f + \frac{1}{2}\eta^c\mathcal{D}_{ab}\psi_c\mathcal{D}^{ab}q\det\mathcal{E}\\ &-\frac{1}{2}\eta^a\mathcal{D}_a{}^b q\;\partial_{bc}\psi^c\det\mathcal{E}.\end{aligned} \tag{B.15}$$

Substituting now the expression for $\partial_{bc}\psi^c\det\mathcal{E}$ from (B.5) and slightly rearranging the terms, we obtain

$$\begin{aligned}\delta_Q\mathcal{L} = &-\eta^c\mathcal{D}_c{}^a q\mathcal{D}_{ab}\,\psi^b - \frac{1}{4}\eta^a\mathcal{D}_a{}^b q\;\psi_b\mathcal{D}^{cd}\psi_d\mathcal{D}_{ce}\psi^e + \frac{1}{16}\psi^2\eta^a\mathcal{D}_a{}^b q\mathcal{D}_{bc}\psi^c\mathcal{D}^{de}\psi^f\mathcal{D}_{de}\psi_f\\ &+\frac{1}{2}\eta^c\mathcal{D}_{ab}\psi_c\mathcal{D}^{ab}q\det\mathcal{E}.\end{aligned} \tag{B.16}$$

Finally, combining the terms in the first line together, we will get the following simple form of the variation of the integrand

$$\delta^*_Q\mathcal{L} = -\eta^c\left(\mathcal{D}_c{}^a q\mathcal{D}_{ab}\psi^b - \frac{1}{2}\mathcal{D}^{ab}\,q\mathcal{D}_{ab}\psi_c\right)\det\mathcal{E}. \tag{B.17}$$

Unfortunately, further simplifications are not possible. The simplest way to be sure that $\delta^*_Q\mathcal{L}$ (B.17) gives zero after integration over d^3x is to find the "equation of motion" for q which follows from the "Lagrangian" (B.17)

$$\frac{\delta}{\delta q}\int d^3x\;\delta_Q\mathcal{L} = 0. \tag{B.18}$$

Clearly, the expression (B.18) has to be identically equal to zero if our action is invariant under unbroken supersymmetry. After quite lengthly and tedious, but straightforward calculations, one may show that this is indeed so.

Thus, our action (11.5.10) is invariant with respect to both broken and unbroken supersymmetries.

Appendix C

In this Appendix we collected some formulas describing the nonlinear realization of $N = 1, D = 5$ Poincaré group in its coset over $d = 3$ Lorentz group $SO(1, 2)$.

In $d = 3$ notation the $N = 1, d = 5$ Poincaré superalgebra contains the following set of generators:

$$\text{N} = 4,\ \text{d} = 3\,\text{SUSY} \quad \propto \quad \left\{P_{ab}, Q_a, \overline{Q}_a, S_a, \overline{S}_a, Z, \overline{Z}, M_{ab}, K_{ab}, \overline{K}_{ab}, J\right\}, \tag{C.1}$$

$a, b = 1, 2$ being the $d = 3$ $SL(2, R)$ spinor indices.[8] Here, P_{ab} Z and $\overline{Z}$ are $D = 5$ translation generators, $Q_a, \overline{Q}_a$ and $S_a, \overline{S}_a$ are the generators of super-translations, the generators M_{ab} form $d = 3$ Lorentz algebra $so(1, 2)$, the generators K_{ab} and $\overline{K}_{ab}$ belong to the coset $SO(1, 4)/SO(1, 2) \times U(1)$, while J span $u(1)$. The basic (anti)commutation relations read

$$\begin{aligned}
&[M_{ab}, M_{cd}] = \epsilon_{ad} M_{bc} + \epsilon_{ac} M_{bd} + \epsilon_{bc} M_{ad} + \epsilon_{bd} M_{ac} \equiv (M)_{ab,cd}\,,\\
&[M_{ab}, P_{cd}] = (P)_{ab,cd}\,,\ [M_{ab}, K_{cd}] = (K)_{ab,cd}\,,\ \left[M_{ab}, \overline{K}_{cd}\right] = \left(\overline{K}\right)_{ab,cd}\,,\\
&\left[K_{ab}, \overline{K}_{cd}\right] = \frac{1}{2}(M)_{ab,cd} + 2\left(\epsilon_{ac}\epsilon_{bd} + \epsilon_{bc}\epsilon_{ad}\right) J,\\
&[K_{ab}, P_{cd}] = -\left(\epsilon_{ac}\epsilon_{bd} + \epsilon_{bc}\epsilon_{ad}\right) Z,\ \left[\overline{K}_{ab}, P_{cd}\right] = \left(\epsilon_{ac}\epsilon_{bd} + \epsilon_{bc}\epsilon_{ad}\right) \overline{Z},\\
&\left[K_{ab}, \overline{Z}\right] = -2P_{ab},\ \left[\overline{K}_{ab}, Z\right] = 2P_{ab},\\
&[M_{ab}, Q_c] = \epsilon_{ac} Q_b + \epsilon_{bc} Q_a, \left[M_{ab}, \overline{Q}_c\right] = \left(\overline{Q}\right)_{ab,c}\,,\\
&[M_{ab}, S_c] = (S)_{ab,c}\,, \left[M_{ab}, \overline{S}_c\right] = \left(\overline{S}\right)_{ab,c}\,,\\
&\left[\overline{K}_{ab}, Q_c\right] = -\left(\overline{S}\right)_{ab,c}, \left[K_{ab}, \overline{Q}_c\right] = (S)_{ab,c}\,,\ \left[\overline{K}_{ab}, S_c\right] = \left(\overline{Q}\right)_{ab,c}\,,\\
&\left[K_{ab}, \overline{S}_c\right] = -(Q)_{ab,c}\,,\\
&[J, Q_a] = -\frac{1}{2} Q_a,\ \left[J, \overline{Q}_a\right] = \frac{1}{2}\overline{Q}_a,\ [J, S_a] = -\frac{1}{2} S_a,\ \left[J, \overline{S}_a\right] = \frac{1}{2}\overline{S}_a,\\
&[J, K_{ab}] = -K_{ab},\ \left[J, \overline{K}_{ab}\right] = \overline{K}_{ab},\ [J, Z] = -Z,\ \left[J, \overline{Z}\right] = \overline{Z},\\
&\left\{Q_a, \overline{Q}_b\right\} = 2P_{ab},\ \left\{S_a, \overline{S}_b\right\} = 2P_{ab},\ \{Q_a, S_b\} = 2\epsilon_{ab} Z,\\
&\left\{\overline{Q}_a, \overline{S}_b\right\} = 2\epsilon_{ab}\overline{Z}.
\end{aligned} \tag{C.2}$$

Note, that the generators obey the following conjugation rules:

$$\begin{aligned}
&(P_{ab})^\dagger = P_{ab},\ (K_{ab})^\dagger = \overline{K}_{ab},\ (M_{ab})^\dagger = -M_{ab},\ J^\dagger = J,\ Z^\dagger = \overline{Z},\\
&(Q_a)^\dagger = \overline{Q}_a,\ (S_a)^\dagger = \overline{S}_a.
\end{aligned} \tag{C.3}$$

---

[8] The indices are raised and lowered as follows: $V^a = \epsilon^{ab} V_b,\ V_b = \epsilon_{bc} V^c,\ \ \epsilon_{ab}\epsilon^{bc} = \delta_a^c$.

We define the coset element as follows

$$g = e^{ix^{ab}P_{ab}} e^{\theta^a Q_a + \bar{\theta}^a \overline{Q}_a} e^{i(\boldsymbol{q}Z + \bar{\boldsymbol{q}}\overline{Z})} e^{\psi^a S_a + \bar{\psi}^a \overline{S}_a} e^{i(\boldsymbol{\Lambda}^{ab}K_{ab} + \overline{\boldsymbol{\Lambda}}^{ab}\overline{K}_{ab})}. \quad \text{(C.4)}$$

Here, $\{x^{ab}, \theta^a, \bar{\theta}^a\}$ are $N = 2, d = 3$ superspace coordinates, while the remaining coset parameters are Goldstone superfields, $\boldsymbol{q} = \boldsymbol{q}(x,\theta,\bar{\theta}),\ \bar{\boldsymbol{q}} = \bar{\boldsymbol{q}}(x,\theta,\bar{\theta}),\ \psi^a = \psi^a(x,\theta,\bar{\theta}),\ \bar{\psi}^a = \bar{\psi}^a(x,\theta,\bar{\theta}),\ \boldsymbol{\Lambda}^{ab} = \boldsymbol{\Lambda}^{ab}(x,\theta,\bar{\theta}),\ \overline{\boldsymbol{\Lambda}}^{ab} = \overline{\boldsymbol{\Lambda}}^{ab}(x,\theta,\bar{\theta})$. These $N = 2$ superfields obey the following conjugation rules:

$$\left(x^{ab}\right)^\dagger = x^{ab},\ (\theta^a)^\dagger = \bar{\theta}^a,\quad \boldsymbol{q}^\dagger = \bar{\boldsymbol{q}},\ (\psi^a)^\dagger = \bar{\psi}^a,\ \left(\boldsymbol{\Lambda}^{ab}\right)^\dagger = \overline{\boldsymbol{\Lambda}}^{ab}. \quad \text{(C.5)}$$

The transformation properties of the coordinates and superfields with respect to all symmetries can be found by acting from the left on the coset element g (C.4) by the different elements of $N = 1, D = 5$ Poincaré supergroup. In what follows, we will need only the explicit form only for the broken $(S, \overline{S})$, unbroken $(Q, \overline{Q})$ supersymmetries, and $(K, \overline{K})$ automorphism transformations which read

- Unbroken (Q) supersymmetry $[g_0 = \exp\,(\epsilon^a Q_a + \bar{\epsilon}^a \overline{Q}_a)]$

$$\delta x^{ab} = i\left(\epsilon^{(a}\bar{\theta}^{b)} + \bar{\epsilon}^{(a}\theta^{a)}\right),\quad \delta\theta^a = \epsilon^a\,,\ \delta\bar{\theta}^a = \bar{\epsilon}^a. \quad \text{(C.6)}$$

- Broken (S) supersymmetry $[g_0 = \exp\,\left(\varepsilon^a S_a + \bar{\varepsilon}^a \overline{S}_a\right)]$

$$\delta x^{ab} = i\left(\varepsilon^{(a}\bar{\psi}^{b)} + \bar{\varepsilon}^{(a}\psi^{b)}\right),\quad \delta\boldsymbol{q} = 2i\,\varepsilon_a\theta^a,\quad \delta\bar{\boldsymbol{q}} = 2i\,\bar{\varepsilon}_a\bar{\theta}^a,\quad \delta\psi^a = \varepsilon^a,\quad \delta\bar{\psi}^a = \bar{\varepsilon}^a. \quad \text{(C.7)}$$

- Automorphism $(K, \overline{K})$ transformations $[g_0 = \exp i\,\left(a^{ab}K_{ab} + \bar{a}^{ab}\overline{K}_{ab}\right)]$

$$\begin{aligned}
\delta x^{ab} &= -2i\left(a^{ab}\boldsymbol{q} - \bar{a}^{ab}\bar{\boldsymbol{q}}\right) - 2\theta^c\psi_c\bar{a}^{ab} + 2\bar{\theta}^c\bar{\psi}_c a^{ab},\\
\delta\theta^a &= -2i\,a^{ab}\bar{\psi}_b,\quad \delta\bar{\theta}^a = 2i\,\bar{a}^{ab}\psi_b,\\
\delta\boldsymbol{q} &= -2i\,a^{ab}x_{ab} - 2a^{ab}\left(\theta_a\bar{\theta}_b - \psi_a\bar{\psi}_b\right),\quad \delta\psi^a = 2i\,a^{ab}\bar{\theta}_b,\\
\delta\bar{\boldsymbol{q}} &= 2i\,\bar{a}^{ab}x_{ab} - 2\bar{a}^{ab}\left(\theta_a\bar{\theta}_b - \psi_a\bar{\psi}_b\right),\quad \delta\bar{\psi}^a = -2i\,\bar{a}^{ab}\theta_b.
\end{aligned} \quad \text{(C.8)}$$

As the next step of the coset formalism, one can construct the Cartan forms

$$g^{-1}dg = \Omega_P + \Omega_Q + \overline{\Omega}_Q + \Omega_Z + \overline{\Omega}_Z + \Omega_S + \overline{\Omega}_S + \cdots. \quad \text{(C.9)}$$

In what follows we will need only the forms $\left\{\Omega_P, \Omega_Q, \Omega_{\bar{Q}}, \Omega_Z, \Omega_{\bar{Z}}, \Omega_S, \Omega_{\bar{S}}\right\}$ which explicitly read

$$\Omega_P = \left\{ \left(\cosh 2\sqrt{\boldsymbol{Y}}\right)_{ab}^{\;\;cd} \Delta x^{ab} - i\left(\overline{\boldsymbol{\Lambda}}^{ab}\Delta\boldsymbol{q} - \boldsymbol{\Lambda}^{ab}\Delta\bar{\boldsymbol{q}}\right)\left(\frac{\sinh 2\sqrt{\boldsymbol{Y}}}{\sqrt{\boldsymbol{Y}}}\right)_{ab}^{\;\;cd} \right\} P_{cd},$$

$$\Omega_Q = \left\{ d\theta^b \left(\cos 2\sqrt{\overline{\boldsymbol{T}}}\right)_b^{\;c} - i\, d\bar{\psi}^b\, \boldsymbol{\Lambda}_b^{\;a}\left(\frac{\sin 2\sqrt{\overline{\boldsymbol{T}}}}{\sqrt{\overline{\boldsymbol{T}}}}\right)_a^{\;c} \right\} Q_c,$$

$$\Omega_Z = \left\{ \Delta\boldsymbol{q} + \left(\overline{\boldsymbol{\Lambda}}^{ab}\Delta\boldsymbol{q} - \boldsymbol{\Lambda}^{ab}\Delta\bar{\boldsymbol{q}}\right)\left(\frac{\cosh 2\sqrt{\boldsymbol{Y}}-1}{\boldsymbol{Y}}\right)_{ab}^{\;\;cd} \boldsymbol{\Lambda}_{cd} \right.$$

$$\left. + i\, dx^{ab}\left(\frac{\sinh 2\sqrt{\boldsymbol{Y}}}{\sqrt{\boldsymbol{Y}}}\right)_{ab}^{\;\;cd} \boldsymbol{\Lambda}_{cd} \right\} Z,$$

$$\Omega_S = \left\{ d\psi^b \left(\cos 2\sqrt{\overline{\boldsymbol{T}}}\right)_b^{\;c} + i\, d\bar{\theta}^b\, \boldsymbol{\Lambda}_b^{\;a}\left(\frac{\sin 2\sqrt{\overline{\boldsymbol{T}}}}{\sqrt{\overline{\boldsymbol{T}}}}\right)_a^{\;c} \right\} S_c, \tag{C.10}$$

$$\Delta x^{ab} = dx^{ab} - i\left(\theta^{(a} d\bar{\theta}^{b)} + \bar{\theta}^{(a} d\theta^{b)} + \psi^{(a} d\bar{\psi}^{b} + \bar{\psi}^{(a} d\psi^{b)}\right),$$
$$\Delta\boldsymbol{q} = d\boldsymbol{q} - 2i\,\psi_a d\theta^a, \quad \Delta\bar{\boldsymbol{q}} = d\bar{\boldsymbol{q}} - 2i\,\bar{\psi}_a d\bar{\theta}^a. \tag{C.11}$$

Here, we defined matrix-valued functions $\boldsymbol{Y}_{ab}{}^{cd}$, $\boldsymbol{T}_a{}^b$ and $\overline{\boldsymbol{T}}_a^b$ as

$$\boldsymbol{Y}_{ab}{}^{cd} = \boldsymbol{\Lambda}_{ab}\overline{\boldsymbol{\Lambda}}^{cd} + \overline{\boldsymbol{\Lambda}}_{ab}\boldsymbol{\Lambda}^{cd}, \quad \boldsymbol{T}_a{}^b = \boldsymbol{\Lambda}_a{}^c\overline{\boldsymbol{\Lambda}}_c^b, \quad \overline{\boldsymbol{T}}_a^b = \overline{\boldsymbol{\Lambda}}_a^c\boldsymbol{\Lambda}_c^b. \tag{C.12}$$

Note, that all these Cartan forms transform homogeneously under all symmetries.

Having at hands the Cartan forms, one may construct the "semi-covariant" (covariant with respect to $d = 3$ Lorentz, unbroken and broken supersymmetries only) as

$$\Delta x^{ab}\nabla_{ab} + d\theta^a\nabla_a + d\bar{\theta}^a\bar{\nabla}_a = dx^{ab}\frac{\partial}{\partial x^{ab}} + d\theta^a\frac{\partial}{\partial\theta^a} + d\bar{\theta}^a\frac{\partial}{\partial\bar{\theta}^a}. \tag{C.13}$$

Explicitly, they read

$$\nabla_{ab} = (E^{-1})_{ab}^{cd}\partial_{cd},$$
$$\nabla_a = D_a - i\left(\psi^b D_a\bar{\psi}^c + \bar{\psi}^b D_a\psi^c\right)\nabla_{bc} = D_a - i\left(\psi^b\nabla_a\bar{\psi}^c + \bar{\psi}^b\nabla_a\psi^c\right)\partial_{bc}, \tag{C.14}$$

where

$$D_a = \frac{\partial}{\partial\theta^a} - i\,\bar{\theta}^b\,\partial_{ab}, \quad \overline{D}_a = \frac{\partial}{\partial\bar{\theta}^a} - i\,\theta^b\,\partial_{ab}, \quad \{D_a, \overline{D}_b\} = -2i\,\partial_{ab}, \tag{C.15}$$

$$E_{ab}{}^{cd} = \delta_a^{(c}\delta_b^{d)} - i\left(\psi^{(c}\partial_{ab}\bar{\psi}^{d)} + \bar{\psi}^{(c}\partial_{ab}\psi^{d)}\right), \tag{C.16}$$

$$(E^{-1})_{ab}{}^{cd} = \delta_a^{(c}\delta_b^{d)} + i\left(\psi^{(c}\nabla_{ab}\bar{\psi}^{d)} + \bar{\psi}^{(c}\nabla_{ab}\psi^{d)}\right). \tag{C.17}$$

The derivatives obey the following algebra:

$$\begin{aligned}
\{\nabla_a, \nabla_b\} &= -2i\left(\nabla_a\psi^c\nabla_b\bar{\psi}^d + \nabla_a\bar{\psi}^c\nabla_b\psi^d\right)\nabla_{cd},\\
\left\{\nabla_a, \overline{\nabla}_b\right\} &= -2i\,\nabla_{ab} - 2i\left(\nabla_a\psi^c\overline{\nabla}_b\bar{\psi}^d + \nabla_a\bar{\psi}^c\overline{\nabla}_b\psi^d\right)\nabla_{cd},\\
[\nabla_{ab}, \nabla_c] &= -2i\left(\nabla_{ab}\psi^d\nabla_c\bar{\psi}^f + \nabla_{ab}\bar{\psi}^d\nabla_c\psi^f\right)\nabla_{df},\\
[\nabla_{ab}, \nabla_{cd}] &= 2i\left(\nabla_{ab}\psi^m\nabla_{cd}\bar{\psi}^n - \nabla_{cd}\psi^m\nabla_{ab}\bar{\psi}^n\right)\nabla_{mn}.
\end{aligned} \tag{C.18}$$

The $d = 3$ volume form is defined as

$$d^3x \equiv \epsilon_{IJK}dx^I \wedge dx^J \wedge dx^K \quad\Rightarrow\quad dx^I \wedge dx^J \wedge dx^K = \frac{1}{6}\epsilon^{IJK}d^3x. \tag{C.19}$$

Transition from the spinor notations to the vector one is set as follows

$$V^I \equiv \frac{i}{\sqrt{2}}\left(\sigma^I\right)_a^{\ b} V_b^a \quad\Rightarrow\quad V_a^b = -\frac{i}{\sqrt{2}}V^I\left(\sigma^I\right)_a^{\ b}, \qquad V^{ab}V_{ab} = V^IV^I. \tag{C.20}$$

Here we are using the standard set of σ^I matrices

$$\sigma^I\,\sigma^J = i\,\epsilon^{IJK}\sigma^K + \delta^{IJ}\mathcal{I}, \quad \left(\sigma^I\right)_a^{\ b}\left(\sigma^I\right)_c^{\ d} = 2\delta_a^d\delta_c^b - \delta_a{}^b\delta_c^d, \tag{C.21}$$

were ϵ^{IJK} obeys relations

$$\epsilon^{IJK}\epsilon_{IMN} = \delta_M^J\delta_N^K - \delta_N^J\delta_M^K, \quad \epsilon^{IJK}\epsilon_{IJN} = 2\delta_N^K, \quad \epsilon^{IJK}\epsilon_{IJK} = 6. \tag{C.22}$$

References

1. S.R. Coleman, J. Wess, B. Zumino, Structure of phenomenological lagrangians. I. Phys. Rev. **177**, 2239 (1969)
2. C.G. Callan Jr, S.R. Coleman, J. Wess, B. Zumino, Structure of phenomenological lagrangians. II. Phys. Rev. **177**, 2247 (1969)
3. D.V. Volkov, Phenomenological lagrangians. Sov. J. Part. Nucl. **4**, 3 (1973)
4. V.I. Ogievetsky, Nonlinear realizations of internal and space-time symmetries. in *Proceedings of the Xth Winter School of Theoretical Physics in Karpacz*, vol. 1, p. 117, 1974
5. D.V. Volkov, V.P. Akulov, Possible universal neutrino interaction. JETP Lett. **16**, 438 (1972)
6. D.V. Volkov, V.P. Akulov, Is the neutrino a Goldstone particle? Phys. Lett. **B46**, 109 (1973)
7. J. Bagger, A. Galperin, Matter couplings in partially broken extended supersymmetry. Phys. Lett. **B336**, 25–31 (1994). arXiv:hep-th/9406217
8. J. Bagger, A. Galperin, New Goldstone multiplet for partially broken supersymmetry. Phys. Rev. **D55**, 1091 (1997). arXiv:hep-th/9608177

9. J. Bagger, A. Galperin, The tensor Goldstone multiplet for partially broken supersymmetry. Phys. Lett. **B412**, 296 (1997). arXiv:hep-th/9707061
10. M. Rocek, Partial breaking of global D = 4 supersymmetry, constrained superfields, and 3-brane actions. Phys. Rev. **D59**, 106001 (1999). arXiv:hep-th/9811232
11. F. Gonzalez-Rey, I.Y. Park, M. Rocek, On dual 3-brane actions with partially broken N = 2 supersymmetry. Nucl. Phys. **B544**, 243 (1999). arXiv:hep-th/9811130
12. E. Ivanov, S. Krivonos, N = 1, D = 4 supermembrane in the coset approach. Phys. Lett. **B453**, 237 (1999). arXiv:hep-th/9901003
13. S. Bellucci, E. Ivanov, S. Krivonos, Superbranes and super born-infeld theories from nonlinear realizations. Nucl. Phys. Proc. Suppl. **102**, 26 (2001). arXiv:hep-th/0103136
14. S. Bellucci, E. Ivanov, S. Krivonos, Partial breaking of N = 1, D = 10 supersymmetry. Phys. Lett. **B460**, 348 (1999). arXiv:hep-th/9811244
15. S. Bellucci, E. Ivanov, S. Krivonos, Partial breaking N = 4 to N = 2: hypermultiplet as a Goldstone superfield. Fortsch. Phys. **48**, 19 (2000). arXiv:hep-th/9809190
16. S. Bellucci, E. Ivanov, S. Krivonos, N = 2 and N = 4 supersymmetric Born-Infeld theories from nonlinear realizations. Phys. Lett. **B502**, 279 (2001). arXiv:hep-th/0012236
17. E.A. Ivanov, V.I. Ogievetsky, The inverse Higgs phenomenon in nonlinear realizations. Teor. Mat. Fiz. **25**, 164 (1975)
18. S. Bellucci, E. Ivanov, S. Krivonos, Towards the complete N = 2 superfield Born-Infeld action with partially broken N = 4 supersymmetry. Phys. Rev. **D64**, 025014 (2001). arXiv:hep-th/0101195
19. M. Aganagic, C. Popescu, J.H. Schwarz, Gauge-invariant and gauge-fixed D-brane actions. Nucl. Phys. **B495**, 99 (1997). arXiv:hep-th/9612080
20. E. Bergshoeff, F. Coomans, R. Kallosh, C.S. Shahbazi, A. Van Proeyen, Dirac-Born-Infeld-Volkov-Akulov and deformation of supersymmetry. JHEP **1308**, 100 (2013). arXiv:1303.5662 [hep-th]
21. S. Bellucci, S. Krivonos, A. Sutulin, Supersymmetric component actions via coset approach. Phys. Lett. **B726**, 497 (2013). arXiv:1306.1115 [hep-th]
22. S. Bellucci, N. Kozyrev, S. Krivonos, A. Sutulin, Partial breaking of global supersymmetry and super particle actions. J. High Energy Phys. arXiv:1309.3902[hep-th]
23. S. Bellucci, N. Kozyrev, S. Krivonos, A. Yeranyan, Supermembrane in D = 5: component action. arXiv:1312.0231[hep-th]
24. T.E. Clark, M. Nitta, T. ter Veldhuis, Brane dynamics from nonlinear realizations. Phys. Rev. **D67**, 085026 (2003)
25. M. Hanneaux, L. Mezincescu, A σ-model interpretation of Green-Schwarz covariant superstring action. Phys. Lett. **B152**, 340 (1985)
26. D. Sorokin, Superbranes and superembeddings. Phys. Rept. **329**, 1 (2000). arXiv:hep-th/9906142
27. S. Bellucci, E. Ivanov, S. Krivonos, O. Lechtenfeld, ABC of N = 8, d = 1 supermultiplets. Nucl. Phys. **B699**, 226 (2004). arXiv:hep-th/0406015
28. A. Achúcarro, J. Gauntlett, K. Itoh, P.K. Townsend, World-volume supersymmetry from space-time supersymmetry of the four-dimensional supermembrane. Nucl. Phys. **B314**, 129 (1989)
29. E. Ivanov, S. Krivonos, O. Lechtenfeld, Double vector multiplet and partially broken N = 4, D = 3 supersymmetry. Phys. Lett. **B487**, 192 (2000). arXiv:hep-th/0006017
30. S. Bellucci, E. Ivanov, S. Krivonos, Goldstone superfield actions in AdS5 backgrounds. Nucl. Phys. **B672**, 123 (2003). arXiv:hep-th/0212295

MIX
Papier aus verantwortungsvollen Quellen
Paper from responsible sources
FSC® C105338

If you have any concerns about our products,
you can contact us on
ProductSafety@springernature.com

In case Publisher is established outside the EU,
the EU authorized representative is:
Springer Nature Customer Service Center GmbH
Europaplatz 3, 69115 Heidelberg, Germany

Printed by Libri Plureos GmbH
in Hamburg, Germany